100 Years of Civil Aviation

100 Years of Civil Aviation

A History From the 1919 Paris Convention to Retiring the Jumbo Jet

Ben Skipper

AIR WORLD

First published in Great Britain in 2023 by
Pen & Sword Aviation
An imprint of
Pen & Sword Books Ltd
47 Church Street
Barnsley
South Yorkshire
S70 2AS

Copyright © Ben Skipper, 2023

ISBN 9781399065962

The right of Ben Skipper to be identified as Author of this work has been asserted by him in accordance with the Copyright, Designs and Patents Act 1988.

A CIP catalogue record for this book is available from the British Library.

All rights reserved. No part of this book may be reproduced or transmitted in any form or by any means, electronic or mechanical including photocopying, recording or by any information storage and retrieval system, without permission from the Publisher in writing.

Printed and bound in the UK by CPI Group (UK) Ltd., Croydon. CR0 4YY

Pen & Sword Books Ltd incorporates the Imprints of Pen & Sword Archaeology, Atlas, Aviation, Battleground, Discovery, Family History, History, Maritime, Military, Naval, Politics, Railways, Select, Transport, True Crime, Fiction, Frontline Books, Leo Cooper, Praetorian Press, Seaforth Publishing, Wharncliffe and White Owl.

For a complete list of Pen & Sword titles please contact
PEN & SWORD BOOKS LIMITED
47 Church Street, Barnsley, South Yorkshire, S70 2AS, England
E-mail: enquiries@pen-and-sword.co.uk
Website: www.pen-and-sword.co.uk

Contents

Introduction		9
Chapter 1	Beginnings	16
Chapter 2	Beyond the Horizon: Pioneers and Adventurers	32
Chapter 3	In for the Long Haul: Opening the Passages to the Masses	49
Chapter 4	The Arrival of the Jet	79
Chapter 5	Living the Jet Dream: Bigger, Farther, Faster	105
Chapter 6	Women in Aviation: The Pioneers and the Pilots	125
Chapter 7	The Infrastructure and Organization of Flight	149
Chapter 8	Civil Aviation Beyond the Holiday	166
Chapter 9	Icarus Has Fallen: War, Disaster and Terror	198
Conclusion	Into the Second Century: The Future is Now	232
Bibliography		235
Index		238

Introduction

The story of flight is one of human endurance, ingenuity and persistence. The early years of the twentieth century saw civil aviation take faltering steps, with the First World War acting as catalyst. In the dying embers of conflict, the leading aviation and law experts of the International Commission for Air Navigation (ICAN) gathered in Paris to produce a Convention Relating to the Regulation of Aerial Navigation, subsequently signed on 13 October 1919. The Convention provided the necessary regulation of safe and secure flight by defining and guiding the principles and provisions concerning flight. It heralded the boom of flight. Within twenty years of its signing, increasingly sophisticated passenger aircraft would be crossing continents and oceans in ever-increasing numbers.

To fly and soar have been one of humanity's greatest desires, fed by tales of flying pioneers, including Daedalus and Icarus, and their wings of wax and feather. It was the ancient Babylonians who first noted that to fly was 'a great privilege. Knowledge of flying ... is a gift from the gods'. Indeed, there exists a great deal of evidence that the ancients achieved flight.

In China, around 2,000 BCE, there are stories of flying emperors and scholars taking to the skies to escape some earthbound calamity or map the lands beneath them. By 400 BCE the Chinese were using kites to help understand flight mechanics and characteristics. This led to some rather interesting experiments, including one by Emperor Wan-Hu who built a chair-like device to which he attached two kites and forty-seven rockets. The emperor then ignited the rockets, dissappearing in the resultant explosion.

To the east was the Persian Shah Kai Kawus whose own attempts at flight with four tethered eagles saw him land, according to legend, in China. The most inspiring tales come from India's sacred texts, the *Samaranga Sutradhara*, which give the reader 200 stanzas regarding the operation of Vimanas, flying machines akin to jet aircraft with regards to propulsion method and overall performance.

In the eleventh century a Benedictine monk in Malmesbury, Britain, known as Elmer, was experimenting with flight. Previous experiments had seen the aspiring aeronaut attach homemade wings to his back. Copying designs from Greek myths, Elmer used a mix of materials to create wings which were then attached to his arms and feet.

Once his creation was ready, he ascended one of the monastery's towers and threw himself out into the air. Miraculously, his design worked; with a flight distance of 755 feet (230 metres) recorded, the smoothness of Elmer's flight was interrupted by a crash-landing likely caused by air turbulence created by the physical presence of the monastery structure. Although he broke both legs, Elmer could understand why he crashed, but not how he could have prevented it.

The next innovation came from fellow monk, philosopher and scholar, Roger Bacon, in the thirteenth century. While not a practitioner like his predecessor Elmer, Bacon considered the theory of successful flight. He made two key developments. The first was the design of a craft known as an ornithopter, which gained flight through flapping wings, much like a bird. Unfortunately, this method of flying, despite repeated attempts, remains unachievable.

The second development was a theory that, like water, the air was fluid. It was in his book *History of Art and Nature* that Bacon concluded a globe filled with 'ethereal air' or 'liquid fire' would give it buoyancy, allowing it to rise. While Bacon's idea of what constituted ethereal air and liquid fire was pure supposition, he was on the right track in theorizing on the existence of gases such as helium and hydrogen as well as hot air – all of which would be proven, nearly 600 years later.

Bacon's work was published before the significant expansion of European creative and scientific consciousness that heralded the age of the Renaissance. It was an age that would develop aeronautical innovations still in use today. One of the era's great thinkers, Leonardo da Vinci, sketched over 500 designs of various flying machines, the most famous of which was the early human-powered helicopter. Da Vinci's collection of sketches also included an ornithopter and wings that were designed and built as a result of his study of birds' and bats' wings. Although many of his designs would never fly due to a lack of understanding of the mechanics behind bird flight, da Vinci's fundamental discovery was the need for a tail to help stabilize any act of flying.

In 1670, a discovery was made by Italian priest Father Francesco de Lana who reasoned that removing the air from a suitably sized cooper sphere would make it lighter than its surroundings and thus rise. Unfortunately, while the science was perfect, the manufacturing processes of the period were not. So, de Lana's theories could not be put into practice.

The discovery of gaseous hydrogen by scientist and philosopher Henry Cavendish in 1766 advanced the prospect of lighter-than-air flight. Hydrogen weighs a tenth of the air around us, with a mass of two grams per 22.4 litres of volume, or seven percent of atmospheric air. Nevertheless, despite Cavendish's discovery, the process of achieving lighter-than-air flight was not immediately seized upon. Instead, that crown went to the hot-air pioneers, the Montgolfier brothers.

Joseph and Étienne Montgolfier, sons of a papermill owner, were spellbound by the idea of flight and all the marvels it might bring. Early experiments used paper envelopes filled with steam to achieve lift. These attempts failed as the paper envelopes became little more than a soggy mess. Unperturbed, the brothers continued their work, using self-refined hydrogen from coal to help achieve lift. But, like the earlier steam experiments, nothing happened initially.

Then, in 1782, the brothers had their first success: they designed a taffeta balloon lined with fireproof alum-treated paper and filled it with hot air. The experiment was a success and spurred the brothers' enthusiasm. Several more experiments followed, with flight distances constantly increasing. Finally, in September 1783, they made their first public breakthrough. The Montgolfiers made a royal demonstration of their balloon in the presence of Louis XVI. The flight was the first to carry passengers: a duck, a cock and a sheep. The flight would also establish whether there would be any ill effects on possible future passengers. Again, the flight was a success, and on 21 November 1783 the first passengers took to the air in a larger balloon fuelled by an iron furnace. Physicist Jean-François Pilâtre and the Marquis d'Arlandes François Laurant ascended to a height of around 2,950 feet (900 metres), travelling 6.2 miles (ten kilometres) before landing safely. The first obstacle to artificial flight had been overcome, setting imaginations ablaze.

By 1785 the first cross-channel flight in a balloon had taken place with François Blanchard and Dr John Jeffries flying from Dover to Calais on 7 January. Later that year, Blanchard would demonstrate the first practical use of a parachute. He would cross the Atlantic in 1793 to continental North America, where he delivered the first piece of airmail to George Washington after a balloon flight from Philadelphia to New Jersey.

The next leap forward occurred in 1886 when another Frenchman, Clémant Ader, designed and made the *Ader Éole*, a bat-wing-inspired aeroplane powered by a 20hp steam engine. Although on the right track, the lack of control meant that most of Ader's flights were little more than wild hops, but the point was made that powered flight was possible. The experimentation continued, and on 31 July 1894, industrialist and inventor of the machine gun, Sir Hiram Stevens Maxim, achieved powered flight, albeit uncontrolled and unintended. A test rig, designed to test wing form, broke free of its shackles and flew for 180 metres with Maxim and three others on board, before crashing.

Then, on 17 December 1903, on the edge of the Atlantic Ocean, the miracle finally happened: a controlled, powered flight took place near Kitty Hawk, North Carolina. For twelve seconds the world must have stood still for Wilbur Wright as he watched his brother Orville control their aeroplane into history.

The twentieth century saw incredible advances in aviation as the Wright Flyer ushered in a new age of innovation and adventure. In 1906, Glenn Curtiss, whose designs would help establish the first commercial air service, met with the Wright brothers to exchange notes on aircraft design and handling. By 1908 Curtiss had designed and built the June Bug, a biplane capable of taking off under its own power and fitted with a wheeled undercarriage. The wings of his biplane curved towards one another and were fitted with wingtip-mounted ailerons. As a result of his work Curtiss became the first licensed US pilot. Any idea of a mutually beneficial air race between the Wrights and Curtiss was short-lived. What should have been a friendly commercial rivalry between all parties and newcomers descended into an almost endless stream of litigation by the Wrights. As a result, the Wrights' many designs languished while European designers and manufacturers flourished.

At the same time, the Europeans were ahead of the curve in developing the aeroplane. Anglo-French aviator, aircraft designer and manufacturer Henri Farman kicked off the almost frenetic innovation of the period by completing a one-kilometre (1,094-yard) circular flight on 8 January at Les Moulineaux in his Voisin-Farman I biplane. The aeroplane, designed and built with business partner Gabriel Voisin, completed the course in one minute and twenty-eight seconds. Farman won the 50,000-franc Grand Prix d'Aviation prize, sponsored by Henri Deutsch de la Meurthe for anyone completing the course. Petroleum businessman de la Meurthe was fascinated by flight and used his considerable wealth to help stimulate aviation innovation by private individuals like Farman. By March that year Farman was completing flight circuits of 1.2 miles (two kilometres), and by October he had completed a cross-country flight of seventeen miles (twenty-seven kilometres) between the towns of Bouy and Reims.

While impressive in terms of how quickly Farman had built up his aeroplane's endurance and his proficiency as a pilot, he was still short of the Wright Brothers' achievements. Before taking their 1905 break, the Wright brothers had achieved vital benchmarks, including flight distances of twenty-five miles (forty kilometres), with an airborne endurance of some thirty minutes. There was also the matter of control, which genuinely defined an aviator's mount as an aeroplane. The Wrights' progress in this field allowed them to perform figures of eight relatively easily.

Initially, these claims of the aeroplane's agility and endurance were met with disbelief by most Europeans. However, the Wrights returned to flying in 1908 after a brief pause. While Orville stayed in the United States to fly their new two-seater aeroplane for the US Army, Wilbur wooed European crowds during the high summer months. Not wasting any time, Wilbur's first outing took him over the Le Mans racecourse, where he demonstrated the agility of his aeroplane to astonished crowds. During the four months he was in France, Wilbur performed over 100 flights, totalling some twenty-five hours. He also carried passengers on more than sixty occasions and on one occasion flew for two continuous hours. Orville's work with the US Army ended with the crash of his Flyer due to the propeller shattering on 17 September. His passenger, army Lieutenant Thomas Selfridge, died of his injuries later that day, becoming the first official aeroplane accident fatality. Orville's wounds, while not life-threatening, would leave him suffering from extreme pain for the next twelve years until X-rays revealed a series of hip fractures and dislocations.

In France, one man in particular had been impressed by Wilbur Wright's deeds, inventor and engineer Louis Blériot. Blériot would design and build the Type XI (Blériot) monoplane in 1909, an aeroplane in which he would successfully make a cross-country flight of twenty-six miles (forty-two kilometres) between Étampes and Chevilly, south of Paris. Perhaps emboldened by his success, Blériot succeeded in making the first crossing of the English Channel (La Manche) by aeroplane on 25 July and, by doing so, picked up the *Daily Mail*'s £1,000 prize.

The flight proved many things, not least Blériot's resolve as he'd made the twenty-seven-minute crossing without a compass in low cloud. It also confirmed that mainland Britain could be reached by aircraft from continental Europe. Some welcomed this significant shrinking of the world; others were alarmed at how the perceived gap between land masses had suddenly exposed vulnerability. Yet, despite misgivings, Blériot received at least 100 orders of his Type XI monoplane.

The Type XI monoplane marked a critical stage in the evolution of aircraft design. Its key features would remain a constant in propeller-driven aeroplane design; these included a crankshaft-driven tractor propeller, a monoplane wing, a covered fuselage structure, rudder pedals and a flexible control stick. The engineers were now making their mark on the development of the aeroplane.

The developments and interest surrounding the aeroplane were now reaching a fever pitch in Europe, with Reims hosting the first air meet between 22 and 29 August 1909. The event attracted thirty-eight aircraft, twenty-eight of which participated in a range of trials and completions sponsored by local champagne producers. As a result of the patronage, the meet also brought an air of unmistakeable glamour.

Joining the likes of Blériot were Glenn Curtiss and Henri Farman. This cosmopolitan collection of aviators would see several records made, including height (508 feet/155 metres) and the distance of 112 miles (180 kilometres) flown in a little over three hours. Nonetheless, there was still a great deal to do, including the design of more reliable aero engines.

1909 was also an unfortunate watershed moment in the history of the aeroplane – the sad beginnings of death related to powered flight. By the end of the year, there had been three deaths. As the pace of development and popularity picked up in 1910, so too did the deaths, with thirty-two more fatalities. One of these deaths was that of Welsh aviation and motoring pioneer, the Honourable Charles Stewart Rolls. Rolls lost his life on 12 July at the Bournemouth air meet when his Wright Flyer broke up mid-air.

1910 also saw a raft of positive developments. The idea of the air meet had spread, with twenty taking place in Europe, the continental United States and North Africa. Innovation continued with the first night flight, an important event in Argentina; the Baroness de la Roche was the first woman to become qualified as a pilot. Then, on his sixth flight, French-American pilot John Moisant, accompanied by his mechanic, Alfred Fileux and a pet cat, Mademoiselle Fifi, as passengers, crossed the English Channel. This flight was a milestone, proving that passengers could quickly be flown to continental Europe and vice versa. In the continental United States, Curtiss biplanes were making headway with ever-increasing long-distance flights, including flights between Albany and New York City, a distance of 158 miles (254 kilometres) and the first use of a radio in an aeroplane.

The following year the aeroplane was well established in the public eye, with its popularity and reliability increasing. French aviator Pierre Prier flew non-stop from London to Paris on 12 April, a distance of 250 miles (402 kilometres) in less than four hours. Air races were becoming more ambitious, with long-distance races becoming prolific. These were completed in stages, with Paris as the main starting point. On the other side of the Atlantic, Calbraith Rodgers completed the first coast-to-coast passage in his Wright Biplane. The 6,437-kilometre (4,000-mile) flight took eighty-two hours and finished in eighty-two stages over forty-nine days. Rodgers' adventure would see him crash no fewer than nineteen times. However, he was supported throughout by his wife and mother, who followed (on the ground) with any necessary spares.

To the south, Randolph Hearst made the 145-kilometre (ninety-mile) flight from Key West in Florida to Havana, Cuba. This was also the year that Curtiss would fly his first practical seaplane. Another first was the carriage of official air mail. On 18 February, French pilot Henri Pequet took a sack containing 6,000 items of mail over the Yamuna River, from Allahabad to Naini in India, in his Humber-Sommer Biplane. While only covering a short distance of five miles (eight kilometres), Pequet's journey was the start of the practical use of the aeroplane in the cargo-carrying role.

1912 continued the trend in innovation and milestones, which saw the first use of the parachute from the aeroplane, made by Albert Berry from a Benoist Headless pusher biplane piloted by Tony Jannus. Jannus would later make history by navigating the first point-to-point journey to carry a paying passenger from St. Petersburg, Florida, to Tampa, with Abram Pheil who paid $400 for the honour. After that, two return flights a day took off from St. Petersburg, except on Sundays. Meanwhile, American pilot and writer Harriet Quimby became the first woman to fly across the English Channel, on 16 April, the day after the *Titanic* sank.

The following year saw the first air crossing of the Mediterranean by

Roland Garros in a Morane-Saulnier G two-seater sports aeroplane, on 23 September. The non-stop crossing from Fréjus-Saint Raphaël in the south of France to Bizerte, Tunisia, took Garros eight hours. A fellow Frenchman, Jules Védrines, made the first flight from Paris to Cairo in an epic journey over ten stages that were not without incident. On one occasion, Védrines was tried in absentia by the German authorities for violating German airspace. At this stage in the development of civil aviation, there was no legal framework for overflight, leaving many aviators to rely on the goodwill of the nations they overflew. As well as being a long-distance aviator, Védrines had held the world airspeed record as the first aviator to break the 100mph barrier. To the north, Norwegian explorer and aviator Jens Tryggve Gran was the first to make the hazardous flight across the North Sea on 30 July 1914 in a Blériot XI-2 monoplane. Tryggve Gran departed from Cruden Bay, Scotland, flying to Stavanger, Norway, in a little over four hours and covering a distance of 317 miles (510 kilometres).

By the high summer of 1914, the foundations for efficient and safe civilian aviation as another means of transport, or 'locomotion' as observed by British politician Lloyd George, were now in place. Technical development had gathered speed, and the opportunities seemed endless. But storm clouds were gathering.

Chapter 1

Beginnings

From Swords to Ploughshares: Laying the Foundations

Before the great calamity of the First World War, passenger aircraft were steadily appearing in the skies. The Russian designer Igor Sikorsky was very much a groundbreaker in the arena of the airliner. His first multi-engine aeroplane, the S-21 Russky Vityaz (Russian Knight), was built in 1911, less than a decade after the Wright Brothers' Flyer took off. His design would prove that the aeroplane could carry heavy loads effectively and safely. The S-21, built by the Russian Baltic Railroad Car Works, seemed little more than a small railway carriage with wings, yet it could carry seven people. Sikorsky's S-21 was a quantum leap forward in providing affordable air travel. The S-21 was followed by a range of Sikorsky aircraft, including the flexible Ilya Muromets.

The First World War saw a proliferation of aircraft on all sides, fulfilling various roles and demonstrating how flexible aircraft had become. With the signing of the Armistice, many of the larger aircraft, mainly bombers, became surplus to requirements. While few in number, they sparked the imagination of those wanting to build a better post-war world. 1919 became a pivotal year for civil aviation, with transoceanic crossings established. The first breakthrough came with the transatlantic flight of Lieutenant Commander Albert Cushing Read in a US Navy (USN) Curtiss tri-engine NC flying boat known as NC-4. NC-4 and sister craft NC-1 and NC-3. All three left Naval Air Station Rockaway on 8 May 1919 with intermediate stops at the Chatham Naval Air Station, Massachusetts, and Halifax, Nova Scotia, before flying on to Trepassey, Newfoundland, on 15 May.

As part of an intricate plan and to provide support, including rescue should it be needed, the USN placed a series of eight ships along the eastern side of continental North America. These ships also provided searchlight illumination to aid the navigation of the three flying boats as they flew northward. Also included in the small flotilla was a recently converted minesweeper that served as a tender, the USS *Aroostook*. The *Aroostook* sailed ahead of the three flying boats and awaited their arrival at Newfoundland. She would supply mechanical and logistical support before crossing the Atlantic to meet the three flying boats in England.

On leaving Newfoundland on 16 May, the three flying boats began the longest, most dangerous part of their journey, crossing the 1,181-mile (1,900-kilometre) stretch of the Atlantic to the Azores. Once again, the route was marked by the USN, with twenty-two ships spaced out along it, using a combination of searchlights and illumination shells to help aid night navigation. Fifteen hours later a single flying boat, NC-4, landed at Faial Island in the Azores; NC-1 and NC-3 were both forced to land due to poor visibility. NC-1 was rescued by the Greek ship SS *Ionia*, towing it until the flying boat sank three days later. For their part, the crew of NC-3 taxied 200 nautical miles (370 kilometres) to reach the Azores, where it was taken in tow by a USN ship.

After three days of rest, the NC-4 took off again. Still, mechanical problems forced the flying boat to land 150 miles (241 kilometres) later at São Miguel Island and await the arrival of spare parts so that repairs could be effected. On 27 May, NC-4 took off once more, with thirteen ships providing light to aid navigation between the Azores and the Portuguese mainland. Ten hours later, NC-4 landed in Lisbon harbour. The venture had taken Cushing Read and his crew ten days and twenty-two hours, a flight time of twenty-six hours and forty-six minutes, to cross the Atlantic.

Cushing Read and the crew of NC-4 then flew on to Plymouth, England, arriving on 31 May, taking twenty-three days for the flight from Newfoundland to Great Britain. For this final leg of the journey, ten more USN ships illuminated the route, bringing the total of vessels used to fifty-three. The adventure had proved transatlantic crossings were possible. In later years, Cushing Read would become something of an aviation prophet, stating that flight above 60,000 feet at 1,000mph (1,609km/h) was possible – an idea dismissed as fanciful by peers, but later proved by the exploits of Concorde.

The adventure of the three Curtiss flying boats and their crews included stops, tremendous logistical support and planning, and proved that transatlantic flight was possible – established within a month by Alcock and Brown's non-stop flight.

Their feat was in response to a challenge set by the *Daily Mail*, which sponsored a range of aviation challenges throughout the twentieth century. In 1913, a transatlantic flight prize was offered by the newspaper, but the First World War intervened. In 1918, the prize was revisited with a £10,000 prize. The prize came with conditions, stipulating that it would be awarded to any 'aviator who shall first cross the Atlantic in an aeroplane in flight from any point in the United States, Canada, or Newfoundland to any point in Great Britain or Ireland, in seventy-two consecutive hours'. The conditions also stated that aviators must attempt the flight in a single aeroplane, which meant changing aircraft would disqualify participating teams and pilots. Interestingly, no mention of the flight being non-stop could have allowed participants to land to refuel.

Pilot John Alcock and navigator Arthur Whitten Brown made their non-stop flight in a modified Vickers Vimy VI heavy bomber, powered by two Rolls-Royce Eagle 360hp engines, taking off on 14 June 1919 from St. Johns, Newfoundland, carrying mail stamped with the inscription 'Transatlantic Air Post 1919'.

What followed the pair and their aeroplane, loaded with a staggering 3,900 litres of fuel, was nothing short of an adventure. Departing at 13:45, the team headed east, and by 16:30, they had crossed the Canadian coastline, heading out over the Atlantic. Unfortunately, within an hour, their wind-driven electrical generator had failed, leaving them without a working radio and no contact with the outside world. It also meant that their internal intercom no longer functioned, and they lost power to the aeroplane's heating systems, including their electronically heated flight suits. As a result, the aviators were exposed to the elements for the flight's duration. A blown-out exhaust pipe further compounded this calamity. The resulting noise left all conversation between the two men almost impossible.

At the same time, the pair were flying through thick fog, preventing Brown from using his sextant to aid navigation. Throughout the flight, Alcock's airmanship was tested. He twice lost control of the Vimy, nearly crashing on one occasion after a spiral dive. He also had to deal with a broken trim control, battling an aeroplane becoming increasingly nose-heavy as it consumed fuel. The result was that the altitude flown was often close to sea level.

Just after midnight on 15 June, Brown got his first glimpse of the stars and was able to use the sextant, confirming that they were on course. Unfortunately, by the early hours the aeroplane was flying through sleet, drenching both men, which only added to their woes. By this point, the aeroplane's instruments had iced up, and the possibility of the aircraft becoming unflyable as a result of ice formation throughout its fuselage and wings was a real possibility. In addition, Brown often exposed himself to the risk of falling out of the open cockpit as he cleared the ice.

Finally, at 08:40 on 15 June 1919, the pair landed at Derrygilmlagh Bog, near Clifden, in County Galway, Ireland, having flown 1,889 miles (3,040 kilometres) in fifteen hours and fifty-seven minutes at an average speed of 115mph (185km/h). Brown would later remark that they could have pressed on to London if the weather had been kinder. As a result of their feat, Alcock and Brown won the *Daily Mail*'s prize and proved beyond doubt that uninterrupted transatlantic flight by aeroplane was possible. The pair was later bestowed the title of Knight Commander of the Most Excellent Order of the British Empire (KBE) by King George V for their heroic feat.

Within a month of Alcock and Brown's achievement, the rigid airship (dirigible) the R34, under the command of Major George Scott, made the first return crossing of the Atlantic by aircraft. The twenty-nine military personnel on board for the journey were a mix of sailors, soldiers and airmen from the newly formed Royal Air Force (RAF) and a stowaway, William Ballantyne, along with Wopsie the cat. The crew included Air Commodore Edward Maitland, who had commanded a Naval Airship Station during the First World War. Also aboard were the Americans Lieutenant Commander Zachary Lansdowne and Lieutenant Colonel William Hensley. They acted as independent flight observers and represented the US Navy and Army. As the R34 was not a passenger airship, slinging hammocks in the keel walkway provided extra crew accommodation needed for the flight. Another important addition was welding a plate to an engine exhaust pipe, allowing the crew to prepare hot food. The R34 left East Fortune, Scotland, on 2 July 1919.

One hundred and eight hours later, on 6 July, the R34 came to rest at Hazelhurst Field, Mineola, Long Island, United States, with her fuel almost exhausted. As none of the crew had the necessary experience in docking a giant rigid-bodied airship, Major John Pritchard descended by parachute to make the necessary arrangements. Pritchard also became the first person to reach American soil by air from Europe. The return flight had the wind behind it, and leaving the United States four days later, on 10 July, the R34 took off on her return journey to RNAS Pulham, Norwich, England. The R34 returned home with mail while the crew was gifted a gramophone and a welcome ration of rum. The R34 arrived safely at RNAS Pulham 75 hours later on 13 July. The R34 adventure rightly surmised that well-appointed airships would have a significant part in long-distance travel, where paying patrons enjoyed comforts usually reserved for premier ground or sea travel journeys.

Not to be outdone by the airship, the aeroplane continued to show its mettle. Between 12 November and 10 December 1919, Australian brothers Ross and Keith Macpherson-Smith completed the first flight from Hounslow, Great Britain, to Darwin, Australia. Along with their mechanics, Jim Bennett and Wally Shiers, the four men covered a distance of 11,293 miles (18,175 kilometres). The brothers were awarded knighthoods and a £10,000 prize from the Australian government. For its part in all these adventures, the Vimy would be converted into a passenger aircraft. It would become the parent aeroplane of the somewhat bulbous Vimy Commercial, becoming one of the first purpose-built post-war passenger aircraft. The Commercial would see use with Imperial Airways and French operator Grands Express Aériens.

It was clear the aeroplane was here to stay, and the initial post-war adventures of daring-do opened the floodgates for the dreamers, the designers and the innovators. With every new aviation adventure, the world shrank a little more. The desire to utilize the aeroplane's potential reached fever pitch. The technological and social leaps and bounds that would take place in the coming years may well have been the stuff of dreams in a world shattered by war. But there was no stopping them. The dreams had wings.

From the Ashes Comes Order

While 1919 was a pivotal year for civil aviation, it was also the year that the legal and political frameworks to support civil aviation were established. In many respects, 1919 was the founding year of a formalized and regulated civil aviation industry as introduced at the Paris Convention. The convention had its genesis in the late nineteenth century at a gathering held in 1880 at Oxford where a multinational group of jurists regarded the impacts and implications of cross-border aviation by aircraft. The Oxford conference discussed flight's civil and military consequences and the possible legal complexities that this would bring. Several unofficial, quasi-legal stances existed on what would become aviation law. The line of thought adopted by aviators, though not necessarily by sovereign nations, was merely to expand the notion that the freedom of the high seas carried over into the aeronautical field as freedom of air. The logic here is that, like the high seas, the air was the common property of humanity and so could not be owned or claimed. To counter this was the theory of complete and exclusive sovereignty, whereby the airspace above a state's territory was an extension of that sovereign territory. A final theory recognized the airspace above sovereign territory as national airspace and, as such, the state would allow access rights through freedom of air.

By the turn of the twentieth century, there was a growing consensus regarding some form of airspace regulation above sovereign territories. The complete and exclusive sovereignty idea was growing but had not yet won the day. French and German jurists sought a mixed access model, which took an 'open skies' approach. French jurist Paul Fauchille was the principal force behind formulating a code for international air navigation. The British pushed for 'a right of sovereignty over aerial space above its soil, saving the right of ... passage for a balloon and other aerial machines'. There was work to be done.

Somewhat surprisingly, it was the rules of war, formulated at the Hague Peace Conference, that would help develop later aviation law. At the first Hague Peace Conference in 1899, a five-year ban was placed on discharging weaponry or explosives from aircraft, in this case, balloons. The ban was further extended at the Second Hague Peace Conference in 1907. Even though many of the articles agreed upon in both conferences were broken during the First World War, they proved that international consensus and cooperation were possible.

The Hague Peace Conferences' approach to regulation would also help shape future aviation laws, using national security as a critical factor in formulating legislation and regulation. The need for this formalized approach to aviation law grew in importance as increasingly sophisticated powered aircraft, which seemed to be developing at a prodigious rate, took to the sky. The meetings of thinkers, jurists and government departments continued throughout the pre-war years. As they did so, more nations became involved, each eager to have their say. By 1910 Fauchille had drawn up a code of air law, published at the International Air Navigation Conference held in Paris. Within the code were two articles, 19 and 20, that dealt explicitly with 'restrictions which states might impose on foreign aircraft in the air space above its territory'. The pioneering notions of free airspace had been brushed aside in the name of security and sovereignty.

A further meeting in Madrid in 1911 saw other changes. Most importantly, states had the right to regulate traffic over land and water but should permit the free passage of all airships [aircraft] of all nations. This ambiguity did little to help, and by 1913 France and Germany had adopted the British stance of maintaining airspace as sovereign territory. The British cemented their approach by enacting the Aerial Navigation Act 1911, confirming that airspace over the British Empire was sacrosanct in its sovereignty.

As part of the post-war discussions regarding the future of air navigation, no doubt heavily influenced by the events of the previous four years, the element of air sovereignty re-emerged, this time to follow the British lines of thinking as defined in the Aerial Navigation Act. As part of the peace process, a new body was established, the Aeronautical Commission of the Peace Conference (ACPC). The ACPC had its roots in the Allied military aviation organizations and helped form a sturdy foundation for the 1919 Paris Convention.

The new convention saw fundamental changes in attitudes, most notably from the French delegation, whose views were changed by four years of war. The French and other European nations wanted total air navigation control over their territories. However, the influence of the Anglo-American stance regarding the national sovereignty of airspace, coupled to free movement, prevailed. This benefitted the Americans and the British by ensuring freedom of movement for their yet-to-be fully established lines of air communication. Moreover, it would protect future commercial interests, especially air mail and passenger travel.

Initially, all sides were at an impasse. Then, on 13 October 1919, the Convention Relating to the Regulation of Aerial Navigation (The Paris Convention) was signed. This signified that all signatories had reached a consensus. Article 1 established the rights of a nation's sovereign airspace, followed by granting the rites of passage for civilian aircraft. However, it was Article 15 that opened the doors to the development of a commercial system of air navigation. It stated, 'The establishment of international airways shall be subject to the consent of the States flown over'. For those countries which were not party to the convention, Article 5 reinforced a nation's right to regulate its airspace with permits and authorization. This meant a state could grant temporary passage to aircraft belonging to or registered in a nonsigning state. The conversation had turned from total freedom of the skies to a more controlled and regulated approach, which was safer for all parties in the long run. So, while not a perfect solution, the Paris Convention was a start. Given the year it was written, the convention was a laudable shot in the right direction.

Most importantly, the convention opened the doors for establishing a global commercial flight network and all it entailed. How this network would be implemented was down to the subtlety of the entrepreneurs, engineers, adventurers, diplomats and state regulators. The convention also sought to establish a series of safeguards, including pilot and crew training and certification, the introduction of navigation and airworthiness standards and aircraft registration.

The convention also saw the establishment of the International Commission of Air Navigation (ICAN), which was part of the fledgling League of Nations. ICAN's role was to monitor developments in the world of civil aviation. ICAN's authority was bolstered by the establishment of the International Air Traffic Association (IATA), whose role was to review the standardization of a range of processes. These processes included the interlining of tickets and pricing to allow travellers to pass through several airlines.

As the world of aviation continued to develop, so did the need to amend and add different laws and regulations. A series of smaller conventions were designed to overcome local variations and differences. The 1926 Ibero-American Convention on Air Navigation was an attempt by Spain to create a treaty with Portugal and the Latin-American states to develop regulations similar to the Paris Convention but between the Iberian nations and South America. This desire to run a virtually parallel convention had come about as a result of Spain leaving the League of Nations after it failed to secure a permanent seat on the council. While in some respects it made sense culturally, politically and more importantly, practically, it was little more than a paper exercise. The Latin-American countries were closer to North America, so it made geopolitical sense for them to operate under Paris Convention rules. Meanwhile, the Iberian nations were hampered by the lack of access to suitably developed long-range aircraft to make the convention worth their while.

In 1928 the Sixth International Conference of American States, also known as the Havana Conference, took place in the Cuban capital Havana. Like the 1926 Ibero-American Convention on Air Navigation, it was designed to address various technical issues, including aviation law. Again, these mirrored the contents of the Paris Convention.

With legalization now in place, civil aviation could grow, commerce would expand and, most importantly, the aeroplane would take its place as the preferred method of travel.

Powering the Dream: The Great Leap Forward

Once the Armistice had been signed, and the armouries closed, post-war Europe had a surplus of aircraft, pilots and support staff. Moreover, the Treaty of Versailles ensured that any future industrial development by the Central Powers was curtailed, including the production of military aircraft. However, the experience the Germans, in particular, had gained from operating the mammoth Zeppelin airships meant they could develop a viable civilian aircraft capable of crossing the Atlantic. This air bridge was vital for civilian and mail passage, especially for those civilians prone to sea sickness. Moreover, the 48-hour flight was considered the height of luxury and certainly faster than the sea crossing, which could take between eight and seventeen days. Nevertheless, the airship's dominion was short-lived, partly due to the highly volatile hydrogen used to inflate gasbags and the development of aircraft capable of transatlantic flight in the period running up to the Second World War.

Initially, many surplus First World War aircraft were converted into short-haul passenger or cargo aircraft. Given the number of pilots available, aircraft use increased. The world of aviation was changing and growing, and those manufacturers who had survived the post-war contraction of aircraft manufacturing could easily exploit the market. Handley Page's twin-engine 'W' series aircraft, capable of carrying up to sixteen passengers, was based on the 0/400 bomber and 0/* series transports. The 'W' series would be joined in the skies by the Vickers Vimy Commercial, with both biplanes helping to grow Britain's commercial aircraft interests at a relatively cheap cost. However, both aircraft were quite clearly conversions of military aircraft, like the DH.4A, and these aircraft were dating rapidly. A new and bespoke passenger platform was required by operators and passengers alike.

The development of new types of civilian aircraft, more suited to safe, comfortable and reliable transportation of paying passengers and essential cargo such as mail, continued. Westland, which had gained considerable experience in aircraft design and manufacture during the First World War, produced the first domestically available passenger aircraft, the Limousine, in 1920. However, despite winning an Air Ministry (Great Britain) prize for design, the six-seat aeroplane saw minimal service.

Likewise, de Havilland began to develop their civil aeroplanes, producing the eight-seat DH.18 plywood-clad biplane in 1920. This was the start of a journey into aviation innovation that would become a de Havilland hallmark. The DH.18 provided de Havilland's engineers and designers with a great deal of experience, which they were able to develop. Next, de Havilland would create a thick-section cantilever-winged monoplane, the DH.29, followed by the nine-seat DH.34 biplane.

While these three aeroplanes were steps in the right direction, they lacked the capacity for longer-haul flights, especially those demanded by the burgeoning airline market. This approach, still seen today, saw the design process driven by the commercial interest of a third party, as much as technological and engineering advances. In this case it was the Imperial Airways desire to link Cairo to Delhi that saw de Havilland deliver the three-engine DH.66 biplane. The DH.66 was bigger than its predecessors, and was specifically designed to carry seven passengers and 13m² (459ft²) of space for air mail.

Not to be outdone, arms manufacturer Armstrong Whitworth entered the fray. Armstrong Whitworth had considerable engineering acumen, and had been at the cutting edge of engineering innovation since the 1850s. In 1926, Armstrong Whitworth introduced the Argosy biplane; like the DH.66, the Argosy was powered by three engines and was capable of carrying twenty passengers. Unlike the aeroplanes fielded by de Havilland, the Argosy was fabric covered, although the cabin floor was wooden, passengers were treated to a toilet and opening windows. Baggage was stored fore and aft of the passenger compartment, with the open cockpit situated behind the nose engine, like the DH.66. The Argosy was also the herald of a new concept, a luxury flight experience known as the Silver Wing, which provided patrons with in-flight refreshments and steward service.

In 1931, the serenely beautiful Handley Page H.P.42 took to the skies. The four-engine H.P.42 was the apogee of biplane design, with a wing area almost double that of the DH.66 and a range of 500 miles (804 kilometres). It featured an enclosed cockpit for the four crew; the twenty-four passengers were seated in a metal-bodied fuselage.

Though de Havilland had shown it could produce larger aeroplanes, its design ethos lent itself to creating a series of smaller multi-engine aircraft. Of these it was the DH.84, first flown in the autumn of 1932 that would become a veritable icon of this age. The twin-engine, six-seat biplane would be built in the United Kingdom and Australia, its art deco-inspired design gave it a light, airy cabin. The Gipsy Major engines, enclosed in sleek wing-mounted nacelles, hinted at smooth power and gave the DH.84 a range of some 400 miles (644 kilometres). De Havilland followed this with the four-engine DH.86, which looked very similar to its smaller predecessor. However, the DH.86 could carry up to sixteen passengers over 760 miles (1,223 kilometres). This made the DH.86 the ideal platform for carrying mail.

The de Havilland biplane designs reached their pinnacle with the DH.89 Dragon Rapide, an aeroplane of almost unrivalled beauty and elegance. It was the epitome of the pre-war de Havilland designs. Fast art deco lines merged with modernist functionality, which remained aesthetically pleasing. The eight passengers were treated to a light and airy cabin, now a feature in the DH.8*'s design, while pilots were appointed an almost-regal flight position, placing them in a wonderfully glazed cockpit, affording them a view any passenger would be envious of. Such was the soundness of the design that the DH.89 Dragon Rapide continues to fly, almost a century after its birth.

While an undoubtedly beautiful-looking aircraft, the biplane was virtually obsolete: the 1930s were the dawn of the monoplane. As such, these fantastic-looking airborne works of art had become, in terms of technology, yesterday's design. To compound the end of the biplane as the primary design arrangement for civilian aircraft, Armstrong Whitworth unveiled their AW.15, known as the Atalanta. The AW.15 was a high-wing four-engine aeroplane that was capable of going faster and farther than the H.P.42. This was the dawn of a global race, where manufacturers and designers worked feverishly to produce the winning design. The 1920s had been a period of civilianizing military designs and developing the ideas and technology to deliver reliable aircraft. The 1930s were very much about refining those ideas. It was a time when the giant leaps were starting. Designers and manufacturers were growing in confidence, pushing the envelope to develop supreme expressions of safe, reliable and stylish passage for travellers. The race was on, and Armstrong Whitworth was keen to exploit their lead.

The AW.15 was followed in 1938 by the AW.27 Ensign, an aeroplane whose potential was let down by poor reliability and overall performance, despite its 180mph cruising speed and 800-mile range. While some aspects of the AW.27 were far from perfect, the fourteen examples built provided sterling work during the Second World War, with some having their original Siddeley Tiger engines upgraded to the more powerful Wright Cyclone.

Armstrong Whitworth was not alone in their pursuit of monoplane perfection. Despite their DH.8* series biplanes gracing the skies, de Havilland was far from resting on its laurels. In 1937, the supremely handsome DH.91 Albatross took to the skies. Like Armstrong Whitworth, de Havilland had produced an aeroplane that was the next generation of both multi-engine flight and airliner. It was almost futuristic in appearance, and like the AW.27 Ensign, owed a great deal of its appearance to aircraft being developed overseas, especially in the United States. Initially designed as a mail carrier, the twenty-two-seat Albatross featured a plywood-laminated balsa fuselage that would be used again to construct the Mosquito fighter bomber. Power came from four air-cooled Gipsy Twelve engines, which gave the Albatross a cruising speed of 210mph (338km/h) and a range of over 1,000 miles (1,609 kilometres), but was capable of 3,000 miles (4,828 kilometres). Like the AW.27, the low-winged DH.91 was an aeroplane which showed that designers and engineers were very much thinking on the right lines. For de Havilland their final pre-war design, the DH.95 Flamingo, was perhaps the closest expression of a contemporary airliner as we recognize it today.

The DH.95 featured a wealth of modern equipment and build features, which put it in competition with the Douglas DC-3. The twin-engine Flamingo was, unlike the Albatross, an all-metal construction, aside from the control surfaces, with retractable undercarriage, and was capable of carrying twenty-two passengers and three crew. It took to the sky late in 1938, the Second World War intervening in what could well have been a severe commercial headache to the American market.

While British manufactures were hitting the ground running, especially in the immediate post-war years, as a result of the desire to bring communications with the Empire ever closer, American civil aviation development stalled. This was mainly due to government interference and its need, like the British Empire's, to develop a secure and reliable American airmail network. The commercial needs of the government and operators were reflected in subsequent aeroplane designs. This lent itself to the production of robust and reliable aeroplanes capable of withstanding many environmental changes that would be encountered across the continental United States. The subsequent mail planes were, like British aircraft, often repurposed military aircraft or imports from Europe, with Fokker designs proving popular.

The focus on producing mail planes in the immediate term was beneficial to domestic aircraft designers and manufacturers, with Boeing building their single-engine biplane, the Model 40. First flown in 1925, the Model 40 was similar in design to the early de Havilland DH.18, featuring a forward compartment capable of carrying 540kg of mail, along with two passengers. It could do so while cruising at 105mph (169km/h) and had a range of 650 miles (1,046 kilometres). The Model 40 would continue to be developed alongside new ventures including the elegant Model 80 Trimotor biplane, which entered service in 1928. The Model 80 could carry up to eighteen passengers over a distance of 460 miles (740 kilometres).

By the 1930s Boeing was beginning to develop their repertoire into the arena of transportation aircraft, a growing market with huge potential. This led to experimentation with the advent of the Model 200/221 Monomail in 1930. The Monomail introduced a range of innovations which included a departure from the biplane format, and the introduction of a low-set cantilever wing. In addition, the Monomail was fitted with a retractable undercarriage, powered by a single Pratt & Whitney Hornet enclosed in a low-drag cowling.

The was followed in 1933 by the Model 247, which became Boeing's first mass-produced all-metal airliner. The Model 247 was exceptionally handsome, sleek and somehow futuristic looking, adopting the now-familiar Boeing 'look'. The twelve-seat Model 247 was powered by two supercharged Pratt & Whitney S1D1 Wasps, giving it a combined power output of 1,100hp. Like the Monomail the Model 247 featured a retractable undercarriage as well as an enclosed cabin, autopilot, trim-tabs and de-icing equipment.

With the experience it had gained from developing its Model 247, as well as its large bomber aeroplane, the XB-15, Boeing unveiled its groundbreaking Model 307 Stratoliner on New Year's Eve 1938. The four-engine Stratoliner featured a pressurized cabin capable of carrying its thirty-eight passengers and five crew, including a flight engineer, a distance of 1,750 miles (2,816 kilometres) while cruising at 20,000 feet (6,096 metres) at a speed of 215mph (346km/h). Unfortunately, the start of the Second World War prevented the Stratoliner from fully realizing its commercial potential.

Alongside Boeing, other American manufacturers busy developing their mail and airliner aeroplanes, was Lockheed, who produced the shoulder-winged Vega in 1927. The single-engine Vega appeared like a child's drawing come to life; its fuselage could carry six passengers a distance of 725 miles (1,167 kilometres) at 165mph (266km/h). It became an extremely popular aeroplane and would fulfil various roles, including carrying Amelia Earhart safely over the Atlantic Ocean during her solo crossing.

During its production, it would see its plywood fuselage replaced by metal. It would also be the progenitor of the parasol-winged Air Express, an aircraft specifically designed for the Western Air Express company. It featured a larger wing mounted on a standard Vega fuselage. The Vega's popularity ensured Lockheed's future, and in 1931 the company produced the high-speed Model 9 Orion. This aeroplane was faster than military aircraft of the time. However, unlike the Vega, the Orion featured a low-mounted wing pulled by a single rotary engine capable of transporting six passengers at 220mph (354km/h) over distances of up to 650 miles (1,046 kilometres).

In 1932, another American concern stepped into the growing market of small airliner and executive aeroplane design: Beech Aircraft Company. They were swift in joining the ranks of Boeing, Douglas and Lockheed, and produced a small, six-seat, two-engine, low-winged monoplane, the Beech 18 in 1937. The Beech 18 would remain in production until 1970, with over 9,000 units built. This small and incredibly flexible aeroplane would go on to fulfil a range of roles from floatplane to ambulance, and in doing so become another icon of the skies.

In 1934, there was a significant change in US government safety standards regarding the manufacture and use of civilian aircraft, especially passenger aeroplanes mounting a single engine. The new rules dictated that single-engine aircraft could no longer carry passengers at night, or over hazardous terrain. In addition, the size of the continental United States significantly impacted design, industry and operators, especially as single-engine aircraft were cheaper to produce.

To add to Lockheed's woes, Boeing's Model 247 was making an impact on its market; in response Lockheed fielded the all-metal, twin-engine Model 10 Electra. The Electra was capable of carrying ten passengers over 800 miles (1,287 kilometres) at 176mph (283km/h). It would become a massive success for Lockheed and would see service globally. The Electra would go on to be the basis for two further models. The Model 12 Electra Junior first flew in 1936 and was slightly smaller than the Electra. The Electra Junior was intended to act as a feeder airliner, or as an early corporate or executive aeroplane, and had the same range as the Electra but was able to cruise and 210mph (338km/h). The final pre-war outing was the 1937 Model 14 Super Electra, which could carry fourteen passengers 850 miles (1,368 kilometres) at a cruise speed of 215mph (346km/h).

Despite the efforts of the major manufacturers, both in the United Kingdom and the United States, the Douglas Aircraft Company was about to raise the stakes considerably. The Douglas Commercial (DC) series aeroplanes would lead to the genesis of the DC-3, which would become the most mass-produced airliner ever built, a legend in every sense. The journey to the iconic DC-3 started in 1933 with the production of a low-winged monoplane, the DC-1. This twin-engine, all-metal aeroplane, capable of carrying twelve passengers and three crew, was built to a Trans World Airlines (TWA) request for an aircraft capable of using any airfield on TWA's routes.

Almost immediately it was recognized that Douglas had created an extremely flexible and capable aeroplane, a fact confirmed by tests over a six-month period. TWA asked for numerous changes, including more powerful engines and an increased passenger carriage of fourteen. The new aircraft would be called the DC-2, of which TWA ordered twenty. Like its forebear, the DC-2, launched in May 1934, was to prove a most revealing platform, the zenith of its accomplishments was taking part in the MacRobertson Trophy Air Race that October. This flight would take it from Great Britain to Australia (also known as the London-to-Melbourne Air Race).

On the back of this achievement the DC-2 started to attract commercial attention from other airlines in the growing market in domestic passenger flights. In response to a request from American Airlines, Douglas redeveloped the airframe into the broader and longer-bodied Douglas Sleeper Transport (DST). This aeroplane would replace older biplane types on transcontinental flights. The new fourteen-sleeping-berths aeroplane flew in December 1935 in its twenty-one-seat configuration. It was known as the DC-3.

The DC-3 soon established market dominance and by the end of 1937 the aircraft type accounted for ninety-five percent of all US airline traffic, as well as being adopted by thirty overseas airlines. By the time production of the DC-3 stopped in 1947 over 13,000, including the military C-47 version, had been built, which made the DC-3 the most numerous airliner ever built. Douglas continued to develop the mark into the DC-4 which featured a tricycle undercarriage, four engines, and could seat forty-four. The commencement and protraction of the Second World War prevented Douglas from fully exploiting the commercial viability of the DC-4.

Other American manufacturers also produced airliners, including the iconic Ford 4-AT Trimotor, a high-wing monoplane with a corrugated body and wing. The 4-AT Trimotor first flew in 1926 and was a development of the Stout 3-AT, the first all-metal-constructed American aeroplane. The 4-AT Trimotor could carry eleven passengers, with the later 5-AT version taking seventeen. It had a modest cruise speed of 107mph (172km/h) and a range of 570 miles (917 kilometres).

Curtiss produced a twin-engine biplane, the Condor II, which was a development of their military B-2 Condor bomber. The Condor II was designed as a passenger luxury-sleeper transport. Operating out of Dallas, these cumbersome-looking aeroplanes flew passengers to locations across the southern regions of the United States. Some Condor IIs found their way across the Atlantic flown by International Air Freight, a British company operating a trans-European service before the Second World War.

While America and Great Britain engaged in a race to perfect the airliner, European designers were busy perfecting their art too, led by Dutch aviator Antony Fokker. Fokker, whose designs had been used by the Imperial German Army Air Service (Luftstreitkräfte), initially had mastery over contemporary Allied aircraft, was continuously developing his designs, often at the expense of existing wartime contracts. Post-war, Fokker returned to the Netherlands and focused on designing civilian aeroplanes, his first being the high-winged, single-engine Fokker F.II in 1920. The F.II carried four passengers in an enclosed cabin and saw use by Dutch and German airlines. In 1924 he unveiled the F.VII. The design was essentially an enlarged F.II capable of carrying eight passengers and featured an enclosed cockpit.

Fokker added two extra engines to the design, placing one under each wing, retaining the single nose engine. The refined aeroplane was an instant success for Fokker, with global sales, licences awarded and production lines established in the United States and across Europe. This trimotor would form the basis of a series of models culminating in the imperious-looking twelve-seat F.XX of 1933. Fokker's last outing into civil aircraft was the four-engine, thirty-two-seat F.XXXVI high-wing monoplane in 1934. Unfortunately, the fabric-covered F.XXXVI was obsolete the moment it flew. As Douglas and de Havilland were making great strides in developing their multi-engine aeroplanes, Fokker called it a day on his civil designs, focusing on military aircraft.

The French continued their pre-war tradition of making elegant aeroplanes with Blériot unveiling their SPAD S.27 single-engine biplane in 1919. The S.27 was merely a development of the S.20 fighter with two passengers enclosed in the fuselage behind the pilot. On the flip-side Farman produced their wonderful art nouveau F.60 Goliath in 1923, a biplane, which could carry eighteen passengers in an enclosed cabin some 500 miles (800 kilometres).

This was joined in 1926 by the Lioré et Olivier LéO 21, a stunning twin-engine biplane that could take, like the F.60 Goliath, eighteen passengers.

Société des Avions Marcel Bloch, known simply as Bloch, had established itself as a producer of military aircraft. In 1930, Bloch began producing civilian aeroplanes, starting with the MB.60 shoulder-wing trimotor mail plane. This was followed in 1932 by the eleven-seat MB.120, designed for French-government work overseas. The MB.120 had a range of 620 miles (998 kilometres) and, like the MB.60, was a shoulder-wing trimotor design.

It was the MB.220 that broke the proverbial mould and mirrored the twin-engine, low-wing designs that the Boeing 247 had started. First flying in 1936, the all-metal MB.220 carried four crew and sixteen passengers. Bloch expanded on the theme of the MB.220, wheeling out the four-engine MB.161 in 1939. The twin-rudder MB.161 was capable of carrying thirty-three passengers 2,000 miles (3,218 kilometres) and was manned by a crew of five. The Second World War intervened in its development, but post-war saw the MB.161 wheeled out as the Sud-Est SE-161, where it flew with the French military and Air France until 1964.

The Treaty of Versailles was a hammer blow for the Germans, with their aviation industry primarily concerned with military aircraft. As a result, German designers had to return to the drawing board and consider their options. From these designs came icons with Junkers Flugzeug und Motorenwerke AG (Junkers) leading the way. Seeing the need to fill the gap in aviation production, especially with civil airliners, Junkers wasted no time in unveiling the low-wing F 13 monoplane, the world's first all-metal airliner.

The F 13 featured the corrugated finish, which became a defining feature of Junker designs. The four-seat, single-engine aircraft was a massive success for Junkers and was followed by more innovation, and in 1924 the fourteen-seat trimotor G.23/G.24 took flight. This aeroplane featured an enclosed cockpit capable of transporting passengers 410 miles (660 kilometres) in relative comfort. The G.23/G.24 established Junkers as the market leader in continental Europe with over 1,000 units produced over a four-year period, and with licensed production also taking place in Sweden.

Junkers continued developing the theme, and in 1929 unveiled the huge four-engine G.38, which was the largest land-based aeroplane at the time of its first flight. The G.38 was capable of carrying thirty passengers and seven crew, including mechanics who were able to access the engines in flight thanks to the large blended-wing body design, a feature later copied by Boeing. The G.38 had a range of just over 2,000 miles (3,219 kilometres), due to the design intended to be a transatlantic mail-carrying seaplane. However, Hugo Junkers changed the design to make the G.38 land-based.

In 1931 Junkers returned to a more conventional format and launched the Ju 52. The first three aeroplanes featured single engines. Subsequent aircraft received two wing-mounted engines, transforming them into the famous trimotor. The Ju 52/3m could carry seventeen passengers and two crew a distance of 620 miles (1,000 kilometres).

The other German manufacturer making airliners was Focke-Wulf who produced a prolific stream of aircraft, from seaplanes to sports aircraft. Its first commercial aircraft, the A.16, was launched in 1924. This unique-looking aircraft was a high-winged monoplane capable of carrying a pilot and four passengers. It was followed by various other small airliners and mail planes, including the seven-seat A.32 Bussard and the Fw 47 Höhengeier meteorological survey aeroplane in 1931.

Focke-Wulf's design team revelled in creating unusual-looking aircraft and were not afraid of experimentation, going on to design a series of autogyros. The Focke-Wulf design team led by Kurt Tank, a man of seemingly infinite creativity, designed their pre-war magnum opus in 1937, the Fw 200 Condor. Manned by a crew of five and capable of travelling 2,200 miles (3,541 kilometres) courtesy of its four engines, the low-wing forty-seat Condor was the last word in contemporary large-aeroplane design. Had the war not intervened, it is possible the Condor would have been able to establish a regular air link between Europe and the continental United States.

This period also witnessed a considerable amount of technical development work in aviation, with engineering refined and design perfected. Manufactures and enthusiasts were able to take advantage of many of the technological leaps that had occurred and apply these to various types of aeroplanes. These included the manufacture of smaller leisure aircraft as well as record-breaking aircraft and those that would symbolize the majesty of flight in almost unimaginably luxurious ways.

Another area most designers agreed was critical to the success of new aircraft types was the development of the radial or rotary engine. These engines kept maintenance costs to a minimum due to their lack of a cooling system – a boon if flying in an area without suitable maintenance facilities and which kept the prices down on the engines themselves. On civilian aircraft, designers also wanted powerplants that were reliable and safe; their primary concern at this stage was not speed. However steadfast the rotary engine was, its dominion would be challenged inexorably by the arrival of the turbojet. The turbojet was a child of the coming calamity of war that would shape and develop civilian aviation beyond the wildest dreams of those early pioneers.

Chapter 2

Beyond the Horizon: Pioneers and Adventurers

Pioneers of Civilian Aviation: Owners, Designers, Manufacturers

Over the past 100 years of civil aviation there have been hundreds of designers, manufacturers and pioneers, some of whom became household names. Many have long and illustrious histories, others are surprisingly short. Yet their brand names instantly conjure up air icons, not all of which were airliners. Within every company sat the quiet genius or flamboyant dreamer, desperate to make their dreams a reality.

For many aircraft manufacturers, the key objective in the years after the end of the First World War was to get a foothold in the highly lucrative post-war market. The First World War opened up the world of aviation. It heralded the dawn of regular passenger flights and services on a new and exciting scale, making the airliner the most lucrative product for many manufacturers. As a result, some invested their profits into the business to develop new and improved models that would woo the fledgling industry and passengers alike. This approach was essential, especially during the depression era of 1929 to 1933, as it ensured manufacturer and brand longevity.

One of the first British manufacturers was Short Brothers plc, formed in 1908 by brothers Eustace and Oswald. The Shorts initially specialized in the manufacture of observation balloons and were soon joined by brother Horace, a turbine engineer. After the First World War, which had seen Short aircraft fulfil a wide range of roles, they developed a range of civilian aircraft. The first was the Short Sporting Type, initially based on an RAF contract for training seaplanes. Short then unveiled the underpowered but groundbreaking S.1 Cockle all-metal monoplane in 1924. Made of corrosion-resistant duralumin, the single-seat Cockle was intended as a sports aircraft. However, while it was anything but sporty, it proved all-metal seaplanes could be built.

Shorts continued to develop their racing aircraft, and in 1927 unveiled the Short Crusader racing seaplane. The Crusader was designed to take part in the Schneider Trophy as a practice aircraft but was lost in an accident on its inaugural flight from Venice's famous Lagoon. By 1931 the Schneider Trophy was permanently in British hands, leaving Shorts to continue developing its seaplanes. Their most remarkable product was the 1937 Short Mayo Composite which saw an S.20 Mercury Seaplane mounted atop an S.21 flying boat. The intention was to provide a reliable aircraft capable of connecting continents.

By late 1938 the Short Mayo Composite had crossed the Atlantic between Ireland and Canada, followed by a momentous two-day flight from Scotland to South Africa. Unfortunately, the only example of the Short Mayo Composite was rendered obsolete by another Short aircraft, the S.26 G-class flying boat. Nevertheless, Short Brothers continued to develop their range of seaplanes and flying boats during the Second World War.

In 1948, Shorts unveiled their first post-war purpose-built aircraft, the SA.6 Sealand amphibian. This small aircraft was designed to take advantage of sheltered bays, lakes, rivers and prepared runways, proving popular with military and civil clients. It would be Short's last purpose-built seaplane.

The post-war years saw Short Brothers experiment with gliders; the S.343 and SA.5, and the first British fixed-wing vertical take-off/landing (VTOL) aircraft, the SC.1. The SC.1 was an experimentation aeroplane fitted with five Rolls-Royce RB108 turbojets, using four engines for vertical flight and one for forward flight. An innovative fly-by-wire system guided the SC.1 flight control surfaces. It marked the zenith of the Short brothers' technological developmental work. By the 1960s, they had established themselves as manufacturers of a range of transport and passenger aircraft. In 1989, Short Brothers was bought by Canadian firm Bombardier Aviation.

A year after the Short brothers formed their company, Glenn Curtiss and Augustus Herring established the Curtiss Aeroplane and Motor Company, initially known as the Herring-Curtiss Company. Curtiss quickly launched a range of groundbreaking aircraft which fulfilled an experimental and training role to help meet the interests of a growing market. These included monoplanes, biplanes, flying boats and racing biplanes such as the R3C, which won the Pulitzer Trophy in October 1925. The Pulitzer Trophy-winning seaplanes also won the Schneider Trophy in 1923 and 1925. Curtiss would continue to develop monoplanes, many with enclosed cabins.

Two critical developments by Curtiss were Autoplane and the Curtiss-Bleecker SX-5-1 helicopter. The Autoplane was the first attempt to design a practical flying car in 1917. In 1929 Curtiss teamed up with engineer Maitland Bleecker to create the Curtiss-Bleecker SX-5-1 helicopter. Bleecker's design, built by Curtiss, featured four oversized rotors, each powered by a propeller driven by a central driveshaft. The design failed and the SX-5-1 helicopter was consigned to history in 1933 as a result of uncontrollable vibration. At that point, Curtiss Aeroplane and Motor Company merged with eleven other companies, including Wright Aeronautical, which had its basis in the original Wright Company, to form the Curtiss-Wright Corporation.

In Italy, the Aeronautica Macchi was founded in 1912 by aeronautical engineer and entrepreneur Giulio Macchi. Initially, it built Nieuport monoplanes under licence for the Italian military during the First World War. However, in the post-war world of aviation adventure, Aeronautica Macchi lost no time putting its experiences to good use and focusing on seaplane design. The spur for Aeronautica Macchi was to win the Schneider Trophy with initial aeronautical designs by the visionary Mario Castoldi, based on single-seat fighter aircraft such as the Macchi M.7bis biplane seaplane. These beautiful aircraft created a visible chart of technical progress, often hampered by mechanical failings and bad luck. Yet they led to the first win for Macchi with the M.39 racing in November 1926. The M.39 would also set the world speed record for seaplanes later the same month, reaching 258.87mph (416.61km/h).

The M.39 would form the design for the M.52 and M.52R, which defended the previous year's victory. However, the three M.52s which entered the race in 1927 suffered from mechanical issues and could not complete the race. But all was not lost, and in March 1928, pilot Major Mario de Bernardi took the single M.52R out into the sky. The R version featured redesigned floats along with a smaller wing. It quickly returned the investment in its design and development by breaking the 300mph (500km/h) speed ceiling, with de Bernardi reaching a staggering 318.62mph (512.77km/h). By 1931 Aeronautica Macchi and, more importantly, Castoldi, had perfected their art and produced the M.C.72.

The M.C.72 was as sublime as it was aggressive. The red-bodied floatplane, powered by a Fiat AS.6 V24 supercharged engine, produced 3,100hp via the contra-rotating three-blade propellers. The elongated fuselage was a mix of metal and wood, with the pilot positioned aft of the trailing edge of the monoplane. Mechanical difficulties frustrated Aeronautica Macchi once again, and the M.C.72 missed what would be the final Schneider Trophy competition in 1931.

For the next three years the fleet of five M.C.72s was tweaked and developed under political pressure to prove that Italian aviation was as technologically capable as any other nation. The Italian government pushed the M.C.72 and its Fiat powerplant to produce results. On 23 October pilot Warrant Officer Francesco Agnello set the highest speed achieved by a piston-engine seaplane, a stunning 440.68mph (709.20km/h). After this the M.C.72 was stored, never to fly again.

Alongside the float and seaplanes, Aeronautica Macchi built the M.C.73 light touring and the M.16 sports biplanes, which got no further than the development phase. After the Second World War Aeronautica Macchi concentrated on light aeroplanes, producing a series of propeller and jet-propelled aircraft.

Supermarine Aviation Ltd, whose floatplanes would take and hold the Schneider Trophy for Great Britain, was formed in 1913. It had initially been called Pemberton-Billing Ltd, established by aviator and publisher Noel

Pemberton-Billing. During the First World War, Pemberton-Billing Ltd built their single-seat scout aeroplane for the Royal Navy Air Service, the P.B.25, and twelve Short S.38 seaplanes. This would give the company the experience it needed in the post-war years as it developed its racing seaplanes.

In 1916, Pemberton-Billing, a Member of Parliament, sold the company to his factory manager, and aircraft and boat designer, Hubert Scott-Paine, who invested in attracting the right talent to help create a successful company. A new chief designer, William Hargreaves, was appointed. The following year Reginald Mitchell was hired as Scott-Paine's assistant to help with technical matters relating to the company. This seemingly innocuous appointment would lead to the development of one of the most famous aircraft ever built, the Spitfire. That year Supermarine gained experience building float- and seaplanes as orders from Shorts Brothers and the short-lived Norman Thompson Flight Company were awarded to meet wartime demands. By the war's end, Supermarine had become an incredibly innovative company focused on flying boats.

Post-war saw Supermarine join the likes of Vickers and Handley Page and convert old war stock into passenger-carrying aircraft. In 1919, Mitchell replaced Hargreaves as chief designer. The following year he would also take on the role of chief engineer. Funded by commercial aircraft sales, Supermarine was able to develop their seaplanes and produce their first single-seat racing flying boat, the Sea Lion. The Sea Lion was a development of the wartime Supermarine Baby biplane, which was intended to take part in the Schneider Trophy that year; however, a collision with debris, or FOD – Foreign Object Damage, a perennial problem for aircraft – saw the Sea Lion sink at the event. Unperturbed, Supermarine continued to develop the aircraft and, in 1922 fielded the Sea Lion II, which won the trophy, thus preventing the Italians from being the early winners of the challenge. However, the luck was not to last, and Supermarine failed to retain their win the following year, leading to Mitchell's dramatic departure from Supermarine design dogma. In 1925 the S.4 monoplane was unveiled in a floatplane configuration. The S.4 was also lost before being able to participate in that year's Schneider Trophy. However, it had set a new world speed record of 226.75mph (364.9km/h) before its untimely demise.

Supermarine returned with the S.5, co-sponsored by the Air Ministry and featuring an all-metal fuselage and floats, unlike its predecessor. However, due to inexperience with metal working on the shop floor, the Supermarine entry did not participate in the competition until 1927, when it achieved first and second place finishes.

The following year, Vickers-Armstrongs purchased Supermarine. In doing so, Supermarine became Supermarine Aviation Works (Vickers) Ltd The union saw many cultural and organizational changes, not all of which were welcomed. However, what followed was the ultimate expression of Mitchell's design vision: the S.6B, powered by the Rolls-Royce R racing engine. This sleek-looking aircraft was the epitome of contemporary racing floatplane design; its pedigree was beyond doubt. Taking to the water in the September 1931 Schneider Trophy competition, the S.6B dominated the race. With it, Great Britain retained the Schneider Trophy for perpetuity.

The Schneider Trophy indeed dominated a great deal of Supermarine's time and resources. Supermarine continued to develop military seaplanes that no doubt benefitted from the company's Schneider Trophy experiences. However, by 1960 the company was little more than a shadow of its inter-war years and became part of the British Aircraft Corporation (BAC).

Società Idrovolanti Alta Italia (SIAI) Seaplane Company of Upper Italy, established in 1915, was another Italian company whose work focused on developing fast military seaplanes. Not long after its founding, the company merged with fellow aircraft company Società Anonima Costruzioni Aeronautiche Savoia. This was followed by a further name change in 1922 when the company became Savoia-Marchetti after the appointment of its long-serving chief designer Alessandro Marchetti. Marchetti was as flamboyant as he was unconventional in his approach to flying-boat design. In 1924, Savoia-Marchetti unveiled its unique, twin-hulled monoplane, the S.55. Despite its odd appearance it was a massive success for Savoia-Marchetti, with almost 250 units produced. The S.55 was also a prolific world-beater, with the prototype breaking no fewer than fourteen speed, altitude and payload world records, with the production models often carrying out staged crossings of the Atlantic.

As the company continued to build its knowledge and experience, it experimented with even more adventurous designs, and in 1928 the S.64 took to the skies. The aircraft featured a 21.5-metre (seven-foot six-inch) wingspan, designed for endurance flights. It proved its worth almost instantly and began accumulating world records. Of these, the most remarkable was the 58 hours thirty-four-minute endurance flight of Arturo Ferrarin and Carlo Del Prete.

Unlike Aeronautica Macchi, Savoia-Marchetti steered clear of the Schneider Trophy and focused on developing touring aircraft such as the SM.80 amphibian. This was the last of the civilian Savoia-Marchetti seaplanes, with the company focusing on designing and building large landplanes and military aircraft. Post-war, the company developed a series of small light aircraft. In 1983, Savoia-Marchetti was absorbed by helicopter specialists Agusta.

Meanwhile, Britain's de Havilland Aircraft Company Ltd was not resting on its laurels. Two years after the company's establishment, it fielded the DH.37, a three-seat sports biplane. The DH.37 was explicitly designed for aviator Alan Butler. He would join de Havilland as chairman a year later. While the DH.37 was a modest aeroplane, it pointed to de Havilland's corporate desire to furnish the sophisticated aviator with stylish and dependable aircraft. This approach reflected the company's founder and chief designer, Captain Geoffrey de Havilland's enthusiasm for flight. Innovation remained very much the watchword at de Havilland, so much so that the company established its technical school in 1928. The DH.37 was followed by the DH.60 series of 'Moth' biplane touring and training aeroplanes, which first flew in 1925 and led to the DH.82 Tiger Moth in 1931. The Moth series culminated in the closed-cabin DH.87 Hornet Moth biplane in 1934, an aeroplane which remains in use today.

While exploiting the possibilities of the biplane layout, de Havilland also experimented with monoplane designs; the DH.53 Humming Bird was its first serial production monoplane. This led to a series of sporting and touring monoplane designs, including the 1930 124mph (200km/h), high-wing DH.80 Puss Moth, which was the highest performing commercially available civil aircraft of the time.

For de Havilland, the next step was to produce a racing aeroplane capable of taking on the arduous 1934 11,300-mile (18,200-kilometre) England–Australia MacRobertson Trophy Air Race with a winning prize of £15,000. For this task de Havilland assigned their recently promoted chief designer Arthur Hagg to create the company's entry.

Hagg's background was in aircraft and nautical design. De Havilland's technical and engineering expertise backed his skills and imagination. Quickly built, the DH.88 Comet was ready for flight a mere six weeks before the start of the race, with three aircraft entered by third parties. The twin-engine, cantilever monoplane design featured stressed plywood skin, an enclosed cockpit for the pilot and navigator and a retractable undercarriage. Of the three aircraft, G-ACSS *Grosvenor House* would finish the race in 70 hours 55 minutes, making it the winner. Five DH.88s were constructed, with the type making a series of record-breaking, long-distance flights between 1934 and 1938.

After the war, de Havilland embraced jet technology, applying it to their latest aeroplane, the DH.106 Comet, which first flew in 1949. However, while the Comet was in the de Havilland spirit, it was also technically flawed, with three high-profile structural failures. By 1954 the Comet had been withdrawn from service to enable de Havilland and the Royal Aircraft Establishment (RAE) to investigate the issue. The failures were found to have been caused by the then little-known phenomenon of metal fatigue. Other factors included poor window-shape design and rivet use.

By the time the issues had been identified and solved, de Havilland's key competitors, Boeing and Douglas, had caught up technologically. As a result, they could apply de Havilland's lessons learned to their designs. But unfortunately, while de Havilland applied the lessons learned to later Comets, these were too late to prevent the company's demise. In 1960, it became part of the Hawker Siddeley Group, and by 1963 all reference to de Havilland had disappeared.

American Speed

Where European manufacturers were limited by space, thus necessitating the development of the seaplane, the Americans were not short of land over which to fly. By the early 1930s, the United States touring and sports landplane market was well established. A series of companies now focused on designing and producing these often-specialist aircraft; the Granville Brothers Aircraft (Gee Bee) were one such concern. Founded in Springfield, Massachusetts in 1929 by the five Granville brothers, the company initially produced a two-seat biplane, the Model A, which sat the occupants side by side. This design hinted at the unorthodox approach to aeroplane design that would become a Granville Brothers' hallmark. The following year the they unveiled the first of the Gee Bee Sportster series of racing aeroplanes designed to compete in the All-American Flying Derby. Toy-like, the low-winged monoplane Sportster offered an open cockpit in a short fuselage with just enough wing surfaces (8.8m² or 95ft²) to get airborne. Fitted with either in-line or radial engines, the Granville Brothers produced six versions, including the Model E, which was considered the definitive model of the type. One of the brothers, Zantford Granville, was killed when his Model E experienced an engine failure and he crashed the aeroplane trying to avoid airfield employees in early 1934.

That incident aside, the Sportster was a successful aircraft and took part in the 1930 All-American Flying Derby piloted by Lowell Bayles; it came second. The following year a redesigned example, the Model D, took part in the Cleveland Air Races. Bob Hall, who would excel in military aircraft design, won the Williams Trophy. In 1931 an enlarged two-seat Sportster was produced by the company known as the Model Y Senior Sportster.

Two Senior Sportsters were produced, with the first model taking part in races flown by Maud Tait, a winner of the 1931 Shell 3-Kilometer Speed Dash. The second aeroplane would be an engine test bed fitted with a powerful Wright Whirlwind rotary engine. The Whirlwind-powered aeroplane was flown by Florence Klingensmith, a founding member of the Ninety-Nines. Disaster struck when fabric detached from an upper right-wing surface in flight, causing the aeroplane to crash, killing Klingensmith instantly.

The next aeroplane delivered by Granville Brothers was the Model Z Super Sportster, designed to win the Thompson Trophy, a low-altitude, high speed, adrenaline-fuelled, closed-circuit challenge. Like previous Granville Brothers' aeroplanes, the Model Z featured a small wing surface and short fuselage with the pilot treated to an enclosed cockpit. Flown by Lowell Bayles, the Model Z won the Thompson Trophy in September 1931. However, while undoubtedly fast, the Model Z was inherently unstable and hard to fly. Yet it won the Thompson Trophy, the Goodyear Trophy and the General Tire and Rubber Trophy.

After the Trophy races, the Model Z was re-engined with a powerful 750hp (560kW) Wasp Senior radial in preparation for an attempt at the landplane world speed record. On 5 December 1931 Bayles took off, but wing failure saw the aircraft crash killing Bayles and earning the Granville Brothers an unenviable reputation for building aircraft that killed their pilots.

What followed was perhaps the ultimate creative expression of Gee Bee racers, the R-Series. The two R-Series aeroplanes featured a teardrop-shaped design by Stanford and Howard Miller. In addition, the cockpit was moved back to sit at the root of the tailfin to give the pilot a better view during racing. The R-1, the first built, was flown by Jimmy Doolittle and won the 1932 Thompson Trophy while simultaneously setting a world landplane speed record of 296mph (476km/h).

However, due to its design, the R-1 had a high stall speed of 110mph (117km/h), and in 1933 pilot Russell Boardman was killed shortly after take-off during a Bendix Trophy point-to-point race. A subsequent rebuild gained the R-1 an extra 460 millimetres (eighteen inches) in fuselage length, while the wings from R-2 were removed and added to the R-1. However, the aptly named R-1/2 crashed on landing not long after. In 1933, before effective repairs could be made, the Granville Brothers were declared bankrupt. Pilot Cecil Allen bought the aeroplane's wreckage, which he rebuilt as *Spirit of Right* with a new wing design and additional fuel tank, but a take-off accident in 1935 killed Allen, and the R-1/2 was never rebuilt.

The final Granville Brothers racer was the R-6 International Super Sportster, a more refined aeroplane for racing and touring requirements. The R-6 featured a tandem cockpit set in a larger fuselage based on the R-Series. This would allow the R-6 to compete in distance competitions, including the MacRobertson Trophy Air Race. The bankruptcy of Granville Brothers saw the unfinished project completed for Jacqueline Cochran, who would later lead the Women Airforce Service Pilots (WASP). However, the R-6's career was unsuccessful, and despite being entered for many distance challenges, she failed to complete a single one. In 1938, the R-6 was bought by aircraft dealer Charles Babb and entered in the Bendix Race, but due to engine failure, it failed to finish. The following year Mexican pilot Francisco Sarabia set a record in the R-6 after buying it from Babb in September 1938, flying from Mexico City to New York City in ten hours forty-seven minutes on 24 May. Sarabia was killed on his return journey after a rag, ingested by the engine's carburettor, led to engine failure on take-off.

Founded in the same year as the Granville Brothers, the Wedell-Williams

Air Service Corporation also produced racing aircraft. However, unlike the Granville Brothers, Wedell-Williams's designs were conventional in appearance and shaped by the experiences of the aviator and flying instructor James Wedell. During the First World War, Wedell, who had taught naval aviators, returned to aviation post-war and made money from rebuilding damaged aircraft, barnstorming and teaching. It was while teaching that he developed a firm friendship with millionaire Harry Williams. The two shared a love of aviation, and Wedell's experiences of barnstorming gave him an edge when he became involved in air racing.

Williams soon built his own factory in Patterson, Louisiana, specializing in low-wing monoplanes, and introduced the Wedell-Williams Model 22 racing aircraft not long after. The Model 22, initially known as 'We-Will Jnr', was built to compete in the 1930 All-American Flying Derby, which took the aviator from Buffalo, New York, to Detroit, Michigan. The Model 22 was the culmination of previous experiences from an aeroplane designed for the 1930 Miami Air Races.

Similar to Granville Brothers' aeroplanes, the Model 22 was designed to be as aerodynamically and mechanically efficient as possible, which was vital given the arduous flight from Buffalo to Detroit. Initially doing well, the Model 22, as was often the case with aircraft of the time, suffered mechanical issues and saw Wedell finish eighth. Another issue was the engine, the Cirrus Hi-Drive, which, although supercharged, remained underpowered. In 1932, Wedell significantly rebuilt and remodelled the Model 22, shortening the wings and fuselage while applying aerodynamic styling to the cockpit, wheel spats and engine cowling. At the same time the aeroplane was officially named the Model 22 and spent the summer competing in air races in the south. The following year the Model 22 received an engine upgrade in the form of a 6-cylinder Menasco B6 Buccaneer air-cooled engine, which was popular with the light aircraft manufacturers of the time. However, the performance of the B6 engine was far from what Wedell expected, and allied to a host of safety risks and poor performance, the Model 22 was sidelined by the Model 44.

The Model 44 was everything the Model 22 was not – aggressive looking, fast and, in the right hands, a record-breaking aeroplane. It was every inch the racer, yet its roots lay in the sister aeroplane of 'We-Will Jr', 'We-Will'. Fitted with a Pratt & Whitney R-1690-S1C3G, the Model 44 took on board lessons learned by Wedell-Williams, from building the 'We-Will Jr' and another aeroplane, the 'We-Winc'. In 1931, the low-winged monoplane Model 44 took to the skies and came second in the National Air Races.

With the experiences of the National Air Races behind him, Wedell began to refine the Model 44, also taking an order to build one for flamboyant aviator Roscoe Turner. Unfortunately, a test flight incident led to Turner's aeroplane losing a wing and crashing, necessitating a radical redesign driven by aeronautical engineer Howard Barlow. This remedial work gave the aeroplane a redesigned wing, which proved exceptional in handling, resulting in two other Model 44s receiving upgrades. The three remaining Model 44s were unstoppable at the United States air race meets, winning trophies galore. In the September 1933 Chicago International Air Race, Wedell set a new world speed record for landplanes at 305.33mph (491.38km/h)

Another 1930s racing circuit icon was the Hughes Aircraft Company H-1 Racer. Designed by Richard Palmer, the H-1 would be the last privately owned world record speed holder, flown by engineer and aviator Howard Hughes. The H-1 had its genesis on the set of the Hughes aviation blockbuster *Hell's Angels*. It was there that Hughes and head of aircraft maintenance Glenn Odekirk discussed the possibility of creating a record-breaking aeroplane.

In 1934, the design process was in full flow, with two sets of low-mounted monoplane wings produced, one set for racing and speed, the other for endurance flights. The aluminium skin was attached to the fuselage via flush-fit rivets, it had a fully enclosed cockpit and was fitted with a Pratt & Whitney R-1535 twin-row 14-cylinder radial engine tuned to produce 1,000hp. The H-1 also featured fully retractable land gear. On its test flight on 13 September 1935, the H-1 piloted by Hughes broke the world speed record for landplanes, recording an average of 352.39mph (567.12km/h) over four passes. Despite a crash-landing due to running out of fuel, Hughes was desperate to push the aeroplane's capabilities. After post-crash repairs, including the addition of longer wings, the H-1 broke the time record for the flight between Los Angeles and New York, with Hughes taking seven hours and twenty-eight minutes to cover the 2,473-mile (3,980-kilometre) route.

Rewarding Aviation Excellence

Behind the burgeoning aviation industry in the early twentieth century were legions of designers and technicians, enthusiasts and dreamers, pilots and pioneers, all waiting and willing to push the envelope. These men and women were spurred on by practical factors beyond the order book, including the financial rewards with prizes funded by newspapers and individuals. These prizes helped push designers, aviators and manufacturers towards more significant achievements in aviation. The accompanying publicity that these key achievements and innovations in flight attracted, especially in the early days, was invaluable. There was also the accolade of simply knowing that their adventures had made their mark in the annals of aviation history, establishing a legacy that remains.

The awards were often specific, targeting innovation in particular areas from seaplanes to gliders to airmanship; each was specific in its intent. They would capture the imagination of aviators and the general public, shining a light on what was happening in aviation. As was often the case, these prizes were incredibly generous, and awards came with wonderfully extravagant trophies that reflected the achievements they represented. The speed and endurance trophies gained the most fame and notoriety, pushing aviators and machines to the physical and mechanical limits, leading to aviation development.

The first major award to be rolled out was the Gordon Bennett Aviation Trophy, instigated by American publisher James Gordon Bennett Jnr in 1909. This was the third of three trophies established by Gordon Bennett and sat alongside his Automobile Cup, which ran from 1900 to 1905 and the ballooning cup, established in 1906 and remains a sought-after prize by aeronauts to this day.

The Aviation Trophy was awarded to those aviators who achieved the fastest time over a measured distance (time trial). Like the Schneider Trophy, each race was organized and hosted by the preceding year's winner's home nation. The similarities did not stop there; the country which won the trophy three times was awarded the Aviation Trophy for perpetuity. In 1920, French aviator Joseph Sadi-Lacointe did just that, covering the 186.5-mile (300-kilometre) course in one hour and six minutes, achieving an average speed of 168.73mph (271.55km/h).

Where the Gordon Bennett Aviation Trophy focused on landplanes and speed, the Coupe d'Aviation Maritime Jacques Schneider, or the Schneider Trophy, was awarded to seaplanes. First offered as a £1,000 prize by French aviator Jacques Schneider in 1912, along with an ornate trophy depicting a winged nymph kissing a reclining zephyr carried by the sea, the award would be hotly contested over the coming years. The main aim of the Schneider challenge was not the achievement of speed per se, although that would be valuable; his primary purpose was to stimulate aviation excellence. From the technological to the human, the point of the Schneider Trophy was to push the boundary of modern seaplane aviation to see technologies developed by competitors spill over into the civil aviation industry.

However, as the competition progressed, the focus of designers became speed and the national prestige that would come with having the fastest seaplane. Eventually, the main competitors were the Italians and British, each vying for the win. Regulated by the Fédération Aéronautique Internationale and the hosting nation's aero club, and watched by crowds, often over 200,000 strong, the aircraft taking part flew around a triangular course. The winner had to win three races in five years to retain the trophy. Aside from the understandable break during the First World War, the competition ran every year between 1919 and 1931, except for 1924. All competitors made serious technological advances throughout the competition, with many technical innovations making their way into the Second World War air forces. The imposing trophy now resides in the Science Museum in London.

At the same time, the British newspaper, the *Daily Mail*, established a series of contests and challenges started by proprietor Alfred Harmsworth. The *Daily Mail* would set specific goals for civil aviators to accomplish, with the first winning a cash reward. Of the early prizes, the most notable awards went to Louis Blériot for making the first cross-channel flight in 1909 and Alcock and Brown for their non-stop transatlantic flight in 1919. The prizes culminated in 1930 with Amy Johnson's successful solo flight from England to Australia, which saw her collect £10,000. In 1969, to celebrate the fiftieth anniversary of Alcock and Brown's crossing, the *Daily Mail* held a Transatlantic Air Race, attracting civilian and military aviators, including the niece of Alcock, Anne Alcock. The winner of the shortest time taken to make the flight was Squadron Leader Tom Lecky-Thompson, who flew in the new Hawker Siddeley Harrier, arriving in New York six hours and eleven minutes after leaving London. Lieutenant Commander Peter Goddard arrived in London five hours and eleven minutes after leaving New York. Both men received £6,000 prize money.

In 1922, George V of Great Britain instigated the King's Cup Race to stimulate development in the light aircraft and engine field. The cross-country race could only be flown by Commonwealth pilots and followed a return race between London and Glasgow broken by an overnight stay. Despite the king's wishes, the cup was never more than an annual sporting event in which entrants took part, more for the experience than anything else, with amateur pilots participating in standard aeroplanes. The racers were handicapped by a time penalty depending on the power of their aircraft. This gave the smaller aeroplanes a reasonable chance against the more powerful ones. It also gave the pilots a challenge to maintain their lead, thus improving the skills of aviators taking part.

A later race, the Jubilee Cup, was established in 1952; this later became the British Air Racing Championship in the 1980s. The championship sees sixteen handicapped races take place across eight venues in a season, with the winner being the pilot who wins the most points over the season. Each race takes part in a predetermined course, which is lapped four to five times and covers between 80 and 100 miles (129 and 161 kilometres).

Many awards were given to the numerous air races, and challenges held in the continental United States that took place under the banner of the National Air Races. These races attracted aviators and spectators in their droves, providing thrills and spills as pilots and machines were put through their paces. As well as the hair-raising air races, these events also included other aviation disciplines, such as glider demonstrations, airship flights and parachute-jumping contests.

The National Air Races were established in 1920 by publisher Ralph Pulitzer and featured a mix of both circuit and cross-country racing taking place at various sites across southern California. The races initially showcased machines flown by aviation legends such as Jimmy Doolittle and Roscoe Turner, with participants vying for the Pulitzer Trophy Race and the Pulitzer Speed Trophy awards. By 1929 the National Air Races had evolved into a ten-day event from late August to early September, managed by aviation promoter Cliff Henderson and attracting the cream of American aviation talent.

The Pulitzer family would also go on to sponsor the Pulitzer Trophy as a reward for the fastest to finish a predetermined course. The trophy itself was the epitome of art deco sculpture which reflected the adventurous spirit of the time that drove many aviators to the skies. Each pilot had to complete four laps of a predetermined thirty-two-mile (fifty-kilometre) course in the fastest time possible. From 1920 to 1925 the trophy attracted mainly military pilots flying DH.4s. In 1921, Bert Acosta, whose antics in the air often drew fines and suspensions, became the only civilian to win the trophy in his Curtiss CR-1 biplane. While the military pilots dominated the Pulitzer Trophy, the event saw the winning average speed increase from 156mph (251km/h) in 1920 to 248mph (399km/h) in 1925.

For many aspiring aviators, especially young women, the Women's Aerial Derby, known as the 'Powder Puff Derby', was started in 1929 with pioneers such as Louise Thaden and 'Pancho' Barnes. The inaugural race also saw the founding of the Ninety-Nines, a mutual-support group to promote aviation to women, a role it continues with to this day. However, while many women held pilot's licences, only forty had the experience to participate in the races.

The entry requirements were exacting, with entrants needing 100 hours of solo flight time, including 25 hours of solo cross-country flight of more than forty miles between points. All pilots also had to hold a valid flying licence issued by the Federation Aéronautique Internationale and an annual sporting licence issued by the contest committee of the National Aeronautics Association. On top of these requirements, all participants had to be able to carry a gallon of water and a three-day food supply. Of the forty who participated in the inaugural race, only twenty stepped forward to participate in the eight-day challenge, with fourteen finishing, inspiringly led by Louise Thaden. By the start of the Second World War women were regularly pitting their skills against men. The organizers' earlier discriminations were considered old-fashioned and fears of feminine frailty completely unfounded.

Post-Second World War, the jet age essentially pushed the piston-powered aeroplane of the National Air Races aside. However, all was not lost, and in September 1964, under the guiding hand of aviator Bill Stead, the Reno National Championship Air Races were instituted. The air races still take place with six classes open to pilots, including sports and biplane aerobatic classes.

Another great spectacle was the Cleveland National Air Show, founded in 1929 by local businessmen Louis Greve and Frederick Crawford. Like the National Air Races, the event was held over a ten-day period between 24 August and 2 September and allowed entrants to compete against one another in time trials and cross-country challenges. Similarly, the opening was extremely well supported, with over 100,000 members of the public watching the feats of daring-do take place in the Ohio sky.

Sponsored by Cleveland manufacturer Charles E. Thompson, the Cleveland National Air Show featured the endurance time trial Thompson Trophy which ran from 1929 to 1961, excluding the war years. The race was ten miles (sixteen kilometres) long, with a fifteen-metre (fifty-foot) tower marking the turns. The course aimed to encourage low-altitude flying and manoeuvrability at high speeds, with this excitement being the main draw for spectators. To add to the thrills, all aircraft taking part in the Thompson Cup took off at ten-second intervals, with a mere 100 feet (thirty metres) between them. As a result, the Thompson Cup became the deadliest air race of the time, with aircraft crashing all too frequently as competitors jostled for the lead, wingtip to wingtip. Yet the prize money, which fluctuated year on year, drew the competitors in, thus earning it the unofficial accolade of being the most prestigious race of the global aviation meets.

The period in which the Thompson Trophy was awarded was divided into two: the pre-Second-World-War period of 1929 to 1939 and the post-war trophy of 1946 to 1961. The post-war trophy was split into two classes, 'R' for piston-engine aeroplanes flown by civilians and 'J' for jet-powered military aircraft. The 'R' races were held between 1946 and 1949, and the 'J' races from 1951 to 1961, except for the 1952 and 1960 races which were cancelled. The Thompson Trophy also contained a clause excluding women pilots from participating in any of the races it hosted.

In 1931 the Bendix Trophy was patronized by industrialist Vincent Bendix after a suggestion by Cliff Henderson, the managing director of the US National Air Races, for a new cross-country air race. In its pre-Second World War iteration, the Bendix Trophy gained a reputation as being a pioneering challenge, attracting the most daring of aviators. Initially, women were excluded, partly due to the death of racer Florence Klingensmith in 1934, but in 1935 the ban was dropped, and women were able to participate. This opened the door to many of America's talented female pilots who were keen to participate. Like the Thomson Trophy, the Bendix Trophy ceased during the war years, restarting in 1946 with the piston and jet classes, with the same distinction between pilots. By 1962 the last cross-country race had been held, and the trophy wound up. It reappeared in 1998 as the Honeywell-Bendix Trophy for Aviation Safety until its demise in 2011.

Air racing continued to be a great draw. However, the jet age presented civilian and amateur aviators with a considerable challenge, not helped by the military fielding increasingly powerful aeroplanes, thus dominating the field.

This led to the development of Formula One Air Racing. Not only did this new racing class reinvigorate the post-war racing scene, it also challenged aviators' technical and mechanical skills. However, aircraft couldn't be powered by engines greater than 190 cubic inches in capacity. Even so, the tiny air racers were still capable of reaching speeds of up to 200mph (322km/h). The first race was held in 1947, with competitors flying a 3.19-mile (5.13-kilometre) oval course in a range of agile monoplane aeroplanes. Now in its sixth decade, Formula One Air Racing continues to draw crowds and competitors.

For the twenty-first-century European community, Red Bull Air Racing offers pilots and spectators an aerial display that captures the feel of contemporary extreme sports. Participants fly a low-level 2,000-metre course which runs close and parallel to the spectators while performing a series of aerobatic manoeuvres. The course is not for the faint-hearted, and the aviators taking part are at the top of their game. The races take place annually, usually with three events per year, with the final race taking place at Reno and the competitors taking part in the Red Bull Air Race World Series Championship.

The sponsors cancelled the Red Bull series of races in response to the combined obstacles of the 2020 COVID pandemic and dwindling corporate interest. However, Red Bull's replacement, the Air Race World Championship, will lead to a new age of air racing. Air Race will build on the aviation legacy of the famous Red Bull races. It will also focus on technological advances in green aviation and urban air mobility, ensuring air racing is relevant in the twenty-first century.

Across-the-world Challenges

While short-distance air races offered aviators and spectators the thrills, there remained other challenges, beyond the cross-country events, that presented the aviator with the possibility of far greater laurels. These were offered by various organizations that sought to push civil aviation's technical and endurance envelope. Alongside Great Britain's *Daily Mail*, which offered a wide range of prizes, other organizations and individuals offered the enterprising aviator a moment of fame and immortality.

The England to Australia Race of 1919 was the brainchild of the then Australian Prime Minister Billy Hughes to help promote aviation and show that Australia was no longer a distant outpost of the Empire. The rules were simple: the first Australians to fly from London to Darwin in a British-built plane in thirty days or less would win the £10,000 prize. Six Australian and one French team took up the challenge; two crews died, and two failed to reach Darwin due to crashes. However, two men were more than capable of taking on the challenge, the Macpherson-Smith brothers: Ross had flown as

T. E. Lawrence's pilot in the Middle East, with Keith also serving as a pilot during the First World War.

Along with their support crew, Sergeants James Mallett-Bennett and Walter Shiers, the four men took off from Hounslow Heath Aerodrome, London, on 12 November 1919 in their Vickers Vimy. Over the next twenty-eight days, the men flew for 135 hours fifty-five minutes, covering approximately 11,123 miles (17,901 kilometres) before making landfall on 10 December 1919. The Macpherson-Smith brothers were made knights on landing, Mallett-Bennett and Shiers were given their commissions and the £10,000 prize was split between the four men.

The Ford Company started the Ford National Reliability Air Tour. This event was held between 1925 and 1931 offering aviators and manufacturers the opportunity to prove their mettle over a 1,900-mile (3,058-kilometre) course, which grew over time, with stops in ten cities over several days. Over six years the event saw an increasing variety of aircraft taking part, with a Pitcairn PCA-2 autogyro participating in the 1930 event. In 2003 a re-enactment event was held; it visited twenty-seven cities, covered 4,000 miles (6,437 kilometres) and featured over thirty vintage aircraft.

In 1934 the MacRobertson Trophy Air Race sought to re-enact the 1919 Great Air Race, instituted by the Lord Mayor of Melbourne, Sir Harold Gengoult Smith, with the £15,000 prize donated by confectioner Sir Macpherson Robertson. Race competitors would take off from RAF Mildenhall, Great Britain, landing at Flemington Racecourse, Melbourne, after travelling approximately 11,300 miles (18,200 kilometres). Unlike the 1919 event, the participants' aircraft had evolved to almost fictional heights of technical development. The rules allowed any aircraft, regardless of size or power, with no limits on crew size, to take part. There were strict instructions which forbade the picking up of a pilot once an aircraft had left the start line, and there were also rules on what technical and survival equipment was carried. At dawn on 20 October 1934, watched by a 60,000-strong crowd, the twenty competitors took off. Just three days later, in a time of 71 hours, the DH.88 Comet *Grosvenor House*, flown by Flight Lieutenant Charles Scott and Captain Tom Black, landed at Flemington Racecourse, winning the race.

Scott would notch up yet another win flying a Percival Vega Gull in the Schlesinger African Air Race of 1936. Here, the challenge set by South African tycoon Isidore Schlesinger was to fly from Portsmouth, Great Britain, to Johannesburg, South Africa. Schlesinger offered a prize of £10,000 which was split between two races, a speed and a handicap. Not only was the race influenced by the MacRobertson Trophy Air Race, but Schlesinger felt it was an excellent vehicle to promote the Empire Exhibition, a celebration of the founding of Johannesburg. Entry was open to aviators and aircraft of the British Empire only, which narrowed the field, and due to a range of circumstances, of the fourteen teams taking part, only Scott and his co-pilot Giles Guthrie completed the race. Taking off from Portsmouth Aerodrome on 29 September, the two men arrived at Rand Airport on 1 October, the flight taking a total of fifty-two hours and fifty-six minutes.

Despite efforts by promoters, little interest was shown in the pre-war challenges after the Second World War, and in 1953 the last great air race took place between London Heathrow and Christchurch, New Zealand. The 12,300-mile (19,800-kilometre) race between the two locations was split into classes: speed for military pilots and commercial for civilian aircraft. The speed class attracted a £10,000 prize and the Harewood Gold Cup. This prize was won by an RAF English Electric Canberra flown by Flight Lieutenant Roland Burton and navigated by Flight Lieutenant Don Gannon in a time of twenty-three hours fifty-one minutes, which still stands. The commercial class (handicap) was won by a KLM DC-6 flown by a Captain Kooper, completing the journey in forty-four hours and twenty-nine minutes.

In the wake of the Second World War came a raft of technological advances and social changes that changed the world forever. The once-remarkable feats of airmanship and engineering prowess that had wooed the masses in those often-heady inter-war years no longer sparked the public imagination as they once did. Transoceanic and -continental flights had become commonplace, and world records were being broken daily with the advent of the jet engine. With its arrival, the exclusive glamour of piston flight was gradually, but not wholly, replaced by the heat of the massed and hurried jet set.

Chapter 3

In for the Long Haul: Opening the Passages to the Masses

The Foundations of Civil Aviation are Laid; an Industry is Born

As of April 2020, the United Nations International Civil Aviation Organization (ICAO) had over 5,000 airlines on its registers. Of these, almost 800 were international carriers with access to 41,700 airports worldwide, of which 13,000 were based in the continental United States alone. By the end of 2019, ICAO figures showed that some 4.7 billion airline passengers had been flown by operators, a steadily growing figure, which dropped by almost three billion, understandably, by the time of the Coronavirus pandemic of 2020. For airlines, the 2019 passenger marketplace was Asia-Pacific-centric, with the region holding a staggering 34.7 percent share of the air passenger market, followed by Europe with 26.8 percent and North America with 22.3 percent. Freight figures, which had been steadily growing, were beginning to plateau in 2019 due to various economic and social trends. Despite this, the World Bank showed that by the end of 2019 operators had carried 221 million tonnes of airfreight in that year alone, globally. All of these passengers and freight had been taken by 39,203,774 registered carrier departures.

These figures resulted from foundations laid a century earlier, in 1919, with the signing of the Paris Convention. From the beginning of regulated flight the development of large aircraft, allied to the carriage of cargo and passengers, was seen as the key to making civil aviation accessible and sustainable by designers, owners and financiers. It was a truism that the larger the aircraft, the more passengers and freight could be carried. So, it stood to reason that the more passengers and cargo flew, the greater the revenue, the greater the income, the larger the investment in development, the larger the aircraft.

There was also another factor: distance. Passengers wanted to visit exotic locations, so aircraft had to be big enough to carry the necessary fuel and be powered by reliable engines capable of long hours of operation. Here designers could look at the progress already made by the long-range dirigibles of Frenchman Henri Giffard and the rigid airships of Count Ferdinand von Zeppelin.

The Ukrainian Igor Sikorsky was the first designer to recognize the possibilities of long-range flight using bigger aircraft, outside the field of dirigible design. His first large aircraft, the S-21, was a quantum leap forward in design a mere decade after the Wright Brothers' first flight. Unfortunately, the First World War and the Russian Revolution stopped further development, but Sikorsky's designs proved larger multi-engine aircraft could fly. Indeed, the development of the concept, allied to the long-distance flight of types such as the Vickers Vimy, was proof-positive that the idea had plenty of room for growth.

Hastened by the newly enacted Paris Convention, the imaginations of designers and entrepreneurs were sparked. Their enthusiasm was fuelled by the achievement of those early transatlantic flights. These flights opened the far-flung corners of the globe for passengers to enjoy new sights hitherto restricted to film, postcards and the picture press.

By the mid-twentieth century, the fragile aircraft of that pioneering generation had been replaced by veritable intercontinental jet-powered behemoths of the sky. Aircraft were filled with passenger numbers that pre-war aircraft designers and operators could only have dreamt of. This new era of flight fed the burgeoning package holiday boom where no destination was beyond reach and, for the airlines, the opportunity to shrink the world further still.

The Empire that Never Slept: Early British Civil Aviation

From the moment the Montgolfier brothers proved that flight was possible on a regular, almost industrial scale, there have been those who sought to conquer the sky one way or another. From the start of regulated civil aviation, as the 1919 Paris Convention recognized, some could see flight's potential uses and advantages beyond speed and excitement. The many prizes on offer in the early days were proof that men and women were willing to invest their fortunes and, more importantly, their reputations, in aviation; such was the strength of belief that flight was predominantly a force for the betterment of humanity.

In the post-Paris Convention world, there was a hint of things to come: the adventurous were quick to take the first steps towards establishing a more formalized approach to flight by starting airlines. Surplus aircraft from the First World War, sold at a fraction of the cost they were bought for, played a huge part in opening the skies, especially in Great Britain. Here an entrepreneur could purchase an Avro 504K training aircraft for a mere £40 from a private sale in those immediate post-war years. The two-seater Avro offered the enterprising investor, keen to promote 'airmindedness' among the masses, with several options, especially if an entire fleet of aircraft could be purchased. Recently demobbed pilots started flying schools, with others offering flight experiences. However, those who provided scheduled flights and the opportunity to travel were perhaps the best prepared for the many changes that would take place over the next twenty years. One can never be sure whether it was foresight or plumb luck, but the men and women who chose to develop the field of commercial air travel struck it rich – most of the time.

One of the first airlines to appear in Europe during the First World War was Aircraft Transport and Travel (AT&T), established in 1916 by the Aircraft Manufacturing Company Limited (Airco). Despite its early founding, AT&T had to wait until August 1919 before it could operate from the aerodrome at Hounslow Heath, near the present-day Heathrow; the aerodrome even came complete with a customs tent. On 25 August 1919 AT&T made their first flights from their Hounslow airfield, bound for Paris. The first flight was the carriage of one passenger in a DH.14 biplane, flown by Bill Lawford; this was the first scheduled international flight between Great Britain and France. This flight was followed later in the day by Major Cyril Paterson and four passengers in an Airco de Havilland 16 biplane designed by Geoffrey de Havilland. Again, their flight would take them from London to Paris.

De Havilland was an aviation pioneer in every sense of the word. He had honed his skills in the automotive world before establishing himself as an aviation entrepreneur in 1909. He understood what an aircraft needed to be. Drawing on his engineering and flying experiences, he created a series of aircraft, designing for the Royal Aircraft Factory before moving on to Airco. He finally set up de Havilland Aircraft Company Ltd on 26 September 1920 with the help of his former boss at Airco, the industrialist and fellow aviation pioneer George Holt-Thomas. De Havilland would create an enviable portfolio of aircraft for the civil aviation market, including the Moth and Dragon Rapide series of biplanes. Another repurposed design was the bulbous and slightly comical-looking Vickers Vimy Commercial. This ten-seat biplane was a popular aeroplane and was also operated by French rivals Grands Express Aériens.

While Lawford had flown the first scheduled international passenger and cargo flight, the laurels of the inaugural international civil flight with passengers was undertaken by Handley Page Transport (HPT). HPT had been formed as an airliner earlier that June and started operations in India as the Handley Page Indo-Burmese Transport. On the same day as the AT&T flights, HTP flew a group of journalists to France from Hounslow Heath before Lawford's AT&T flight took off. HTP would not establish a regular flying schedule to France until September. The race to provide services to the paying passenger was on. Not one to be left behind, the following year saw AT&T link up with Dutch airline KLM to provide a service between London and Amsterdam on 17 May 1920.

Two other airlines, quickly founded in the wake of the Paris Convention, were Daimler Airway, part of the Birmingham Small Arms Company (BSA) and Instone Air Line (IAL), founded by Sir Samuel Instone. Instone was one of the first men to become an aviation entrepreneur and innovator. He introduced uniforms for his flight crews and made the first telephone call in a flying aircraft.

By the end of 1920, the growth of domestic airlines in Great Britain was faltering as a result of French airlines taking advantage of the freedoms that the 1919 Paris Convention had awarded them. British airlines were competing against one another and against subsidized continental companies for a small market. On 28 February 1921 all four British airlines ceased to trade due to a lack of finance, clearly indicating air travel was going to be expensive, a constraint that yolks the industry to this day.

This collapse led to the British government granting subsidies to HTP and IAL, with service resuming in late March. By the end of the 1922 financial year, cross-channel flights had carried 11,042 passengers, with British airlines taking 5,692 alone. The following year, further pressure was applied to HTP and IAL, with Daimler Airway beginning their scheduled, subsidized cross-channel flights in April 1922. IAL started a London–Brussels route in May, clearly sensing a crowded field. Even this didn't help, and by October revised routes were put into operation for each airline, with Daimler flying to Amsterdam via Manchester and London, IAL flying the London–Brussels–Cologne route and HPT maintaining their London–Paris passage.

This arrangement worked well and continued to inspire others to establish airlines. In August 1923 the British Marine Air Navigation Company Ltd was founded, a joint venture between Supermarine Aircraft and Southern Railways, who owned Southampton Docks. The docks were equipped with customs and immigration facilities. They would serve as the departure point for the first scheduled flying boat services to and from the Channel Islands and France, with flights aboard a Supermarine Sea Eagle starting on 14 August, a good month before HTP, who began their London–Paris–Basle–Zürich service.

Despite this early innovation, the British government remained dissatisfied with the performance of the subsidized airlines. So, on 31 March 1924, it merged British Marine Air Navigation Co. Ltd, HTP, IAL and Daimler Airway, to form Imperial Airways. Although a pilots' strike prevented any services until 26 April, Imperial Airways Ltd (IAL) could field a good fleet of aircraft, mostly from Airco, Handley Page and de Havilland. The fleet incorporated civilianized aeroplanes such as the DH.4A and the larger Handley Page W8, based on the Type 0/400 bomber capable of carrying fifteen passengers. It also fielded the Geoffrey de Havilland-designed civil biplanes, the DH.16 and DH.34.

The advent of IAL heralded a new dawn in British civil aviation, focusing on providing a long-haul service to all areas of the British Empire. IAL enabled business and civil service, alongside the fortunate adventurers' and tourists' experiences. As well as recognizing the advantages that air travel had over the longer sea voyages, IAL was quick to capitalize on the potential aircraft offered in establishing often hazardous routes.

IAL followed the European model of the patronage of national airlines, where the government provided air travel subsidies paid to the airline provided it met its mileage quotas. In making Croydon Airport the centre of its operations, IAL set about providing only scheduled international flights. As a result, all domestic flights were cancelled, except for charter flights. The international flights that connected mainland Britain and the continent were the first to be strengthened, and by 1924 scheduled flights to major cities in Central Europe were taking place.

In November 1924, IAL took delivery of its first new airliner, the Handley Page W8f biplane. The following year IAL began providing in-flight entertainment for passengers and became the first airline to show an in-flight film, a staple of long-haul flights in contemporary civil aviation.

However, Empire called, and between November 1925 and March 1926, aviator Sir Alan Cobham charted the London–Cape Town route for IAL. Cobham had learned to fly during the First World War; his first flying role had been as a test pilot for de Havilland and then first pilot for the de Havilland Aeroplane Hire Service. Flying a de Havilland DH.50 floatplane, Cobham set off on his adventure, passing through no fewer than twenty-eight locations, including Athens and Khartoum. No sooner had he returned than he was preparing to set forth again, testing the UK-to-Australia route, a journey of 32,000 miles (52,000 kilometres). Leaving Rochester, Kent, on 30 June 1926, he arrived in Melbourne on 15 August. After resting for a few days, Cobham traced his route back from Melbourne, starting on 26 August and arriving, somewhat fittingly, at the Palace of Westminster on the River Thames on 1 October.

These treks south were followed by a further route-proofing to the jewel of the British Empire, India. On 27 December, a DH.66 Hercules seven-seat biplane left Croydon bound for Delhi, arriving safely on 8 January 1927. On the return flight, the aeroplane was directed home via Cairo, another crucial future route, to ensure this was feasible. 1927 would be a pivotal year for IAL as they established Middle Eastern flights between Cairo and Baghdad, replacing the previous RAF mail route.

Further developments were made in securing a London–Karachi flight, which took place on 30 March 1929. This route would take British aircraft over Persia; the overflight permission required for establishing the air route would take diplomats two years to negotiate. This alone demonstrated how unfit the Paris Convention had become since its original drafting a decade before.

By the end of the year, the route was extended to stop at Delhi. However, the seven-day trip was punctuated by a train journey. Despite route changes there would be a rail link between Basle and Genoa for the next few years as aeronautical technology sought to catch up, to allow aircraft to safely fly over the alps.

The 1930s saw further developments for IAL and, in April 1931, the airline started with an airmail route between London and Sydney via the Dutch East Indies, taking a mammoth twenty-six days. Despite this, passenger flights started the following year on 1 October using the recently established airport in the Trucial State of Sharjah, now part of the United Arab Emirates.

The air routes to the east of the Empire both near and far, were vital to IAL for its commercial survival and for maintaining communications. On 28 February 1931 a route between London and Mwanza on Lake Victoria in Tanganyika began. The route was extended to Cape Town in early December as a temporary measure for airmail delivery, followed by an airmail-only service from London to Cape Town starting on 20 January 1932, before expanding to passenger service on 27 April.

As these new routes opened, IAL introduced the imposing twenty-four-seat Handley Page H.P.42 quad-engine biplanes. Yet, within a year it would also introduce the high-wing Armstrong Whitworth Atalanta. The Atalanta was a purpose-built aeroplane for IAL capable of carrying up to seventeen passengers and would prove to be a successful workhorse able to fly 640 miles (1,030 kilometres), which was an improvement over the H.P.42's 500-mile (800-kilometre) range.

One of the first tasks of the Atalanta was to refine the long London–Brisbane route, and on 29 May 1933, IAL began to plot a new route with Herbert Brackley, the airline's air superintendent in charge of the flight. Brackley was an experienced pilot who had personally proofed many of the routes flown by IAL. The Atalanta arrived at Brisbane on 23 June after taking a mammoth thirty-one-stop flight path, which saw it traverse the critical points of the British Empire.

At the same time, the London–Calcutta route was started on 1 July followed on 23 September by the start of the London–Rangoon flight. Finally, IAL established their London–Singapore service on 9 December, to finish the year. A little under a year later, the London–Brisbane route, with Qantas covering the Singapore–Brisbane route, opened up, initially only taking mail – passengers would follow from April 1934.

Queensland and Northern Territory Aerial Services (Qantas) was established by Wilmot Fysh and Paul McGinness on 16 November 1920, keen to put their experiences, friendship and shared interests to good use. Fysh and McGinness had flown together in the Palestinian Campaign during the First World War and were joined by fellow veteran Fergus McMaster. By 1930 Qantas, which had moved location three times, finally settled in Brisbane, Queensland, where it continues to maintain a hub. Qantas joined with IAL in December 1934 to form Qantas Empire Airways Limited (QEA) and continued to service the Singapore–Brisbane route.

In addition to QEA, IAL also established the Railway Air Services (RAS) with the Big Four railway companies of the United Kingdom in March 1934: London Midland & Scottish (LMS), London & North Eastern Railway (LNER), Great Western Railway (GWR) and Southern Railway (SR). The RAS was a domestic service that operated within mainland Britain connecting Croydon to Glasgow via Birmingham, Manchester/Liverpool and Belfast. This route enabled passengers to connect directly with IAL's international flights from Croydon. A further spoke to the hub was added when a service was established at Cardiff Municipal Airport, which covered the southern and eastern areas of mainland Britain, including flights to Plymouth and Liverpool. IAL's eastern expansion continued, and 14 March 1936 saw the beginning of the London–Hong Kong route, barely a month after establishing the transAfrica route, which would link Khartoum to Lagos by the end of the year.

By the late 1930s the Short Empire had arrived at Southampton's IAL hangars and moorings. The desire to tap into the air tourist market was growing, and for many, flying boats were the epitome of pre-war elegance. Flying routes to South Africa, the Far East, Australia and India, the twenty-four-passenger Empire was hugely successful. By March 1939, there were three scheduled flights to Sydney from Southampton and three flights per week to Durban. Flights took ten and six days, respectively.

New air routes were traced out across the Atlantic and to New Zealand, but world and domestic political events overtook IAL plans, and these routes would not be established for the airline. By the start of the Second World War sentimentality has given way to pragmatism and the British government established the Air Navigation (Restriction in Time of War) Order 1939 on 1 September, effectively grounding all civil aviation activities without a permit or good reason. The government gathered aircraft and administration from IAL and the recently formed British Airways Limited (BAL) at Whitchurch Airport in Bristol. There they were administered by the National Air Communications, a department of the Air Ministry. 1 April 1940 saw IAL and BAL merge to form a new company, the British Overseas Airways Corporation (BOAC).

British Airways Ltd (BAL) was a latecomer to the inter-war British civil aviation industry, formed in September 1935 by the merger of three separate airlines. The first was Spartan Airlines (SAL), founded as a subsidiary of the Spartan Aircraft Company to showcase its Cruiser ten-seat Trimotor monoplane. SAL flew short-haul flights between the UK mainland and the isles of Wright and Man. The second airline was United Airlines Ltd (UK) (UALUK), a merger of Jersey Airways, Guernsey Airways and Highland

Airways in April 1935. It also flew the air services of three of the four major UK rail companies of the time, which had formed the Channel Islands Airways Ltd UALUK began a series of scheduled short-haul flights covering the north of the United Kingdom. As well as Spartan Cruisers, UALUK also flew DH.89 Rapides. The final airline to join BAL was Hillmans Airways, established in November 1931, that provided services between Romford and Paris as well as scheduled domestic and airmail flights. Hillmans operated the high-winged DH.80A Puss Moth monoplane and the DH.83 Fox Moth biplane to provide flights between Romford and Clacton.

BAL had initially been called Allied British Airways Ltd and by December 1935 it was a public company. As it was publicly owned it was able to attract investment to help maintain the thirty-seven aircraft it held. Over the next year, BAL slowly refined its fleet and services, including a merger with transcontinental short-haul airline British Continental Airways (BCA) on 1 August 1936. BAL operated from Croydon, transporting passengers to locations throughout Northern Europe and Scandinavia.

BAL moved its airport hub to help streamline operations to Gatwick Airport in West Sussex in May 1936. BAL continued to refine and update its fleet to the modern mono-wing designs, buying Fokker and Junkers trimotors. In March 1937 the Lockheed Model 10 Electra joined the BAL fleet; in September 1938 the bigger Lockheed Model 14 Super Electras were also added.

As well as fleet growth, BAL continued to exercise its route influence. On 12 August 1937, it gained a 50 percent share of the newly formed Scottish Airways Ltd. Growth brought about by new airmail contracts, combined with concerns over poor runway maintenance and an increase in local aerial congestion saw BAL move its services, aside from night mail flights, from Croydon, to Heston Aerodrome. It was at Heston, at the steps of a BAL Lockheed 14 on 30 September 1938, that Prime Minister Neville Chamberlain announced that he had achieved 'peace for our time' with the signing of the Munich Agreement.

The following year BAL continued to grow its continental network with flights to Lisbon, Warsaw, Berlin and Budapest established in 1939. However, the outbreak of war on 1 September 1939 saw BAL merge with IAL to form BOAC.

European Expansion

Like the British, the French quickly established post-war airlines and air services to connect with its colonies in the Far East and the Caribbean. The first aeroplane to be used for scheduled international flights was a Farman Goliath biplane, carrying twelve passengers between Paris and Brussels, starting 22 March 1919. On 18 April, the airline Compagnie des Messageries

Aériennes (CMA) was founded by French aviation innovators Louis Charles Breguet, Louis Blériot, Louis Renault and René Caudron. CMA would provide a series of air services, first offering internal flights between Paris and Lille and from 16 September between Paris and London. At the same time, rival airline Compagnie des Grands Express Aériens (CGEA) was founded but did not run scheduled flights between Paris and London with Farman Goliaths until the following year on 29 March. Unlike Great Britain, France was closer to her colonies, especially in North Africa, and soon had an enviable network of routes. For the French, the dream remained to link its territories in the Caribbean with France via a practicable air bridge.

Pierre-Georges Latécoère, an enterprising aeronautical engineer, had gained a considerable amount of experience in the First World War producing aircraft for Salmson via Groupe Latécoère, the company he formed in 1917 that is still in business today. This experience fed his fascination with aircraft, especially seaplanes, the design of which would make him famous. In 1918 Latécoère established the Société des Lignes Latécoère (SLL) with the sole aim of opening the air bridge between France and South America. However, the technology to achieve such a long-distance flight remained tantalizingly out of reach. While he sought solutions to set up the transcontinental air bridge, Latécoère would use SLL to stretch out its routes southward using experienced pilots such as Antoine Saint-Exupéry. By June 1925, SLL had established a route to Senegal, with airmail going by steam to South America, where it would then be flown by SLL operating in Brazil. It was not until 12/13 May 1930, that a Latécoère 28 mail plane flown by Jean Mermoz made the 3,058-kilometre (1,900-mile) flight between Dakar and Natal, Brazil.

As well as establishing the air bridge to South America, the French were determined to connect with their colonies in the Orient. To this end, 23 April 1920 saw the founding of Compagnie Franco-Roumaine de Navigation Aérienne (CFRNA). By 3 October, CFRNA was operating a Paris–Constantinople route, flying over some of Europe's most inhospitable terrain and weather. On New Year's Day 1925 CFRNA became Compagnie Internationale de Navigation Aérienne (CIDNA). CFRNA pilots flew in highly hazardous environments and perfected safe night flying, soon in use across the French aviation industry. Also using night flying was Air Union, a merger airline created by CMA and CGEA in 1923, in their scheduled cross-channel night flights in April 1929.

The French used a vast range of aircraft alongside the Farman Goliath to fulfil their flights and air service agreements, including the elegant Blériot-SPAD S.56 six-seat enclosed-cabin biplane. Also in service was the veteran Salmson 2 biplane, which had seen wartime service as a reconnaissance aeroplane, with some examples modified as the Limousine model, featuring an enclosed cabin capable of carrying up to four passengers.

In Germany, Deutsche Luftschifffahrts-Aktiengesellschaft (DELAG) was established in 1909 by Alfred Coleman as an offshoot of the manufacturer Luftschiffbau Zeppelin GmbH. As the world's first airline, DELAG operated a fleet of Zeppelin airships in the air-shipping role. Coleman was an aviation engineer and the general director at Zeppelin who saw the potential for using his designs for travel. It can be argued that Coleman was also the first genuinely aviation-orientated entrepreneur, helping to make Luftschiffbau Zeppelin GmbH a self-sufficient company. He did this by bringing as many of the processes as he could in-house, including the manufacture of the hazardous hydrogen gas produced by Zeppelin Hydrogen and Oxygen Company (ZEWAS). Zeppelin-Hallenbau GmbH, which constructed the vital airship hangars, and Holzindustrie Meckenbeuren GmbH, which supplied crucial timber, were also incorporated. Apart from building a large aviation business, Coleman sought to make the Zeppelin brand socially aware. In 1913, to honour the seventy-fifth birthday of Ferdinand Graf von Zeppelin, the founder of Zeppelin and developer of rigid airship construction, he established Zeppelin-Wohlfahrt GmbH. This affordable housing company remains in operation.

DELAG was soon transporting commercial passengers due to a funding deal brokered between Coleman and Albert Ballin, the general director of the intercontinental shipping line Hamburg-Amerikanische Packetfahrt-Action-Gesellschaft (HAPAG). In the summer of 1910, DELAG made its first foray into passenger services, initially carrying passengers on pleasure flights. By the eve of the First World War DELAG was making inter-city flights between Baden-Baden, Frankfurt, Berlin and Düsseldorf.

Initial airship size restrictions, as laid out by the Treaty of Versailles, would have prevented Coleman and the Zeppelin company from developing large aircraft; however, by 1925, these restrictions had been lifted by the Allies. Dr Hugo Eckener, who had succeeded Graf von Zeppelin as the company manager, sought to expand DELAG's travel portfolio beyond the national boundaries of Germany. On 18 September 1928, the Eckner-designed LZ 127 *Graf Zeppelin* made its maiden flight, followed by its first transatlantic flight between Friedrichshafen, Germany and Lakehurst Field, New Jersey on 11 October 1928. This started a route flown by DELAG between 1928 and 1937, mainly between Germany and South America, making the South Atlantic crossing 136 times.

In 1935 DELAG's fleet of airships became part of the state-sponsored Deutsche Zeppelin-Reederei (DZR). However, after the *Hindenburg* disaster of 6 May 1937, Zeppelin's fortunes waned, and by 1940 Zeppelin flights were no more.

Despite DELAG's apparent public dominance of early German commercial air travel, Deutsche Luft-Reederei (DLR) entred the civil aviation arena in 1917. Due to the ongoing war, its first scheduled service would not occur until 5 February 1919, when an aircraft carried mail and newspapers from Berlin to Weimar. Founded by Walther Rathenau, chairman of the board of AEG, DLR aimed to continue to develop the AEG aircraft brand, which had been established during the First World War. By 1921 DLR's air routes had grown to include the Netherlands, Scandinavia and the Baltic Republics. DLR would be Europe's first regular and sustained scheduled passenger service on its Berlin–Weimer route using heavier-than-air aircraft. DLR operated this route mainly using war-surplus LVG C.VI reconnaissance biplanes, some of which were modified with closed cabins with room for two passengers. These worked alongside other war-surplus aircraft, including the twin-engine Friedrichshafen G.III heavy bomber repurposed as a civilian aeroplane and mainly operating as a transporter. DLR would be reorganized as Deutsche Aero Lloyd AG (DAL) in 1923.

DLR was joined as an airline by Junkers in March 1919, who flew to Weimar from Dessau with two modified J 10 ground-attack monoplanes, featuring a canopy-enclosed cabin encased in the corrugated-finished fuselage now familiar to Junkers aircraft. The J 10 was very clearly the first all-metal aeroplane in service as an airline and, in that respect, was a pioneering aeroplane in itself. In 1921, Junkers formalized the airline by setting up the Junkers-Luftverkehr AG (JLA) airline, which also acted as a vehicle to promote its aircraft, a move which would later be echoed by Boeing.

JLA would be the exclusive operator of the F 13, an all-metal monoplane featuring an enclosed cabin for four passengers which made its first flight on 25 June 1919. After that, Junkers would proliferate, despite an early attempt to form a merger proposed by aircraft manufacturer Sablatnig Flugzeugbau GmbH (Sablatnig). This short-lived airline offered government-subsidized services between Berlin and Bremen. Nevertheless, Junkers pushed on, maintaining its independence, and by 1925 it had established and extended its trans-European flight network, operating from several hubs. Its aircraft could easily carry a passenger from Helsinki to Constantinople over its prolific network of routes, often involving night flights.

January 1926 saw the formation of Deutsche Luft Hansa A.G. (DLH), which would become Deutsche Lufthansa in 1933 and Lufthansa in 1953. DLH was another national merger of commercial airlines, with DAL and JLA at its heart, and with these airlines came a formidable trans-European flight network.

Other German airlines came and went in those immediate post-war years. These were often small operators, who would fall foul of costs, competition and the rapid technical advances in aviation, but also of restrictions set by the Treaty of Versailles that were rigorously imposed in the early years.

Other European nations also built up their civil aviation sector in the immediate post-war years, keen to put the hell of war behind them and advance into a glorious new age led by technology. In Belgium the Syndicat National d'Étude des Transports Aériens (SNETA), founded by Georges Nélis on 31 March 1919, began practical studies into the viability of a national airline. A mix of aeroplanes was used, including a Rumpler C.IV reconnaissance biplane. Initially, services were located in the Belgian Congo through a subsidiary, Comité d'Étude pour la Navigation Aérienne au Congo (CENAC), flying to Matadi, Léopoldville and Stanleyville using a Lévy-Le Pen amphibious biplane.

A fire on 7 September destroyed a hangar and seven of SNETA's twenty-three aeroplanes. SNETA was absorbed by Societé anonyme belge d'Exploitation de la Navigation aérienne (Sabena), which came into being on 23 May 1923. The first commercial flight took place on 1 July between Brussels and London via Ostend. By 1 April 1924 Sabena had launched air routes to Rotterdam and Strasbourg and by the end of the year, services were extended to Amsterdam and Basle, Switzerland. In 1929 Sabena added Hamburg to its destination list, via Antwerp, Düsseldorf and Essen.

Sabena also provided the first long-haul and, more importantly, profitable air route to Léopoldville on 12 February 1925 using landplanes rather than the amphibian craft, favoured by CENAC. The flights also bridged the gap left by Ligne Aérienne du Roi Albert (LARA), a short-lived civilian airline serviced by six amphibians established in the colony. In LARA's short service period, it established a route from Léopoldville to Kisangani using the Congo River to navigate and land on. LARA was the first overseas airline established by a European colonial power and operated between 1 July 1920 and 7 June 1922.

As Sabena only operated landplanes, an aerodrome-building project was instigated, and in 1926, it began internal flights along the Boma–Léopoldville–Élisabethville route. This 1,422-mile (2,288-kilometre) route flew over the dense jungle in the three-engine, ten-seat Handley Page W.8f biplane. By 1931 Sabena was flying forty-three aircraft in the region and was soon cooperating with Air France and DLH.

Air France was established on 7 October 1933 by the merger of Air Orient, Air Union, Compagnie Générale Aéropostale, Compagnie Internationale de Navigation Aérienne (CIDNA) and Société Générale des Transports Aériens (SGTA), which had been the first commercial French airline to be founded by the aeronautically talented Farman brothers on 8 February 1919. Such tie-ins between airlines would rapidly become commonplace, benefitting both airline and consumer. In 1935, Sabena made the first long-haul flight to the Congo, which took five-and-a-half days in a Fokker F-VII/3m Trimotor high-wing monoplane capable of carrying eight passengers. The following year Sabena purchased the larger low-wing, eighteen-seat Savoia-Marchetti SM.73 Trimotor airliner. The Savoia-Marchetti was also a faster aeroplane capable of making long-distance flights to the Congo in four days. However, the SM.73 was dogged by disasters and showed how dangerous flight remained in the early years.

In Europe, Sabena continued to develop its flight routes and, by 1931, had established routes into Denmark and Sweden. These were followed by flights to

Berlin in 1932. For these routes, Sabena used the famous Junkers Ju 52/3m, introduced in 1930 and which made regular flights with airways from March 1932. The Ju 52 was designed by visionary aeronautical designer Ernst Zindel, who had begun working with Hugo Junkers, who had founded the Junkers Luftverkehr AG airline in 1920 at the Junkers Research Institute. In 1925, Zindel was made head of design at the design office of Junkers Flugzeugwerk AG (JFA), where he managed the development of the Ju 52 from 1928. By 1933 Zindel had become the design director, where he applied his aeronautical knowledge to dominate German aviation in terms of flight performance and technical development.

Sabena flew the Ju 52 from its main base at Brussels Haren Airport and utilized a series of hubs throughout the country for internal flights. In 1938, it took delivery of the faster, though smaller Savoia-Marchetti SM.83. The ten-seat SM.83 trimotor was designed by founder Alessandro Marchetti. Unfortunately, despite its speed, its overall performance was considerably poorer than its somewhat accident-prone predecessor.

As well as Belgium, the Netherlands quickly established a national carrier in the wake of the war. On 7 October 1919, Koninklijke Luchtvaart Maatschappij N.V. (KLM) was founded and is now the oldest continuously operating airline. KLM was founded by a group of eight businessmen, including entrepreneur Frits Fentener van Vlissingen, and established its operating hub at Amsterdam Airport Schiphol. Its first scheduled flight took place on 17 May 1920 when KLM's first pilot, Jerry Shaw, took off from Croydon Airport to Amsterdam using a leased AT&T DH.16 four-seat biplane. This was followed by scheduled air services starting in 1921. KLM gained experience over the next couple of years. On 1 October 1924 it made its first intercontinental flight to Batavia (Jakarta) in the Dutch East Indies using a Fokker F.VII Trimotor designed by Walther Rethel. By September 1929, regular service between Amsterdam and Batavia was taking place, making this scheduled route the longest undertaken by civilian aircraft until the beginning of the Second World War.

In Europe, KLM was quick to establish itself, and by 1926 it was offering flights throughout Northern Europe via its Amsterdam hub, flying the five-seat Fokker F.II and Fokker F.III high-wing monoplanes. By 1930 business was becoming brisk, and brand sentimentality for Fokker was pushed aside when KLM purchased the new, all-metal Douglas DC-2. The DC-2 was a game changer for many airlines, including Swissair and Iberia Líneas Aéreas de España, S.A. (LAPE).

Designed to respond to the revolutionary Boeing Model 247, the DC-2 was larger, capable of carrying fourteen passengers and had a range of 1,000 miles (1,600 kilometres) compared to Boeing's 745 miles (1,199 kilometres). This was a sensible investment for KLM, especially for its long-haul flights to the Dutch East Indies. In 1934, KLM put the DC-2 into service on its Amsterdam–

Batavia service and its Batavia–Sydney route. The same year it began its Amsterdam–Curaçao service, a Dutch island in the southern Caribbean Sea. A Fokker F.XVIII Trimotor twelve-seat serviced this particular route.

In 1936, it received its first DC-3s, which replaced the DC-2s in the Dutch East Indies. Two years later, these were joined by the high-wing DC-5s, also used in the West Indies. In 1938, KLM started flying to the newly established Manchester Ringway Airport, becoming the first to do so.

Despite Dutch neutrality, KLM was affected by the outbreak of the Second World War, and its fleet was painted orange to prevent confusion with military aircraft. Flight routes had to be tailored accordingly, and before German forces invaded Dutch territory, routes were servicing Scandinavia, Belgium, the UK and Portugal. After the German attack on 10 May 1940, KLM aircraft flew under the BOAC banner for the remainder of the war in Europe until 1943. In South East Asia, services continued for a further year until Imperial Japanese aggression came to the Dutch East Indies. At that point, KLM crews worked hard, evacuating refugees from the area. While the war restricted flying in Europe and Asia, KLM continued to operate in the Caribbean, expanding its air route network there.

The first airline to be established in Scandinavia, Det Danske Luftfartselskab A/S (DDL) was very much an early starter. Although founded on 29 October 1918, DDL did not start scheduled services until 7 August 1920 with a two-seat Flugzeugbau Friedrichshafen 49c floatplane acquired from DLR. DDL also used a range of chartered aeroplanes, including the high-winged Fokker-Grulich F.III and the Dornier Komet. 1926 saw the arrival of the beautifully appointed Farman F.120 four-engine, nine-passenger, high-wing aeroplane. The F.120 operated out of Copenhagen, flying on to Amsterdam, a hub for flights to London and Paris.

The exotic-looking F.120 was not the success it was thought to be, and by 1929 all four aircraft had been withdrawn from service. The more capable Fokker F.VII replaced them, and in 1933 these were supplanted by Fokker F.XII sixteen-seat aeroplanes. In 1938 DDL made the then seemingly extravagant purchase of two Focke-Wulf Fw 200 Condor 26s. One would be destroyed during the war, while the other would survive hostilities and remain in DDL until damaged beyond repair in 1946.

Public Joint Stock Company Aeroflot – Russian Airlines (Aeroflot) – is one the oldest airlines in service, founded on 3 February 1923 in the Russian Soviet Federative Socialist Republic. It was conceived in 1921 in the early years of communism when the Soviet Union established a shared airmail service with the German operator, Deruluft-Deutsch Russische Luftverkehrs A.G. (DDRL), which flew between Königsberg, East Prussia and Moscow. This seemingly innocuous venture gave the new Soviet authorities experience in international air travel.

By 8 March 1923, the Central Committee of the Communist Party of the Soviet Union established the Enterprise for Friends of the Air Fleet (ODVF), followed a little over a week later by the founding of the Dobrolyot Society. The Dobrolyot, or the Russian Society of Voluntary Air Fleet, was founded as a voluntary organization to help grow the domestic civil aviation fleet and experience. The remit for volunteers of the Dobrolyot was vast. They were essentially building a nationwide asset from the ground up and doing so quickly. Cargo and passenger services were established, with the necessary infrastructure built to facilitate aviation growth. These activities took place alongside developing the domestic aeronautical industries and utilizing aircraft to help further post-revolution economic, social and political growth. Such actions would include air surveys, propaganda missions and facilitating the all-important air mail.

Very rapidly, the Dobrolyot traced routes across Europe and Central Asia, connecting the new Soviet republics using a range of aircraft from Fokker and Junkers. State funding was supplemented by public subscription, which allowed big investors the opportunity to have access to a fleet of aircraft for personal use. At the same time, other airlines, which followed similar lines of organization and funding, were established: Zakaria, which covered Georgia and Azerbaijan, and the Ukrainian airline Ukrpovitroshliakh would merge in 1925.

As the Dobrolyot grew, its remit and area of operations expanded beyond Soviet territorial borders, and by 1925 flights were covering a network of 3,100 miles (5,000 kilometres), carrying 14,000 passengers a year. As a result of the success of the project, Dobrolyot became an all-union enterprise in 1926 after a decree by the Councils of People's Commissars, also known as a Sovnarkom Resolution. In 1928, the Dobrolyot merged with Ukrpovitroshliakh.

As the Soviet Union strengthened its infrastructure, the need to further align the Dobrolyot with a centrally organized body became clear and, on 25 February 1932, all civil aviation activities in the Soviet Union, including those undertaken by the Dobrolyot, came under the control of the Chief Directorate of the Civil Air Fleet. This significant change was followed a month later by adopting Aeroflot's name for the Soviet Civil Air Fleet. The following year the Communist Party Congress set out a five-year plan to use aviation as the primary method to link all the main cities of the union. The programme aimed to develop domestic aeronautical infrastructure further and strengthen aircraft production across the board.

By 1937, the aviation network had grown beyond recognition, including the Soviet Far East, with the route covering 22,000 miles (35,000 kilometres). With the Soviet-German airline, the DDRL, ceasing operations on 1 April 1937, Aeroflot could now provide international flights, starting with Stockholm and Riga. The Aeroflot fleet was deliberately small in type to help keep costs down and allow for ease of maintenance, using aeroplanes such as the eight-engine Tupolev ANT-20, the largest aircraft in the world at the time of its maiden flight on 19 May 1934, to service its routes. Despite having access to this futurist colossus, Aeroflot wisely managed their resources; they also operated the PS-84 (Lisunov Li-2), a licence-build DC-3 featuring Soviet engines that entered serial production in 1939. The PS-84 would rapidly become the workhorse of the Aeroflot fleet on its main routes. In 1938 a third aviation five-year plan was announced with strengthened infrastructure and more routes; by 1940 scheduled flights were operating on over 330 local routes covering 91,000 miles (146,000 kilometres) of the USSR.

On 23 June 1941, the day after the invasion of the USSR by Axis forces, Aeroflot was placed under the command and control of a People's Commissariat, and staff mobilized and transferred to help in the war effort.

Crossing a Continent: The Birth of American Civil Aviation

While most European airlines were helped by the desires of subsidizing governments to develop their links within the empire and beyond, the continental United States was in a very different situation. Its sphere of influence was focused on its neighbours and the Pacific. There were also territories gained due to victory in the Spanish-American War of 1898, including the Philippines and Guam.

As a result of having such an extended sphere of administrational influence, American civil aviation was focused primarily on delivering mail. By the time the 1919 Paris Convention had been signed, the United States Post Office had begun establishing a network of routes crisscrossing the continental United States. Perhaps more vividly than anywhere else, it was here that the network of air route hubs and the smaller airports provided a transcontinental service that was second to none. Initially using the Standard JR1B mail biplane, the Post Office soon supplemented these aircraft with war-surplus DH.4s, facilitating an effective coast-to-coast service.

By 1922, the transcontinental airmail network was well established. However, given the expanses of often featureless terrain, work began on developing ground-based lighting systems along the airmail routes to aid night navigation. The mammoth task, known as the Air Mail Service Beacon System, saw aerodromes along the routes fitted with beacons, with the importance of a site indicated by the strength of luminescence.

The tower-mounted oscillating beacons, which rotated at 6rpm, were sited between three and five miles (five and eight kilometres) apart and fitted with a one-million candlepower bulb which could be seen up to ten miles (sixteen kilometres) away. The towers were then mounted on top of large yellow-painted concrete arrows pointing in the flight direction. The system also used coded

A wood cutting by Albrecht Dürer (1471–1528) showing Daedalus watching as Icarus falls from the sky.

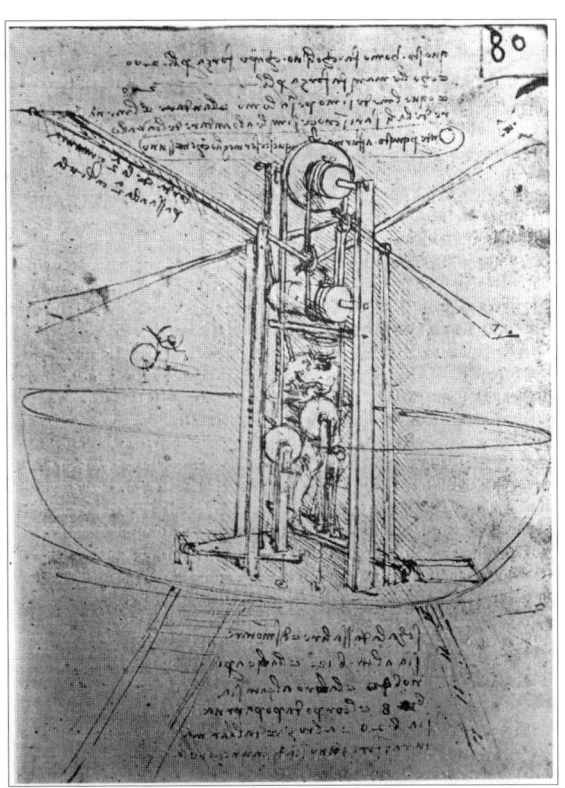

A da Vinci design for a flying machine, 1487. (Leonardo da Vinci)

Above left: A 1697 engraving of Francesco Lana de Terzi's aerial ship. (Unknown)

Above right: An 1851 engraving of Henry Cavendish. Cavendish's work would lead to advancing the idea of lighter-than-air flight. (George Wilson)

A fine illustration of Dr John Jeffries from *Wonderful Ballon Ascents*. (Camille Flamarrion)

Elmer of Malmesbury about to launch himself into the ether, watched by fellow monks. (Unknown)

A moment in time. The Wright Flyer makes its historic flight on 17 December 1903. (Library of Congress)

A steely-eyed Glenn Curtiss in the 'June Bug', 4 July 1908. Note the bow to the wings. (Alexander Graham Bell family)

Wilbur Wright with the Spanish King Alfons XIII at Pau, France, 20 February 1909. (Agence Rol. Agence Photographique)

French aviator Henri Farman (1874–1958) with his wife. (George Grantham Bain Collection, Library of Congress)

Louis and Alice Blériot with his Type XI after landing at Dover. This flight would bring the possibilities of aviation into sharp focus for many in government. (George Grantham Bain Collection, Library of Congress)

Above: Calbraith 'Iron' Rodgers alighting at the finish of his astonishing transcontinental flight. (George Grantham Bain Collection, Library of Congress)

Left: John B. Moisant and Mademoiselle Fifi. (Library of Congress)

Wreckage of the crash that killed Charles Stewart Rolls on 12 July 1910, at Southbourne, England. Rolls would be the first aviation fatality on British soil. (*Illustrated London News*)

Left: Pilot Tony Jannus and Captain Albert Berry with the Benoist-built biplane they used when Berry became the first person to parachute from an airplane on 1 March 1912. (Missouri History Museum)

Below right: An oil-stained Roland Garros with aeroplane. (Gallica)

Centre left: Jules Védrines, second left, in Cairo with Paillard, Jules Munier and Henri Mosseri. (Unknown)

French cavalry observe an army aeroplane fly past, 1916. (Unknown)

Harriet Quimby in her Blériot monoplane, 1911. (George Grantham Bain Collection, Library of Congress)

A Sikorsky Russky Vityaz with Czar Nicholas II (on the balcony, behind the ladder) speaking with aircraft designer Igor Sikorsky. (San Diego Air and Space Museum)

Below: An NC-4 on dry land during maintenance work which shows how short her hull was. (Library of Congress)

A thoughtful-looking Lieutenant Commander Albert Cushing Read aboard the NC-4. (National Photo Company)

Vickers Vimy aircraft of Captain John Alcock and Lieutenant A. W. Brown ready for flight, at Lester's Field, St. John's, Newfoundland, 14 June 1919. (Library and Archives Canada)

The rear gondola of the R34. Note the workman being lifted off his feet. (*Topeka State Journal*)

Hydrogen gas supplies for the R34, at Roosevelt Field, Mineola, Long Island. (Library of Congress)

Above left: Sir Ross and Keith Macpherson-Smith. Of note is Sir Keith wearing his army pips with his new RAF uniform. (State Library of South Australia)

Above right: Navigator Arthur W. Brown (left) and John Alcock, pilot of the Vickers Vimy Rolls biplane pictured at St. John's, Newfoundland on 28 May 1919. (Smithsonian)

Centre left: A Graf Zeppelin airship. The 1919 Paris Convention would open the skies for the passage between continents for these behemoths of the sky. (Unknown)

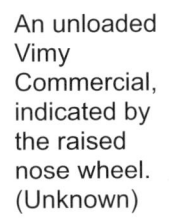

An unloaded Vimy Commercial, indicated by the raised nose wheel. (Unknown)

A Westland Limousine promotion shot for the French aeronautical magazine *Revue L'aéronautique*, May 1920. (Unknown)

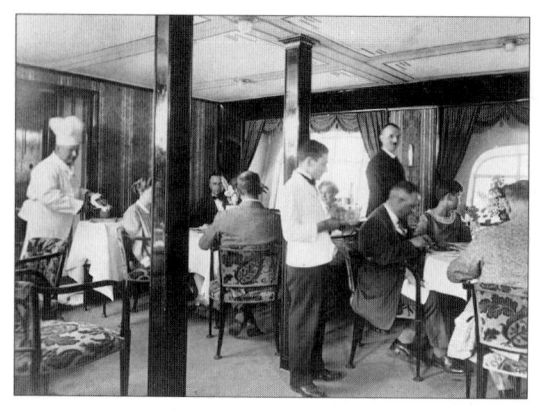

Heinrich Kubis, the first flight attendant (standing) looks at the camera while supervising the assistant steward aboard the Graf Zeppelin, August 1929. (German Federal Archives)

A civilianized de Havilland DH.4 awaiting its next task. These converted bombers would fulfil a variety of roles in the immediate post-war years of the 1920s. (Unknown)

The Handley Page H.P.42 was the acme of biplane airliner design, seeing service across the globe. (San Diego Air and Space Museum)

The Armstrong Whitworth AW 154 Argosy trimotor was capable of carrying twenty passengers and provided in-flight refreshments and steward service. (German Federal Archives)

Below: One of the first generation of specifically designed airliners: the de Havilland DH.18B. (British Flight Testing, Tim Mason)

Above: The de Havilland DH.91 Albatross was another aircraft whose potential was short lived because of the Second World War. (Unknown)

Left: The development of the Armstrong Whitworth AW.27 Ensign was curtailed by poor performance and the advent of the Second World War. (San Diego Air and Space Museum)

Boeing's Model 221 Monomail would prove to be an excellent platform upon which new innovations could be trialled. (Unknown)

A Shell Oil Company Fleet Lockheed Vega in Oakland, 1939. (Bill Larkins)

A Beech 18 of the Morrison-Knudsen Company, San Francisco, 1952. Note the company logo on the nose and tail. (Bill Larkins)

A Swiss Curtiss AT-32C Condor at Boden at Dübendorf Airport. (ETH-Bibliothek)

Farman F.60 Goliath taking a break from its daily London–Paris service. (Unknown)

Well-wishers wave off a Bloch MB.220 in Boden at Dübendorf Airport. (ETH-Bibliothek)

A Junkers F 13 of ABA. AB Aerotransport serviced the night airmail flights between Stockholm and London. (Scandinavian Airlines)

The huge Junkers G.38 was a world away from the F 13, offering a truly unique flying experience. (Meyers Blitz-Lexikon)

A Junkers JU 52/3m at Budaörs Airport, Budapest, 1937. Note the DC-3 in the background. (Fortepan)

A wonderful view into the interior of the Focke-Wulf A.16 on display at Airport Bremen Flugtag, 2009. Note the vase. (Politikaner)

A DDL Focke-Wulf Condor Fw 200 Dania. The military version would be the scourge of Allied shipping in the North Atlantic theatre of the Second World War. (Scandinavian Airlines)

BOAC's Short Empire S.33 C-Class Flying Boat circling Durban, South Africa, after flying the horseshoe route from Sydney, Australia. (IWM)

James H. Doolittle, with his Curtiss R3C-2 Racer, the plane in which he won the 1925 Schneider Trophy Race. (Unknown)

A view of the Curtiss-Bleecker SX-5-1 helicopter showing its four oversized rotors. (NASA)

A Savoia-Marchetti S.55 out of the water shows off its unique twin-hull configuration. The aircrew sat in a pod beneath the engine nacelle between the two hulls. (San Diego Air & Space Museum)

Testing the engine of the Macchi M.39. Such tests would help the M.39 establish itself as a record breaker. (Unknown)

The first S.5 at Calshot preparing for the 1929 Schneider Trophy. (Unknown)

The Supermarine S.6B. The pause before the fury. The S.6B would dominate the 1931 Schneider Trophy, winning it for perpetuity. (Unknown)

Granville Brothers' Gee Bee Model R-1 Super Sportster was a notoriously difficult aeroplane to fly, but one Jimmy Doolittle was able to tame it. (Charles M. Daniels Collection)

French aviator Joseph Sadi-Lacointe broke numerous records during his flying career. He died on 15 July 1944 of injuries sustained while a prisoner of the Gestapo. (Gallica)

Below: Jacques Schneider (right) with M. Bienaimé competing for the 1913 Gordon Bennett Aviation Trophy. Schneider would go on to establish the Schneider Trophy. (Bibliothèque Nationale de France)

The record-breaking Wedell-Williams Model 44 Racer was the sum of hard work and determination. (Charles M. Daniels Collection)

Howard Hughes was drawn to aeronautical excess; where the H-1 gave him the exhilaration of speed, the H-4 Hercules flying boat delivered only disappointment. (Federal Aviation Administration)

Roscoe Turner with his pet lion cub Gilmore in fine form, 1930. Gilmore would accompany Turner on his flights until 1935 when he grew too large. (Associated Press)

'Pancho' Barnes (fourth left) and the Powder Puff Derby at Long Beach, California circa 1930/1. (Unknown)

Louise Thaden, Cliff Henderson and Blanche Noyes at the 1936 Bendix Trophy. (San Diego Air and Space Museum Archive)

Above: Flying the Laird Super Solution, Jimmy Doolittle won the inaugural 1931 Bendix Trophy Race. (Unknown)

Right: Eddie August Schneider and Nancy Hopkins at the 1930 Ford National Reliability Air Tour in Chicago, Illinois. (Nancy Hopkins collection/ International Women's Air & Space Museum Archive)

Above: Charles Scott and Giles Guthrie in front of their Percival Vega Gull after winning the Schlesinger Air Race. (Unknown)

Centre left: DC-2 'Uiver' of KLM at the 1934 MacRobertson Trophy Air Race finished second in both speed and handicap sections. (Mayse Young Collection/ Northern Territory Library)

Bottom left: The KLM DC-6 flown by Captain Kooper (centre) won the commercial class (handicap) Harewood Gold Cup, known as the Last Great Air Race, in October 1953. (KLM) green and red lights, which flashed the exact location of the airfield in Morse Code, another valuable aid for night navigation. Every twenty-five miles (forty kilometres), an emergency intermediate landing field was built. The project delivered eighteen hub airfields, eighty-nine emergency airfields and operated more than 500 beacon lights.

With the completion of the postal service beacons, airmail could be flown from San Francisco to New York City in thirty-four hours, while westbound mail could make use of the prevailing winds to cross the country in twenty-nine hours. In 1926, the system was given to the US Department of Commerce to control due to the airmail service being handed to contractors via the 1925 Contract Air Mail Act, also known as the Kelly Act. The Department of Commerce continued to expand the system, installing towers on spur routes every ten miles (sixteen kilometres), fitting these new installations with one-million candlepower bulbs. The light from these bubs could be seen up to forty miles (sixty-four kilometres) in clear weather.

By 1933 there were numerous advances in navigation, such as the introduction of improved navigational techniques and instrumentation, including the fitting of radios capable of picking up the new Low-Frequency Radio Range (LFRR) system. However, by the time the United States had entered the Second World War, the system was largely redundant, and the Department of Commerce began slowly removing the system, recycling critical materials for the war effort.

While the federal government established a cross-county airmail route, American commercial operators were busy echoing what was happening in Europe. On 19 April 1925 the Ryan Airline Company (RAC) began operations between Los Angeles and San Diego. Headed by aviation and aerospace engineer Tubal Claude Ryan and Ben Mahoney, RAC initially provided flights in three specially converted four-seat Standard J-1 biplanes known as a Ryan-Standard. These were later joined by a single Douglas Cloudster biplane, designed to make the first coast-to-coast flight. A mechanical issue prevented the Cloudster from achieving its design goal. However, Ryan had employed an exceptional engineer, William Bowlus, who was responsible for the Ryan-Standard conversions. He was tasked with converting the Cloudster into a ten-seat passenger aeroplane. Bowlus would later become superintendent of construction on Charles Lindbergh's aircraft, the *Spirit of St. Louis*, which Ryan would make.

In 1925, Bowlus would complete a design from a set of incomplete blueprints bought by Ryan to make the parasol-winged Ryan M-1 mail and passenger plane. Despite the design's successes, RAC would see Ryan and Mahoney part ways in November 1926, although Mahoney would keep the Ryan name and incorporate it into the Ryan Aeronautical Company. The new company would build the Ryan NYP, also known as the *Spirit of St. Louis*, designed by aeronautical engineer Donald Hall, who would later be part of the team that discovered it was a sonic boom, and not material colliding, that was responsible for the crack of a bullwhip sound during flight.

As with many countries, the United States struggled to keep up with its inter-war industrial boom. Domestic legislation had to reflect this, especially as airmail was the key commercial driving force behind the US civil aviation boom. In May 1926, the Air Commerce Act established airways for both airmail and passenger flights, organized air navigation, and licensed aircraft and their pilots. On 31 December 1926 the Act came into force and what followed was a period of complexity and chaos with entrepreneurs, large and small, vying for the lucrative airmail contracts known as Contract Air Mail Routes (CAMs), of which there would be thirty-four in total. When the Act came into force, fewer than twelve airlines operated as feeders to the main Post Office transcontinental route.

The Ford Air Transport Service (FATS) offered the first scheduled private air courier parcel service between Detroit and Chicago on 3 April 1926 using their Stout 2-AT high-wing, all-metal monoplane. Operating CAMs 6 and 7 that covered the states of Illinois, Michigan and Ohio; FATS would start passenger services in August 1926.

Further south, Robertson Aircraft Corporation (RACorp), with chief pilot Charles Lindbergh operated the CAM 2 route between Missouri and Illinois. RACorp flew mail and reconfigured and refurbished military aircraft for civilian use, including aircraft from de Havilland, Waco and Curtiss. RACorp also operated CAM 29, which linked Nebraska to Illinois. By 1928 RACorp was offering a daily passenger and express service flown by its Ford Trimotors.

To the west Varney Airlines (VAL) covered CAM 5 of the tristate area of Washington, Idaho and Nevada. Founded by Walter Varney, who would later establish United Airlines and Continental Airlines, Varney had learned to fly in US Army Signal Corps during the First World War. VAL would soon become one of the predominant airmail contractors and operate the Swallow J-5 biplane, which could also carry two passengers. Another airline operating in the west was Western Air Express (WAE), which flew between California and Utah, covering CAM 4 from April 1926, offering mail and passenger services with its Douglas M-2 biplanes.

CAM 1 covered New York State to the Commonwealth of Massachusetts on the East Coast, with the routes served by Juan Trippe's Colonial Air Transport (CAT). Trippe would become a pioneering figure in commercial aviation, leading several key Boeing designs, including the elegant 314 Clipper flying boat, the Stratoliner and the iconic 747. To the south, Philadelphia Rapid Transit Air Service (PRTAS) operated CAM 13 between the Commonwealth of Pennsylvania and Washington D.C. Flying a Fokker F.VIIa-3m, which carried passengers and mail, PRTAS ran a safe and reliable service, helped by a former KLM employee, André Priester. Unfortunately, despite operating 3 services daily, the route was short-lived. By November 1926, PRTAS had ceased operations due to financial losses as a result of putting passenger carriage before the more lucrative mail carriage.

National Air Transport (NAT) is an interesting airline as its roots lay in the farsightedness of its founder, the financier Clement Keys. Keys gained aviation experience very early on with the Canadian-American aeronautical research group, the Aerial Experiment Association (AEA), founded in 1907 by Dr Alexander Graham Bell. In 1916 he became the vice-president of the Curtiss Aeroplane and Motor Company, which Keys would merge with Wright Aeronautical in 1929 to form Curtiss-Wright and become president of the new company by doing so. The later venture would see Keys involved in the establishment of North America Aviation and Trans World Airlines. Such was his influence in civil aviation that he has been referred to as the 'father of commercial aviation in America'.

NAT was founded in May 1925 and secured the sought-after New York State–Illinois route, CAM 17. The fledgling aviation industry considered this a prestigious route, and NAT initially concentrated on establishing its airmail service rather than passenger bookings. This approach ensured the service was viable and kept the company's focus on delivering its binding contractual obligations. Ten Curtiss Carrier Pigeon biplanes initially serviced the route, which were soon supplemented by eighteen Douglas M-2s from the old Post Office fleet and eight Travel Air 6000 six-seat, high-wing monoplanes. By 1930 scheduled air passenger flights between New York and Chicago were taking place with NAT using the Ford Trimotor as its passenger aircraft.

The final key service provider was Boeing Air Transport (BAT), which covered the California–Illinois route, CAM 18. Unlike NAT, BAT started providing a joint passenger and mail service using their Model 40A two-passenger mail biplanes. The Model 40A was an extremely capable aeroplane, first flying on 20 July 1925 and making its in-service debut on 1 July 1927. Boeing had created BAT to secure the California–Illinois route by providing a lower tender than regional rival Western Air Express (WEA).

However, WEA were far from beaten and were awarded the CAM 4 route between California, Nevada and Utah. Founded by Harris Hanshue, a racing car driver, WEA's first flight used Douglas M-2s, on 17 April 1926, with passenger flights taking place a month later. WEA also operated a seaplane route from Hamilton Cove Seaplane Base on Catalina Island, California, and built two airports, one at Vali and the other at Alhambra.

This new network would provide aviators, entrepreneurs and aircraft manufacturers with immense experience in aeronautical engineering, flying long distances, and, most importantly, generating revenue. Domestically, the first indicator of the advance of American aeronautical art was Ford's all-metal Trimotor high-wing monoplane which made its first flight in June 1926. The Trimotor introduced a level of sophistication hitherto missing from the domestic civil aviation market, and operators welcomed it with open arms. Another key event was Lindbergh's crossing of the Atlantic in an American aeroplane, proving the home of powered flight could still produce marvels. However, Lindbergh's New York-to-Paris flight did something else: it reignited public interest in aviation.

By the end of the decade the mergers which had been prevalent in European civil aviation were starting stateside, with BAT buying Pacific Air Transport (PAT) in January 1928. PAT was established in January 1926 by Vern Gorst. He ran an Oregon bus company and saw the opportunity to extend his business into aviation with the advent of the Kelly Act. PAT ran a service along the Pacific coast from Seattle to Los Angeles, which covered CAM 8, using Ryan M-1 high-wing monoplanes to transport mail and passengers. Echoing the work done by the Air Mail Service Beacon System, Gorst persuaded the Standard Oil Company to paint the names of towns on the roofs of its buildings. He also established lights and beacons along his aeroplanes' route to aid safe navigation. Unfortunately, despite all his preparations, Gorst was not making the money he had envisaged and, in January 1926, sold the company to BAT.

BAT now dominated the West Coast and had soon built their passenger traffic up to such a level that a change in aircraft was required. In October 1928 BAT introduced the Model 80 trimotor biplane into service. Although capable of carrying twelve passengers, BAT soon replaced the early Model 80s with larger eighteen-seat Model 80As.

To the southeast, the Pitcairn Aircraft Company (PAC), which autogyro pioneer Harold Frederick Pitcairn had established, began operating CAM 19 which covered the route between New York City and Georgia. This service began utilizing the Air Mail Service Beacon System to run a night service. A latecomer to the CAM system, PAC started their mail service on 1 May 1928 and took over Florida Airways' (FAW) CAM 10 routes by December of that year. FAW had been co-founded by First World War aces and air transport pioneers Eddie Rickenbacker and Reed Chambers in 1923. As well as mail, FAW also carried passengers, though the revenues were always low, leading to FAW struggling to maintain a profit. Such was its desperation that FAW added stamped bricks to mail bags to add to the weight.

In 1928 Transcontinental Air Transport (TAT), another airline formed by Clement Keys, began planning a transcontinental route serviced by Ford Trimotors. TAT had two unique selling points. The first was that they would provide passengers with meals; the second would be an agreement with the railways to offer an air–rail transcontinental service. The service began on 7 July 1929, with travellers initially travelling by rail from Pennsylvania Station, New York, to arrive at the epically built air and rail terminal at

Port Columbus in Ohio. They would fly from there to Waynoka, Oklahoma, before carrying on by rail to Clovis in New Mexico. From there, passengers would take off and continue to head west by air to Los Angeles. The journey would take passengers forty-eight hours to complete. November 1929 saw TAT purchasing Maddux Air Lines (MAL), founded by car salesman Jack Maddux in September 1927. It also operated the Ford Trimotor, which gave TAT access to nine Trimotors.

In 1929 a new Postmaster General was appointed, Walter Folger-Brown. Folger-Brown was an experienced lawyer and politician, so his appointment heralded enormous changes for an industry increasingly becoming wayward in its approach to its contractual agreements. He was keen to tighten the fractured network of CAMs to ensure value for taxpayers' money. The rationalizations soon started. American Airways (AA) was formed as a merger of the Robertson Aircraft Corporation and Colonial Air Transport on 25 January 1930. RAC had been established by brothers William, a former military aviator, and Frank Robertson in 1918, while a small airline started services in March 1923. Both airlines carried mail and passengers, covering the Tennessee and Texas CAM 33 air routes. American Airways would soon grow, and by 2019 it was the largest airline in the world.

On 16 July 1930, Folger-Brown enforced the merger of TAT and WAE to form Transcontinental and Western Air (TWA). TWA started the first all-air coast-to-coast flights between New York and Los Angeles on 25 October 1930 using Ford Trimotors. Even with a night stop at Kansas City, Missouri, the new flight took thirty-six hours. The following airline to emerge was United Air Lines (UAL) in July 1933, created by merging of BAT, NAT, PAT and VAL.

A fourth merger, not driven by Folger-Brown, yeilded Eastern Air Transport (EAT). EAT resulted from Keys purchasing PAC, cementing his hold on the northeastern CAM routes. EAT joined the three government-merged airlines into what would become known in the United States as the Big Four.

While Key's merger was generated by commerce, Folger-Brown was forcing unions through the 1930 Air Mail Act. This, in turn, led to a meeting with airline executives known as the 'Spoils Conference'. This saw the Contract Air Mail Routes divided among the larger carriers at the expense of the smaller airlines and the emergence of the Big Four (American, Eastern, TWA and United). However, the meeting would not be forgotten, and in 1934 the Air Mail Scandal broke.

As well as the lucrative CAM routes, there were those pioneers who were keen to establish passenger-only lines, with Ludington Airline (LAL) being one of them. Despite being unsuccessful in gaining a CAM route, LAL pushed on regardless. The airline began operations on 1 September 1930, flying hourly between Newark, New York State, and Camden, Philadelphia. LAL used Stinson Trimotor high-winged monoplanes. Within two years, LAL had carried more than 133,000 passengers and had begun to expand its network and establish subsidiary companies. Despite this, LAL ran at a loss, and in 1933, on the verge of bankruptcy, it was bought by EAT. This acquisition would see the start of the investigation leading to the Air Mail Scandal.

Despite being a powerful tool that kick-started domestic civil aviation and aeronautics development, it was clear the Kelly Act had created a proverbial free-market monster. This left the federal state little choice but to intervene as rumours of mismanagement and corruption began to surface. The subsequent Senate investigation into the behaviour of all concerned began in 1934. It resulted in a citation of Contempt of Congress awarded against William MacCracken Jr for refusal to attend any hearings. MacCracken had been the first US Assistant Secretary of Commerce for Aeronautics and had drafted vital safety standards and regulations which were enforced as part of the 1930 Air Mail Act. Resigning from the government in 1929, MacCracken was called back by Folger-Brown to chair the 'Spoils Conference', doing so in a manner that called into question his neutrality in the affair. The investigation soon uncovered a raft of questionable corporate behaviours among those involved, including Boeing, whose holding company, United Aircraft and Transport Corporation (UATC), was deemed anti-competitive. UATC combined every aspect of the aeronautical business, from propellers to airmail carriage to providing a one-stop shop for the consumer, thus virtually eliminating smaller competitors. As a result of Boeing's behaviour, new antitrust laws were passed, which forbade airframe or engine manufacturers from having interests in airlines.

Due to MacCracken's conviction, President Theodore Roosevelt cancelled all CAM contracts on 7 February 1934, and the Army Air Corps was forced to step in to provide the service. This resulted in numerous crashes and the loss of thirteen airmen, which in turn led to a new Air Mail Act and contracts issued on 1 June 1934. Despite the reputational damage to all concerned, the affair led to improvement in technology, passenger operations and investment in the Air Corps.

Despite the changes that were taking place in 1933, February saw the introduction of the revolutionary Boeing Model 247, a low-wing, all-metal passenger aeroplane. Used exclusively by UAL, the Model 247 would become the blueprint of all future airliner designs. Unable to use the Model 247, TWA approached twelve manufacturers, including Douglas, to produce a similar aircraft. Initially, a reluctant aeronautical engineer and company owner, Donald Douglas, designed an aeroplane similar in appearance to the Model 247. Finally, in July 1933, the new aeroplane, the DC-1, made its maiden flight.

Once the dust had settled on the drama of the Air Mail Scandal, the new Postmaster General, James Farley, called the airlines together, seeking to restore service and faith in the national airlines' capacity to deliver an effective airmail service. The meeting was vital for a variety of reasons, the most important being the survival of the domestic civil aviation network at all levels. As a result, the newly organized Big Four airlines were awarded thirty-two new CAM routes contracts drafted by the Roosevelt administration.

The new routes were serviced by increasingly modern aircraft, and which were used to significant effect in helping to build passenger numbers. As aircraft design progressed with experience, Lockheed unveiled its wooden high-winged Vega monoplane in the summer of 1927. The Vega looked more akin to a racing aeroplane than a passenger aircraft, which is not surprising, given that it was designed in the heyday of air racing. Capable of carrying six passengers, the Vega was used by a range of airlines, including TWA and Braniff. Braniff was established in 1928 by brothers Thomas and Paul Braniff. Paul shaped the airline industry by launching a route between Chicago and the Mexican border, expanding passenger and airmail services throughout the continental United States.

Despite the Vega's popularity among operators, it would be the next-generation DC-1, with its memorable thirteen-hour coast-to-coast dash, which took place on 19 February 1934, that would open the doors to successful commercial aviation. A single DC-1 was produced, which TWA extensively tested over 200 flights. TWA's feedback led Douglas to develop the slightly larger DC-2, which it put to work on its transcontinental routes from May 1934. These included the 780-mile (1,255-kilometre) non-stop Newark-to-Chicago route.

The non-stop, cross-county, long-haul flight had arrived, and American Airlines (AAL) began using the Curtiss Condor biplanes for their transcontinental sleeper services. However, the Condor's shortcomings soon became apparent and they were relegated to the short New York–Boston flights; their replacement was a sleeper version of the DC-2. This enlarged aeroplane was fitted with fourteen sleeping berths and became the Douglas Sleeper Transporter (DST). AAL initially purchased ten DSTs, which made their first flight on 17 December 1935 but subsequently changed the order to eight DSTs and eight twenty-one-seat DC-3s, known as dayplanes, which had also made their maiden flight on 17 December.

DST deliveries began the following June, with the first DST starting service as a dayplane on the New York–Chicago route; the DC-3s followed in August. On 15 August AAL began its mammoth sixteen-hour transcontinental skysleeper service named the 'American Mercury'.

As Douglas began developing the DC range of aeroplanes, Lockheed had been watching the developments of Boeing and Douglas and unveiled their all-metal Electra in February 1934. One of the first customers of the Electra was Northwest Airlines (NWA), founded in 1926 by Colonel Louis Brittin with help from a consortium of investors, including Henry Ford.

Developments continued, and before the Second World War, a four-engine Douglas, the DC-4E, was developed after funding by the Big Four and Pan American (Pan Am). At the same time, Pan Am received financing from US Army Air Corps (USAAC) officers Henry Arnold, Carl Spaatz and John Jouett. This was used to counter the influence of German-backed operator Sociedad Colombo Alemana de Transportes Aéreos (SCADTA), which threatened to make headway into the continental United States and presented a risk to national security. As a result, Pan Am was awarded the Florida–Cuba CA. However, due to not having landing rights in Havana, Pam Am formed a triumvirate with the Aviation Corporation of the Americas (ACA), owned by Juan Trippe, and American International Airways (AIA), a small seaplane service operating to Cuba from Key West, Florida. Once established in the Caribbean, Pan Am quickly established routes throughout Latin America.

Pan Am's growth was such that by March 1936 the need for a long-range and reliable aircraft was a top priority, so they joined the DC-4E project. The DC-4E was designed to have a 2,200-mile (3,500-kilometre) range, almost 700 miles (1,127 kilometres) more than the DC-3. However, testing showed that the DC-4E, despite its many innovations, including tricycle undercarriage and pressurized body, was incapable of fulfilling the requirements demanded by the airlines. As a result, the project was cancelled by the summer of 1939, despite a large order from UAL, and the sole example was sold to Imperial Japanese Airways, who purchased the DC-4E to study American aeronautical technology.

While Douglas was distracted by the DC-4E, Boeing had designed and built the Model 307 Stratoliner. Despite its shorter range of 1,750 miles (2,820 kilometres), the Stratoliner made history by becoming the first pressurized airliner to enter service. Pan Am and TWA both bought examples. TWA later introduced it on their transcontinental route on 8 July 1940, while Pan Am used their Stratoliners on their Latin American routes.

While growth within America remained strong and grew steadily, the Pacific Ocean was the most significant barrier to commercial development. The first transpacific crossing, made in three stages, was by Australian Charles Kingsford-Smith, along with four others between 31 May and 9 June 1928 between Oakland, California, and Brisbane, Australia. Flying a Fokker F.VII/3m Trimotor, the crew made a series of visual and audio recordings documenting their flight. The crew consisted of fellow Australian pilot Charles Ulm and Americans James Warner and Harry Lyon, who acted as radio operator and navigator/engineer respectively.

In July 1931, Charles Lindbergh and his wife Anne Marrow, the first woman to gain a US glider-pilot licence (1930) traced a route to Japan. The pair flew northward via Alaska, Siberia and the Kuril Islands, but, political issues prevented the planned route from being used. However, operators were keen to push for a transpacific crossing, so an island-hopping route was devised, using United States territories as stepping stones. The necessary infrastructure was built at Guam and Wake islands, which were used to link San Francisco to Manila via Honolulu. This provided operators with a four-stage journey covering a staggering 8,754 miles (14,088 kilometres).

Pam Am was the first to seize the commercial opportunity that lay before them, wasting no time in issuing the specifications for an aircraft with a range of 2,500 miles (4,023 kilometres), capable of carrying a crew of four and an airmail load of at least 136 kilograms. The Martin Company responded to Pam Am's request with their M-130 flying boats, which could also carry thirty-six day passengers and eighteen in a sleeper configuration. However, the range varied, with a mail-only flying boat capable of achieving a range of 4,000 miles (6,437 kilometres) compared to passenger flights with a range of 3,200 miles (5,150 kilometres).

The aircraft started operations on 22 November 1935 with the M-130 *China Clipper* making the first flight, carrying airmail from Alameda, California, to Manila, making the journey in almost sixty hours. This was followed by the first paying passenger flight, on 21 October 1936, which carried a crew of between six and nine, including cabin stewards.

The Martin M-130s were joined by other flying boats, and at the beginning of 1937, the Sikorsky S-42B entered service with Pam Am. The first task of this new aircraft was to survey a route to Auckland, New Zealand, which would be extended to Hong Kong via Manila. The new service, which began at the end of April 1937, connected Pan Am's flight to the China National Aviation Corporation (CNAC) Hong Kong–Shanghai service. CNAC was a state-owned airline established through the merger of China Airways and the Chinese National Aviation Corporation (CNAC). China Airways was a Sino-American company established by Clement Keys in June 1929, running airmail and passenger services in competition with CNAC. Both airlines were further pressured when Wang Po-Chun, the Minister of Communications, set up his airmail service run by the Ministry of Communications.

With three airlines running the same services, it was only a matter of time before problems arose, and by 1933, Keys had left CNAC which American William Langhorne Bond now managed. Keys' replacement, Thomas Morgan, took over the management of Keys' forty-five percent share in CNAC, promptly selling this to Pan Am. This made organizing the new service easier for Pan Am, and soon the M-130s could complete the San Francisco–Manila round trip in fourteen days.

By the end of 1937, Pan Am had also established their South Pacific San Francisco–Auckland service, via Hawaii, Kingman Reef and Samoa, which started on 23 December. This was marred by disaster, and Pan Am lost the

S-42B on its second flight along with all on board. This led to the service being suspended until 12 July 1940, when a Model 314 Clipper took off on the first fortnightly airmail service. Flying to Auckland via Hawaii, Canton Island and New Caledonia, Pan Am started its passenger service on 13 September 1940.

Despite the growth in aviation, the Great Depression left its mark on the industry. The result was a lack of investment in infrastructure, design and training, which impacted the sound operation of civil aviation in several ways. In 1933, the newly elected President Franklin Roosevelt instigated the New Deal Program. Under this programme of national investment, the Bureau of Air Commerce was established within the commerce department, and the federal government took charge of the nation's air-traffic-control centres. Still mindful of the Air Mail Scandal of four years earlier, the federal government sought further regulation of aviation and, on 23 June 1938, the Civil Aeronautics Act became law. This new act extended the reach of federal regulation into civil aviation and transferred responsibility to an independent Civil Aeronautics Authority (CAA) and a three-member Air Safety Board. The CAA's remit was to encourage, develop and regulate air transportation for the improvement of mail service, as well as national defence, along with supporting commercial activities within aviation. It was also awarded powers of rate regulation and approving proposed airline routes.

Alongside these tasks, the CAA also performed several functions to improve aviation and air commerce. These included using air traffic surveys to determine passenger flow and identify new routes; it was also responsible for scrutinizing the airmail rates charged by private carriers to assess their fairness and examine price complaints from passengers. Another key role was the authorization of new overseas routes alongside providing representation in any international discussion regarding aviation regulations.

A further role of the CAA was the improvement of air safety through the Air Safety Board, which investigated 2,668 accidents between August 1938 and October 1939, making appropriate recommendations as a result of its studies to prevent similar incidents. In parallel to this new role, the CAA also began to implement and enforce new safety standards initiated via the Bureau of Safety Regulation. These included the certification of aviators and mechanics, aircraft inspection certificates and aeromedical education and examination. The Act and the work of the CAA were broadly welcomed by the industry, with Thomas Braniff calling the Act 'the Magna Carta of Aviation' as it revolutionized the entire United States aviation and aeronautics industry reorganizing it into a cohesive whole.

Another aspect of the new act meant that airlines operating in the United States were required to be issued a Certificate of Public Convenience and Necessity (CPCN) to prove that their service was viable in terms of the nation's air travel needs. Pan Am and Panagra, whose services were complementary to one another's, were immediately issued with their CPCNs, as both carriers operated a 'circle' of services throughout Latin America, covering the Atlantic and Pacific coasts along with the Caribbean. It was also recognized that their commercial operations were crucial in maintaining influence and presence in these regions.

Alongside the safety measures, the CAA also ran the Civilian Pilot Training Program (CPTP), a vocational programme to introduce college-aged students to flying. The CPTP was divided into two parts: seventy-two hours of ground-school training where the basics of aviation and aeronautics were taught, followed by up to fifty hours of flying instruction. The CPTP started in 1939 with thirteen colleges and 330 students taking part; within five years 1,132 CPTP educational institutions were delivering the course with 435,165 graduates. The project's impact would be felt the most in the immediate post-Second World War years. The CPTP had trained more pilots than any programme previously run in the United States and, in doing so, it met one of the remits set by the New Deal Program: easing unemployment.

In 1938, the CAA was split into two new agencies: the Civil Aeronautics Board (CAB), which regulated routes and carriers, and a Civil Aeronautics Administration in the Department of Commerce to handle safety certification. The CAB would be closed due to the Airline Deregulation Act of 1978, while the latter would see its functions transferred to the Federal Aviation Agency (FAA), a new independent agency, in 1958.

The storm clouds of war broke on 7 December 1941 and America quickly found herself at war and needing all the aircraft she could muster to help evacuate civilians from the Far East territories, as well as supporting operations in Europe and North Africa.

Over Latin Skies

German nationals, and money, dominated the airline industry in Central and South America. Sociedad Colombo Alemana de Transportes Aéreos (SCADTA) was founded in 1919 by Hellmuth von Krohn, flying Junkers F 13 floatplanes along Colombia's Magdalena River and coast with easy access to its neighbours and Caribbean islands such as Cuba. Using the river as a route, SCADTA F 13s flew between the Caribbean port of Barranquilla and Girardot, situated eighty-eight miles (142 kilometres) southwest of Bogotá.

The route between the two towns was immense: a total of 650 miles (1,450 kilometres), that would generally take a steamer up to ten days to travel. It took the F 13 a mere seven hours. Over the next twenty-one years, SCADTA grew, and passenger and airmail routes were added, including international flights to the United States. Then, in 1940, under pressure from the Columbian and US governments, principal shareholder Peter Paul von Bauer sold his shares to Pan Am to prevent the Nazi regime from acquiring the airline. After the Japanese attack on Pearl Harbor SCADTA ceased operations; its assets were absorbed by the Columbian national carrier, Aerovías Nacionales de Colombia (Avianca), which had been formed the previous year by the merging of SCADTA with the then national carrier Servicio Aéreo Colombiano (SACO). Avianca, despite two bankruptcies, continues to service Columbia and the Americas.

In September 1925, Lloyd Aéreo Boliviano S.A.M (LAB) was established in Bolivia by businessman Guillermo Kyllmann with flight operations using Junkers F 13 landplanes between the cities of Cochabamba and Santa Cruz. This journey, which would typically take up to three days, could be completed in three hours. LAB was the first airline to operate the Junkers Ju 52/3m, commandeered for military operations in the Gran Chaco War between Bolivia and Paraguay in the early 1930s.

In 1930, LAB founded a route between La Paz and Corumbá, Brazil, working with Brazilian operator Syndicato Condor, a Deutsche Luft Hansa subsidiary, providing a service between Corumbá and Rio de Janeiro. In May 1941 Pan American-Grace Airways, a joint venture between Pan American World Airways and Grace Shipping Company (Panagra), which had been formed in September 1928, took over the administration of LAB on behalf of the Bolivian government.

Another German-influenced airline, also based in Brazil, was Sociedade Anônima Empresa de Viação Aérea Rio-Grandense (Varig). Varig was founded in May 1927 by First World War pilot Otto Ernst Meyer-Labastille who operated the ten-seat parasol-winged Dornier Do J Wal flying boat from the coastal city of Porto Alegre. By the time the United States had entered the Second World War, Varig had built up several routes and was operating the DH.89 Dragon Rapide, which serviced its first international route to Montevideo, Uruguay. However, in 1942, Meyer-Labastille resigned as the manager director and Érico de Assis Brasil was voted to take over by the company's board. De Assis Brasil was killed in an air crash, and Ruben Martin Berta, Varig's first employee, took over the role, leading the company until his death in 1966.

Peru saw the Compañía de Aviación Faucett founded by a group of Peruvian businessmen, including Paul Winer and Ernesto Ayulo, led by American Elmer Faucett, in September 1928, and in doing so, became the country's first commercial airline. Faucett had arrived in Peru as a representative of the Curtiss Export Company in 1920, where he established the Stinson aircraft factory. The factory produced the aircraft Faucett would use in the airline, and Faucett, the pilot, proved his skills as an aviator by completing the first trans-Andean flight between Lima to Iquitos. The early years for Faucett were spent gaining experience flying in a country featuring a vast mix of terrain, much of it challenging. By 1937 the airline had absorbed Cia de Aviacion Peruanas

SA from Panagia and was operating seven aircraft to cover a range of internal destinations from its Lima hub. The airline would continue to run into the late 1990s, overcoming Marxist insurrection and political strife before its weak financial structure led to its bankruptcy and collapse in 1999.

The military initially managed Chile's air service; however, Línea Aeropostal Santiago-Arica (LASA), founded by Commodore Arturo Merino Benítez, the founder of the Chilean Air Force in 1929. Initially carrying airmail, then passengers, LASA rebranded itself in 1932 as Línea Aérea Nacional de Chile (LAN). Benítez was careful to maintain corporate dominance over the domestic routes he had established with LAN and restrict external competition, mainly from Panagia. This would result in the airline being unable to source US-built aircraft until the Second World War. These would take the form of Lockheed Electras and Lodestars as well as DC-3s, which replaced French Potez 560 monoplanes and Junkers Ju 86Bs, parts for which were virtually impossible to source. Despite Benítez's desire to retain sovereignty over domestic routes, LAN worked with LAB and Faucett in providing international flights to and with its neighbours.

In November 1933, Viação Aérea São Paulo (VASP) was established by the state government of São Paulo, designed to operate an air route between São Paulo and Rio de Janeiro known as the Ponte Aérea (Air Bridge). VASP provided a staggering 370 flights a week in both directions serving the route with the General Aircraft Monospar ST-4 utility aeroplane, often operating from improvised landing strips. As VASP used only landplanes, it was felt prudent to construct a landing strip, and in 1936, VASP built Campo da VASP, now known as Congonhas Airport, in São Paulo. In 1939 VASP bought Aerolloyd Iguassú S.A., established in 1933, and gave VASP an extended reach into the south of Brazil.

In 1931, New Zealand Royal Flying Corps veteran Lowell Yerex founded Transportes Aereos del Continente Americano (TACA), based in Honduras. TACA soon offered international flights to Panama, Guatemala, Nicaragua and El Salvador, echoing Yerex's desire to establish an airline in every Latin American country. Soon a network of TACA subsidiary airlines was operating. But unfortunately, despite his hard work and building up a unique system of international routes, Yerex could not realize his dream.

That said, he would shape the fortunes of other Latin American and Caribbean airlines, founding BWIA West Indies Airways Limited in 1939 at the invitation of Lady Margaret Young, the wife of Trinidad and Tobago's governor, Sir Hubert Young. This was followed in 1942 by the founding of Empresa de Transportes Aéreos Aerovias Brasil S/A as the national airline of Brazil.

In Mexico, the aviation industry was quickly established post-war, and in 1921, Compania Mexicana de Transportación Aérea (CMTA) began services between Mexico City and Tampico, where it used Lincoln Standard L.S.5

biplanes to transport wages to the workers in the Tampico oilfields. In 1924, William Mallory and George Rihl, who had established the Cia Mexicana de Aviación (CMA) in August 1921, bought CMTA. The following year Sherman Fairchild purchased a twenty percent stake in the company and in 1928 introduced the Fairchild FC-2 high-winged monoplane to the fleet.

In 1929, Trippe of Pan Am bought a majority share in the company, which opened up an international route from Mexico to the United States using the Ford Trimotor to fly the route, which Charles Lindbergh opened. CMA soon established routes throughout Latin America and expanded and strengthened domestic routes. CMA was the first foreign airline to provide a service to Los Angeles, making its inaugural flight on 3 January 1936.

By the beginning of the 1940s civil aviation had established routes, hubs and an ever-growing infrastructure, with numerous mergers and, in some cases corporate monopolies, being established. Aircraft were increasingly all-metal types that were flying further and faster. The skies were opening up, and electronic navigation aids were making long-haul flying safer than ever. Commercial aviation had overtaken the numerous short-lived flying circus and madcap races to become the lynchpin of civil aviation, with the aid of those early pioneers, providing the aeronautical and allied service industries with stability. However, the onset of war would deliver new challenges and innovations that would advance aircraft design and open the skies for even cheaper exploration.

Chapter 4
The Arrival of the Jet

Rewriting the Rules: Legislating for the Post-War Boom Years

The post-war era of flight and rebuilding what would become a truly global business would be helped by changes in legislation. The first essential step toward supporting an industry working in a post-world war market was the signing of the Chicago Convention on International Civil Aviation (Chicago Convention). Fifty-two states signed this refinement of the 1919 Paris Convention on 7 December 1944, all members of the newly established United Nations. Such was its success and importance that today the convention has 193 signatories agreeing to its articles and annexes. The 1944 Chicago Convention became the framework upon which effective international air law was written, including the all-important rules regarding air navigation.

The first of the ninety-six articles was the most striking. The use of precise language made the convention's intent unmistakeable. 'Every state has complete and exclusive sovereignty over airspace above its territory.' This statement was the definition the 1919 Paris Convention had initially sought to establish and be accepted by the League of Nations.

The Second World War had provided jurists, diplomats and industry lobbyists with more than enough examples of why this simple statement was vital for the post-war world aviation industry. This article also allowed aircraft from other contracting states to make non-scheduled flights into or over the territory of another signatory state. Aircraft could also make a landing for non-traffic purposes, that is, landing for any purpose other than taking on or discharging passengers, cargo or mail. They could do this without prior permission, but such acts would remain subject to certain limitations. This would prove to be a helpful clause in the advent of an emergency experienced by an aircraft en route between two points, technical or otherwise.

The 1944 Chicago Convention was a little more restrictive for scheduled international flights, reassuring many, if not all, signatories. It also ensured the skies above land or sea territories would not become a dangerous and unregulated free-for-all. Scheduled international flights would require permission to overfly the territory of another state except where special consent or other agreement with that state was in place. Such bilateral agreements are critical to any shared air-traffic management system that uses the mutual exchange of commercial rights for aircraft operation by named operators, such as airlines.

An Air Transport Agreement (ATA) or an Air Service Agreement (ASA) are such formal arrangements. The first bilateral agreement between two countries was the Bermuda Agreement, signed by the United States and the United Kingdom in 1946. This was replaced on 24 June 2010 by the more comprehensive, encompassing European Union–United States Open Skies Agreement, to which the United Kingdom remains a member for the time being. However, renegotiations have started on a future United States–United Kingdom ATA due to the United Kingdom leaving the European Union in 2020.

A further piece of legislation adopted by the 1944 Chicago Convention was the International Air Services Transit Agreement (IATA), also known as the 'Two Freedoms' agreement. The IATA is an integral part of legislation for the world of civil aviation as it allows aircraft of a signatory state to fly as part of scheduled international air service and gives them the freedom to overfly other signatory states' territory without landing. It also allows aircraft landing rights for non-traffic purposes such as refuelling and maintenance. Although similar to Article 1 of the 1944 Chicago Convention, under the IATA, states can impose remuneration for services rendered to visiting aircraft, which can occasionally result in controversy due to the amounts charged.

The International Air Transport Agreement (IATA), or the 'Five Freedoms' agreement, was another vital element of aviation legislation introduced with the 1944 Chicago Convention and acted as the trade association for the world's airlines. The IATA works in parallel with the IATA's 'Two Freedoms' and grants aircraft operators the right to commercial privileges. In this instance the aircraft can put down passengers, mail and cargo from either the country where the aircraft is registered or can pick up or put down passengers, mail and cargo from or to any state which is a signatory to the agreement. Such agreements tend to be bilateral in nature, but as trading blocs became more politically aligned through member treaties, such as the European Union, these bilateral agreements were revised to reflect the new geopolitical status of members. In total, the 1944 Chicago Convention created nine recognized 'Freedoms of the Air'.

Another critical aspect of the 1944 Chicago Convention was the establishment of the International Civil Aviation Organization (ICAO) in April 1947, a body to which all 193 signatories of the Chicago Convention belong. ICAO took over the role previously led by the International Commission for Air Navigation (ICAN), which worked exclusively for the League of Nations. ICAN was founded in 1922 and advised the League of Nations on air navigation and air law until 1945.

ICAO is a specialized UN agency with a core function to maintain an administrative and expert bureaucracy that supports diplomatic interactions. It researches new air transport policy and standardization innovations as directed and endorsed by governments through its governance and management structure. As a UN agency, the ICAO's autonomy and impartiality are guaranteed, and it has universal authority over its involvement with aviation. As well as a headquarters office in Montréal, Canada, there are eight additional offices worldwide. The key aims of ICAO are to develop the principles and techniques of air navigation and to ensure the safe and systematic growth of civil aviation and aeronautics.

Within ICAO are the assembly, an elected council, and four separate committees responsible for a range of tasks relevant to the safety and efficiency of civil aviation: the Multilateral Aviation Convention and International Aeronautical Body, Technical Standards and Procedures, Provisional Air Routes and the Interim Council. In addition, within each commission is a series of sub-committees. These include Air Navigation Principles, Air Traffic Control Practices and Registration and Identification of Aircraft. Each committee and sub-committee perform a vital role in enabling civil aviation to take place regardless of how dry the subject appears. The standards it sets, from passport design to airport code allocation, allow for the transit of some five billion passengers and 70 million tonnes per year.

As ICAO designs the standards deemed appropriate for safe air travel, all the Chicago Convention signatories must apply these standards, often in collaboration with other signatories.

The final critical international body of civil aviation is the International Air Transport Association (IATA). Founded in April 1945, the IATA replaced the association of the same name founded in 1919. Its initial work focused on the technical elements of aeronauts and aviation. Much of its expertise helped to establish ICAO. In addition, as the 1944 Chicago Convention did not address the economic regulation of the post-war airline industry, it was felt that IATA could fill this void.

IATA established itself as the airline industry's economic regulator, ensuring prices that did not leave airlines out of pocket. This role has led to accusations that IATA is a cartel favouring the industry's interests over that of the consumer. However, many airlines are subject to domestic competition laws; tariffs and rates are subject to federal government approval, not IATA. This means that IATA cannot legally be considered a business cartel. It also sets the standard for universal air transport documentation and flight safety, championing more environmentally considerate practices within the industry.

New Horizons and Wide Vistas: Passing the Baton

Aside from the bounty of technological benefits, including radar, the Second World War would provide a significant developmental boost to the airline industry: sustained and efficient long-range flight. By the war's end, there was a surplus of military aircraft, including transport aircraft, often powered by the most up-to-date and fuel-efficient engines. At the same time, new jet technologies were being developed at a fearsome pace as the aviation and airline industries recognized their potential, and development was pushed through as the benefits for civil aviation became increasingly apparent.

Surplus military aircraft were now in reach of the spirited entrepreneur who had the means to purchase and fly aircraft to help build their new business empire. DC-3s (C-47s), whose pre-war price was more than $120,000, were now selling to the public for $800 at disposal sales organized by the War Assets Administration (WAA) and the Reconstruction Finance Corporation (RFC). Such was the rush for military surplus aircraft for business use and memorials, allied to the need to shift redundant airframes that some head-turning purchases were made, including a Boy Scout Troop who paid $300 for a B-17.

Despite the number of surplus aircraft, 175,000 in the United States alone, manufacturers were keen to apply the lessons they had learned during the war and incorporate them into new aircraft designs. Airlines, new and old, were eager to use this new technology alongside their recently acquired military-surplus stock, which had filled gaps in air fleets damaged during the war. This would help usher in a new era of affordable tourist travel, both long- and short-haul. Alongside this surplus stock, there was also an abundance of trained flight crews, engineers and support staff, with experience and expertise that would have been seen as fantastical before the war. Despite post-war austerity, the world was the canny investor's oyster.

This new world was further aided by new airfields, especially in the British Isles, which serviced transatlantic flights. In turn, flights were aided by a range of hyperbolic radio navigational systems, including the Gee and Loran (long-range navigation) systems which were high-powered radio signals used by shipping and aircraft crossing the Atlantic and Pacific oceans. As the pre-war routes were re-opened, or as a result of geopolitical changes, such navigational aids became invaluable.

Further refinement of the system resulted in the Very High-Frequency Omnidirectional Range (VOR), unveiled in 1946. VOR was developed from the pre-war LFR, which in turn was developed into the Visual Aural Radio Range (VAR). LFR had helped pilots successfully navigate the often-dangerous transcontinental flights in the United States. By the late 1930s, an improved system was sought by the Bureau of Commerce which, in 1937, unveiled VAR. Unfortunately, wartime shortages of suitable VHF equipment meant the system could not be fully rolled out until 1944. With the consequent unveiling of VOR, the VAR rollout did not have a chance to fully replace LFR, showing how rapidly ground-based control and navigation technology were advancing.

VOR was designed to provide a 360-degree bearing to and from the transmitting station as a navigation aid for aircrew. VOR would act as an electronic pathway feeding two separate indicators carried by aircraft: the Course Deviation Indicator (CDI) and the Horizontal Situation Indicator (HSI). These signals would help define the air route the aircraft was taking. Initially, VORs were mechanically rotated to provide a lighthouse signal effect which the CDI and HSI then interpreted to give the aircraft its correct bearing from the issuing station. Soon this system was replaced by electronic Doppler rotation, thus removing the physical maintenance commitment to ensure beacons worked. This system is also vastly superior to the mechanical VOR and helped provide a more accurate pathway for receiving aircraft to follow. By the 1960s, the electronic solid-state VORs had replaced the LFRs and VARs that remained in service.

Soon the system was in place worldwide, providing aircraft with two unique routes: air highways operating a route for aircraft flying below 16,404 feet (5,000 metres) and jet routes, operating above this altitude. As the VOR system works using VHF, it remains Line of Sight (LOS), limiting the system's range to the horizon. However, it remains reasonably accurate, offering aircraft an accuracy of ninety metres (295 feet). A further feature of the VOR system is the Distance-Measuring Equipment (DME), which utilizes radio signals to give receiving aircraft the slant range, the LOS distance between two points, further aiding navigation.

With the introduction of such systems now slowly fading out in favour of a Global Positioning System (GPS), airlines were able to offer increasingly safe global air travel. As the system originated in the United States, aircraft such as the Boeing Model 377 Stratocruiser, introduced in 1947, and Lockheed's Constellation, which had entered service in 1943, were able to use these new navigational systems with ease. However, Douglas, whose DC-3s would dominate the skies in those immediate post-war years, would form the foundation for many domestic airlines. However, its joint weaknesses of short range and low cargo capacity prompted Douglas to push its revisited DC-4 design, introduced in 1942, into the civil aviation market. The DC-4 could seat eighty-four passengers using a high-density seating plan, which led to the newly established charter and scheduled airlines taking advantage of these assets and the DC-4's range of 3,300 miles (5,300 kilometres).

Customers such as Pan Am, Braniff, KLM and Air France soon purchased the DC-4. However, the new aircraft, superior to its now-surplus predecessors, was not the vast sales success it should have been. BOAC chose the DC-4 to replace its Avro Tudors; these DC-4s, known as Argonauts, featured a great deal of British equipment, including engines, which helped to keep costs down for BOAC and Douglas.

Another British derivative of the DC-4 was the ATL-98 Carvair converted by Aviation Traders, a company run by Freddie Laker, a British airline entrepreneur who pioneered the no-frills business model of airline service, now used by Ryanair and easyJet. The Carvair could carry five cars and twenty-two passengers a distance of 2,300 miles (3,700 kilometres) and replaced Bristol's wartime-designed Type 170 Freighter, whose range was a mere 820 miles (1,320 kilometres).

The DC-4 also had the laurels of being the first landplane to make a commercial transatlantic flight, on 23 October 1945, between New York and Hampshire, UK, via Gander, Canada and Shannon, Ireland. However, technology rarely sleeps, and this was undoubtedly the case when Douglas revealed the DC-6 in February 1946, another four-engine design. Despite early issues, the DC-6 was the most economical of the post-war piston-engine airliners. A larger aircraft, the DC-7, followed and could accommodate up to 105 passengers and fly a staggering 5,635 miles (9,069 kilometres). Airlines were quick to utilize the power that also came with the new aeroplane engines, an amazing 3,400hp per engine, over double that offered by the DC-4. These features gave operators the perfect transatlantic, transcontinental airliner; Douglas had struck gold. Pan Am was quick to purchase the new airliner, and, finding their hands forced, BOAC, KLM and other European airlines were quick to follow. Launched in 1953, the DC-7 would be Douglas's last propeller-driven aircraft.

While Douglas had the advantage afforded to it by its wartime production of the DC-3, Boeing turned their experience of bomber design, including the groundbreaking B-29, toward civil aviation. The Model 377 Stratocruiser, with its distinctive double-bubble fuselage, was based on the B-29 and utilized Boeing's experience producing long-range aircraft. This design was revolutionary in terms of passenger comfort, offering the traveller two levels, a pressurized cabin and hot food. Boeing was going squarely for the luxury market, and the cocktail-bar-equipped Stratocruiser would provide the glamour it needed to sell the dream to both airline and consumer.

Again, Pan Am, attracted by the aeroplane's glitz and impressive range of 4,200 miles (6,800 kilometres), led the way with orders, followed once more by BOAC and Scandinavian Airlines (SAS). SAS were one of the first merger airlines to emerge in post-war Europe. The new company brought together the national airlines of Norway, Det Norske Luftfartselskap AS (DNL), Sweden, Svensk Interkontinental Lufttrafik AB (SILA) and Denmark, Det Danske Luftfartselskab (DDL) on 1 August 1946. In addition, SAS started offering transatlantic intercontinental flights between Stockholm and New York.

The production run of the Stratocruiser was short-lived. After fifty-six aeroplanes were built, Boeing ceased production in 1950, the same year that de Havilland unveiled the Comet, the world's first jet-powered airliner.

Boeing concentrated on harnessing the jet engine for future projects and was far from a threat to Douglas's domination in the immediate post-war civil aviation scene. On the other hand, Lockheed remained a commercial danger with their Constellation, which had its roots in a 1939 design brief issued by the enigmatic Howard Hughes, the owner of TWA at the time. Hughes, whose involvement in aviation would span almost fifty years, gained numerous achievements, including the award of the Harmon and Collier trophies. Hughes envisaged a pressurized aircraft that would become the 'airliner of the future'. The Lockheed design team, with Hughes, worked to produce his dream.

The resulting aircraft, which first flew on 9 January 1943, was long, sleek, pressurized and fast, featuring a distinctive triple tail. With the onset of war, the first batch of Constellations were used as military transports, with the first actual civilian version flying on 19 October 1946, eight months after the first civilian flight by TWA. This flight took passengers from New York to Paris on what was to be the first of TWA's scheduled transatlantic crossings. The Constellation was an instant hit, with Pan Am, BOAC, KLM and Air France all purchasing this fantastically exotic-looking aeroplane.

The Martin Company also released new aircraft, the most notable being the 4-0-4; however, the two engines were a little too late in coming, and Martin was up against competing companies. Meanwhile, in Europe, the picture among manufacturers remained muted. The industrial giant that had been awoken in the United States by the war now switched its attention to focus on new airliners to meet the needs of the industry.

Britain had designed some fantastic aircraft during the war. However, it was now struggling to get back on track with its civil aviation industry. Most early post-war airliners were merely civilianized bombers. The one design that emerged from the war, which was reasonably practical, was the Avro York, a transport aircraft based on the Lancaster heavy bomber, capable of carrying fifty-six passengers. The York was undoubtedly a hit with many British and Commonwealth airlines, filling an immediate post-war gap, initially by the Tudor. Again, Avro based a transport on one of its bombers, another Roy Chadwick design, the Lincoln. Although the Tudor was Britain's first pressurized airliner, it was released to a market keen for bespoke designs that reflected the new era. A tailwheel undercarriage-equipped aeroplane based on a bomber was not the answer to the domestic aviation market's woes.

Another interesting anachronism of BOAC's wartime operation was the continued use of flying boats. The use of this aircraft type, while everyone else was ditching theirs, led BOAC to introduce the Short Solent in 1948 to fly the Southampton–Johannesburg route.

Hadley Page entered the fray in with their HP.81 Hermes. Despite having pre-war form for airliners, the Hermes was, again, too little too late. It failed to gain a foothold in a rapidly changing market. The Hermes was joined by

Airpseed's Ambassador, an airliner designed to meet the threat posed by the DC-3. Yet this utilitarian-looking, high-wing aeroplane, designed in wartime, and widely used, was not going to be able to compete in an industry that was eying up the future.

One manufacturer that did make the successful transition into propeller-power airlines was Vickers, fielding the small Viking, which could trace its heritage back to the Wellington bomber and the larger Viscount. Although both aircraft met differing needs, the Viking was a straightforward short-haul airliner. At the same time, the turboprop-powered Viscount was not only the first airliner to be so powered, it was also a success story for Vickers with the type purchased by Air France, Trans Australia Airlines and Air Canada.

But it was de Havilland that pulled out the stops. Their post-war DH.104 Dove and DH.114 Heron airliners launched in 1945 and 1950 were short-haul airliners that met specifications called for by the Brabazon Committee. The Brabazon Committee, established in 1942 to meet the post-war airliner market, was led by Lord John Moore-Brabazon, a keen aviator with much flying experience. He turned his experiences in the air into nautical development, designing the world's first and only autogyro boat. Moore-Brabazon was eager to make Britain's post-war aviation industry a world leader, and the giant Bristol Brabazon, the largest aircraft ever made in the United Kingdom, was testimony to his desire. Despite being a costly failure, it showed that the aviation and aeronautical industries retained the desire to produce aircraft worthy of note.

Various specifications were issued, which the Dove and Heron met. For their party piece, de Havilland had very much saved the best till last and, in doing so, would produce a stunning and world-changing airliner: the turbojet-powered DH.106 Comet.

Breath of the Gods: Harnessing the Jet Age

The age of the jet-powered aircraft was heralded by the arrival of gas turbine-powered engines. The turbojet, a three-stage engine, was first patented in 1921 by French agricultural engineer Maxime Guillaume. Sadly Guillaume's design was never realized due to metallurgical limitations of the time. On 12 April 1937, Royal Air Force engineer Frank Whittle powered up his first Power Jets WU (Whittle Unit). Whittle's design was an axial-flow turbine engine, very similar to that designed by Guillaume. Whittle's engine drew the three stages together: compression, combustion and turbine.

The compression stage, also known as the cold section, draws air into the engine through a series of blades, feeding the air into the combustion stage. Here the air is joined by ignited fuel. This heated air is drawn out by the low- and high-pressure turbine stages, which form the engine's hot section. These sections extract the energy released by combustion, forcing the heated jet of hot air through the exhaust. Designers and engineers developed two types of turbojets: centrifugal and axial.

The centrifugal-type engines are similar to turbocharger designs, resulting in a shorter, more bulbous engine such as the Rolls-Royce Nene. The shortcomings of the centrifugal-type engine led designers and engineers to develop axial-flow engines. Instead of pushing the air over the top of the shrouded shaft drive, the air is passed directly along the shaft. This design allows the air to pass through several smaller fans, increasing the compression of the air before being combusted, as opposed to passing through a single sizeable centrifugal impeller fan. Such engines were by design smaller in diameter, but longer, to help enclose the banks of fans in their various compressor stages.

War sped up the development process so, that by the war's end, the axial-flow turbine engine was on its way to powering the next generation of military aircraft. However, the noisy and fuel-hungry attributes of the turbojet meant that the only real customers for such engines would largely remain the military. However, Concorde and the Tupolev Tu-144 would use the engine's simple construction and low weight-to-power ratio. In addition, the turbojet allowed for a higher rate of climb. It was only at high speed that the turbojet produced a higher propulsive efficiency. These aspects made the turbojet, despite its large size, an extremely suitable engine for an aircraft like Concorde, travelling at speeds above 800mph (1,287km/h).

A further development which helped understand ducted airflow was the work done by Italian aircraft manufacturer Società de Agostini e Caproni and engineer Luigi Stipa. Stipa's design, the cartoon-like Stipa-Caproni, had its engine and propeller entirely enclosed by the aircraft's fuselage. This turned the Stipa-Caproni into a ducted fan with wings. While not taken beyond the initial concept stage, the Stipa-Caproni was technically challenging to fly due to its high aerodynamic drag and low speed. However, the Stipa-Caproni did inspire the development of the turbofan.

During the war, the turbofan engine was developed from the turbojet engine. The critical difference between the turbofan and the turbojet is the addition of a shrouding duct between the outer engine nacelle and the three stages of the traditional turbojet engine. At the front sits a sizeable low-pressure spool, also protected by the shroud, feeding air through the duct and the low-pressure compressor. This passes air into the high-pressure compressor which is used in the same manner as a turbojet engine. The air passing through the surrounding duct, which acts as a cold nozzle, mixes the cold air with the hot air of the exhaust. The large fan helps increase the mass flow of the engine, producing a more efficient – and a must for civilian use – quieter engine.

Turbofan engines are rated as either low- or high-bypass types. A high-bypass engine is defined as having a more significant portion of air flowing through the cold air ducts than through the turbojet engine, typically a 2:1 ratio. The high-bypass engine is now the industry standard for those used on airliners, with examples made by Rolls-Royce, General Electric and UEC-Aviadvigatel. Although the military uses the low-bypass types for high-speed use, these are often coupled with an afterburner for increased power. The critical disadvantage of the turbofan engine is its inability to operate at higher altitudes compared to the pure turbojet engine.

Turboprop and turboshaft engines use the exhaust jet to turn a rotating shaft, which turns an auxiliary mechanism via a gearbox to provide thrust. The thrust generated by the propeller, or in the case of a turboshaft, the main rotor, draws its power initially from the turboprop engine. The thrust generated by the propeller or rotor then joins with the thrust of the turbojet to help produce a greater thrust for use in the air. As a rule, the turboprop has better performance than turbojets or turbofans at low speeds due to high propeller efficiency. Still, turboprops become increasingly noisy and inefficient at higher speeds due to propeller vibration and lack of turbojet power.

The Comet was a pivotal moment in civil aviation history when it took to the skies on 27 July 1947. Despite its small size and inability to make transatlantic flights, the first Comet broke a series of distance records. The Comet would make the world's first scheduled jet-powered service on 2 May 1952, flying from London to Johannesburg. De Havilland had struck proverbial gold, and despite the small size of the early Comet, the order books soon filled up, buoying a new era for the British aircraft industry.

On 2 May 1953, a Comet leaving Calcutta, India, inexplicably broke up, followed by two similar catastrophes the following year in Rome. The recovery of the wreckage soon identified fatigue stresses, a little-known and understood phenomenon. De Havilland ground the remaining Comet fleet, notifying other manufacturers of the cause of the aeroplane's loss, applying the lessons learned to the Comet 2 and modifying the original Comets. Despite these efforts and further developments, including a larger 101-seat Comet 4B, the damage was done, and the British aircraft industry would struggle to recover. However, BOAC could pip Pan Am to the post by introducing the first transatlantic crossing with the Comet 4 on 4 October 1958. This was a coup for BOAC and de Havilland as the Comet had beaten the newly introduced Boeing 707 in making the crossing by three weeks. Regardless of the early mishaps, the Comet would continue to fly until 1997, and the military version known as the Nimrod would finally retire on 28 June 2011, over sixty years after the Comet's maiden flight.

The 1950s would see further developments among airlines. Qantas would take over pre-war Australia–North America routes from British Commonwealth Pacific Airlines (BCPA). BCPA was formed by an agreement between the British, Australian and New Zealand governments in September

1946 to service the long-haul transpacific Australia–North America route from Sydney to Vancouver. The route servicing was undertaken by Australian National Airways (ANA) using DC-4s. ANA was founded in 1936 by the merger of several smaller domestic flight providers and remained solvent during the Second World War by providing support service to the American forces based in Australia for the duration.

After the war, ANA flew DC-3s and DC-4s, which it would use on its long-distance, high-capacity services. Despite funding pressures, it would eventually purchase DC-6Bs. These aircraft would also compete with Trans Australia Airlines' (TAA's) Viscounts, especially for overseas routes. To ease commercial tensions between the ANA and TAA, the government of Australia agreed to grant a duopoly in 1952. This agreement, known as the Two Airline Agreement, allowed airlines to operate flights between state capitals and nominated regional centres. The policy also prevented new entrants from making inroads into Australia's domestic aviation market. The agreement also ensured that both airlines carried approximately the same number of passengers, charged the same fares and had similar fleet sizes and equipment.

In 1957, ANA, along with its position in the Two Airline Agreement and overseas routes, was bought by Ansett Airways (AAW) as a result of the death of ANA's owner, Sir Ivan Holyman. AAW was founded by Sir Reg Ansett in February 1936 and was rapidly built into a strong airline. However, Ansett later offered to sell to ANA after establishing the government-backed Trans Australia Airlines (TAA) in February 1946, a direct competitor to both ANA and AAW.

In the end, the sale of AAW to ANA did not occur. However, Ansett remained a shrewd operator and, in 1947, began operating a package holiday serviced by Short S.25 Sandringham and Consolidated Catalina flying boats. The flying boats would take passengers to the paradise of Hayman Island, off the coast of central Queensland. The Hayman Island resort was developed to appeal to the post-war generation, who now had the leisure time and the money to spend on holidays in exotic locations.

Ansett continued to concentrate on building routes out of Melbourne and purchasing smaller domestic airlines to increase his commercial and political influence. These flights were made by adapted DC-3s, with narrower seats installed, so twenty-eight instead of the usual twenty-one passengers could be carried. Added to the low ticket cost and virtually no in-flight service, Ansett's ideas soon caught on, with TAA offering cheaper flights on its Viscount fleet. Not long after purchasing ANA, word reached Ansett that TAA was considering buying the Sud Aviation Caravelle.

The Sud Aviation SE 210 Caravelle was the first jet airliner to be produced by the French. It included some de Havilland designs and components developed for the Comet. But rather than having four engines placed between the fuselage and wings, the Caravelle featured a rear-fuselage engine-mount configuration for the two Rolls-Royce Avon engines. It was larger than the Comet and could carry up to ninety-nine passengers compared to the Comet I and II's forty-four and the Viscount's seventy-five. It was a considerable leap forward technologically and commercially.

Ansett lobbied against any such purchase, as he lacked the capital and, more importantly, the infrastructure to buy jet-powered aircraft for ANA. Furthermore, any such purchase by TAA would automatically place AAW at a commercial disadvantage. Ansett's lobbying to push for the purchase of turboprop-based aircraft was a success. TAA purchased more Viscounts, along with the new four-engine Lockheed L-188 Electra. The L-188 was the first American-built turboprop airliner, making its maiden flight in December 1957. Its giant propellers and small wing surface area gave it exceptional short take-off and landing (STOL) capacities, which were ideal for operating in Australia. Despite his victory, Ansett had merely delayed the arrival of the jet and, in 1964, the Boeing 727-100 began service on Australia's domestic air routes.

In 1979, Peter Abeles, the founder of the logistic company TNT, along with media mogul Rupert Murdoch, wrested control of AAW from Ansett. After numerous mergers, a series of bad business investments and controversial policies, including banning HIV-positive passengers in 1984, AAW, known as Ansett, went into liquidation in July 2002 after negotiations for a consortium rescue package collapsed.

In 1948, BCPA took delivery of four DC-6 sleeper aeroplanes to service the transpacific route, and by 1952 it had ordered six de Havilland Comets. In 1953, BCPA expanded the order to three Comet 2s, which featured the more powerful Rolls-Royce Avon engine, a larger wing and increased range. In October 1953, the three government shareholders approached Qantas with a view to the company purchasing BCPA. The following year the sale was agreed upon, and Qantas took over the Australia–North America routes, as well as the order for the three Comet 2s.

In Europe, developments in post-war aviation continued. For Germany, this meant rebuilding the domestic civil aviation industry and its national carrier. The country, now divided between the Western Powers and the Soviet Union, had its former national carrier, Deutsche Luft Hansa (DLH), dissolved after the war due to its connections to the Nazi Party. The first step in rebuilding a national carrier came on 6 January 1953 when Aktiengesellschaft für Luftverkehrsbedarf (Luftag) was established in Cologne, West Germany. Despite West Germany not having sovereignty over its airspace, Luftag began rebuilding its fleet and infrastructure, including employing former DLH employees.

Despite not knowing whether civil aviation would be allowed to start again, Luftag set up its first hub at Hamburg airport. It also ordered four short-range, two-engine Convair CV-340s capable of carrying forty passengers, and four 106-seat Lockheed L-1049 Super Constellations for its future long-haul flights, which had yet to be agreed to by the occupying powers. These steps in themselves were a huge gamble for Luftag. Still, the new company's self-belief paid off the following year when, on 6 August, it was allowed to purchase and use the Deutsche Lufthansa Aktiengesellschaft (DLH) name and logo, with the name shortened to Lufthansa.

Lufthansa's confidence paid off again on 1 April 1955 when it made its first domestic post-war flight in West Germany, followed on 15 May by the start of international flights, initially within Europe. On 1 June, a Lufthansa Super Constellation made the first transatlantic crossing to New York. In August the following year, the first South Atlantic crossing was made. These flights would lead to well-established transatlantic air routes to the Americas by the end of the decade, as well as routes to the Middle East. However, the new airline also pushed its marketing and public relations departments hard to promote the airline, its new Frankfurt hub and West Germany in general.

To the east, in 1955, the German Democratic Republic tried to use the Lufthansa name for its national carrier, Deutsche Lufthansa (DLH). The West German Lufthansa immediately started legal proceedings against the East German company. It was clear that the original company would win any legal battle. In 1958, the East Germans set up Interflug Gesellschaft für Internationalen Flugverkehr m.b.H. or Interflug as a backup company. Operating out of East Berlin's Schönefeld Airport, which remained closed to Lufthansa until the 1990 German reunification, Interflug provided charter and scheduled international and intercontinental flights. By 1963, Interflug had replaced DLH as the national carrier's name. As a result of the reunification, Interflug was wound up, with its final commercial flight between Berlin and Vienna taking place on 30 April 1991. Former staff acquired five of the company's Ilyushin Il-18 turboprop aeroplanes, establishing Berlin-Brandenburgisches Luftfahrtunternehmen GmbH or Berline, which operated chartered cargo and passenger flights, flying old Interflug routes. Berline was not a success, and on 31 March 1994, the company filed for bankruptcy.

Meanwhile, Lufthansa continued to push itself as a forward-looking company reflecting post-war values and keen to build toward its future. In keeping with this ethos, Lufthansa ordered four of the new Boeing 707s in late 1958, which had first flown on 10 December 1957. The four-engine 707 was capable of carrying over 200 passengers and was most definitely the right aeroplane at the right time.

The post-war boom of package holidays saw many operators start tourist charter flights. While the Comet was very much cutting-edge, the 707 introduced a more critical factor: space. As with all technology, the initial developments were never cheap. The banks were taking a massive gamble with an initial 707 unit price of $4.3 million compared to the Comet 4's $1.7 million cost. Yet the 707 had been through an exacting development process, aided by information kindly shared by de Havilland in light of the early Comet disasters. Boeing also realized that passenger capacity had to be raised to make their aircraft viable purchasing options for those airlines operating on a budget.

The 707 was a huge success and is credited for introducing a viable jet aircraft into civilian air service. With 865 units built, 707s flew with airlines worldwide, providing comfortable long-distance travel. From the 707, Boeing would go on to develop the short-range 720, which made its first flight in November 1959 and which would also see service with Lufthansa.

Using the 707, Lufthansa steadily replaced its Constellations on the transatlantic routes and established routes to the Far East and South Africa. This was the start of a beneficial partnership between Boeing and Lufthansa, with the carrier going on to introduce the tri-engine 727, developed to land at smaller airports on its Frankfurt–Tokyo route in 1964. This was followed in 1968 by Lufthansa making the first overseas order of the twin-engine 737 for servicing their short-haul flights.

To the east, Tupolev had introduced their first jet-powered aeroplane, the twin-engine Tu-104, which made its first flight on 17 June 1955. This medium-range aeroplane was designed to meet the requirements of the national carrier, Aeroflot, which was keen to build a post-war service that was every bit as technologically advanced as in the west. Capable of carrying up to 115 passengers, the Tu-104 filled a unique place in civil aviation history: it was the second jet-powered airliner to enter service after the Comet. Moreover, in 1957 it was operated on the first jet-only route, operated by Czechoslovak Airlines between Prague and Moscow.

After the Second World War, Aeroflot, which had continued to fly in support of the war effort, was keen to repair the considerable damage done to its infrastructure. This was driven, in part, by the desire to reconnect with the many socialist republics that fell under the Soviet sphere of influence. By 1945, Aeroflot had carried an astonishing 537,000 passengers, mainly in pre-war Lisunov Li-2s. In addition, in 1947, the two-engine Ilyushin Il-12 entered service, providing flights throughout the union.

Another aeroplane which came into service in 1947 to provide general aviation services was the Antonov An-2 single-engine biplane. The An-2 would become a success story for Antonov. Although designed primarily as a service aircraft for supporting agricultural needs, the airframe was exceptionally adaptable. As well as being fitted as a crop duster, the An-2 could also fulfil other roles, including firebombing, air ambulance and twelve-seat light passenger aeroplane. By the time production ceased, over 18,000 An-2s had been built.

The investment in infrastructure soon paid off. By 1950 Aeroflot was servicing a network of 183,600 miles (295,476 kilometres) while carrying over 1.6 million passengers alongside 30,500 tonnes of mail and 151,000 tonnes of cargo. 1951 saw the instigation of the new five-year plan, which included further route expansion and the introduction of night flying. In 1954, the four-engine Il-14 entered service, capable of carrying thirty-two passengers across an ever-growing network. 1956 saw Aeroflot challenged to carry an astonishing sixteen million passengers by 1960. To do this, the airline needed a modern and capable airliner to utilize jet power and fly the flag for Soviet aeronautical engineering. The new aeroplane would use the engines, wings and tail surfaces of the Tu-16 but featured a wider pressurized fuselage.

The Tu-104 made its first international flight from Moscow to Prague on 12 October 1956. Despite the arrival of jet power, Aeroflot developed its fleet of turboprop aircraft, which would serve many less-than-perfect runways that still existed. These included the exceptionally sturdy four-engine Il-18 and the Tu-114, which also held the reputation of being the world's largest airliner upon entering service. Featuring a swing-wing design, the Tu-114 could carry 224 passengers or 170 sleepers at speeds that equalled some jet-powered aircraft.

New jets were introduced as confidence and experience in civilian jet-powered aircraft were gained. By the early 1960s, the short-haul fleet received the fifty-six-seat Tu-124 jet airliner. It also received the An-24 high-wing, twin-engine general aviation aircraft, which, as well as a fifty-seat passenger aeroplane, was used as a search-and-rescue (SAR) and ice-reconnaissance aircraft.

The end of the 1960s saw a considerable increase in aircraft types, including the Il-62, capable of carrying 200 passengers and bearing a remarkable similarity to the Vickers VC10, and the Yakovlev Yak-40, a thirty-two-seat short-haul jet airliner. The Il-62 would fly between Moscow and New York from July 1967, with Aeroflot sharing the route with Pam Am. This led to a less-than-harmonious business relationship with accusations that the two airlines were sabotaging the other's passenger bookings.

After its establishment on 1 April 1940, due to the merger of several airlines, British Overseas Airways Corporation (BOAC) became the United Kingdom's national carrier. During the Second World War, BOAC aircraft flew a variety of often hazardous missions, including transporting valuable ball bearings from Sweden and utilizing its flying boat fleet to ferry military and government staff around the globe. One significant role fulfilled by the airline was returning ferry pilots to Montréal from Prestwick, Scotland, using Consolidated LB-30 transport aeroplanes, which were civilianized B-24s. These flights established the first regularly sustained transatlantic crossings by landplanes, paving the way for future crossings by airlines post-war.

In 1946, BOAC was split into three separate airlines, one maintaining the BOAC title and two others, BEA (British European Airways) and BSAA (British South American Airlines). In 1946, the airline made its first purchase of six Constellations, which were bought to service the financially lucrative transatlantic crossing.

It soon became apparent that more Constellations were needed. In 1948, BOAC purchased Constellations from the Irish national carrier Aer Lingus, which had purchased the aircraft in 1946. Aer Lingus had established a separate transatlantic service flying under Aerlínte Éireann in 1947. A change of government and a new period of national austerity saw the project cancelled by newly elected Prime Minister John Costello. He felt that neither Ireland nor Aer Lingus should have such ostentatious dreams as transatlantic flight. BOAC soon put the new Constellations to use, and by 1949, they were flying the United Kingdom–Australia route. BOAC also purchased the Boeing Stratocruiser to supplement the Constellation transatlantic fleet. The decision to buy and use American-made aircraft resulted from the British aeronautical industry lagging in developing a suitable long-range pressurized airliner. For BOAC, this sensible commercial decision would dog relations with the government, the domestic aeronautical sector and, more importantly, the trade unions.

Despite this, BOAC was the first airline to operate a passenger jet service, with the Comet making its first flights to South Africa and the Far East in 1952. However, these flights ceased in 1954 after the catastrophic loss of four Comets led to the fleet being grounded for investigations. To fill the gap left by the Comet's absence, BOAC took delivery of Bristol Type 175 Britannias. The Britannia was a medium- to long-range turboprop aeroplane capable of carrying 139 passengers. Introduced into service in 1956, the Britannia covered the transatlantic routes, flying against Pam Am and TWA, alongside the now expensive-to-run Stratocruiser, which continued to make the transatlantic crossings. In October 1958, the Comet returned and made the first transatlantic civil jet crossing. Regardless, Boeings 707 had now filled the market gap. In 1960, with the BOAC transatlantic fleet ageing, the first 707s arrived in BOAC livery. The new 707s were powered by Rolls-Royce RB.80 Conways, the first turbofan engine. In addition, they featured a redesigned tail to meet rigorous British aviation safety standards.

In 1964, Sir Giles Guthrie became chair and chief executive of the airline. Guthrie had learned to fly at sixteen and, along with Charles Scott, had won the Schlesinger African Air Race between Portsmouth and Johannesburg in 1936. During the Second World War, Guthrie flew with the Royal Navy Fleet Air Arm, where he would eventually become a test pilot. This aviation experience would stand Guthrie in good stead, and his insistence in buying Boeing aircraft saw BOAC begin to make a profit by the end of the decade.

This preference for Boeing products did not sit well politically and after a confrontation in Parliament regarding a cancelled order of the new Vickers VC10, Parliament forced Guthrie to order 17 units of the new aeroplane in the spring of 1967. The VC10 was, if nothing else, a distinctive-looking aircraft, large with four engines mounted in twin nacelles to the rear of the fuselage. First flown in 1962, the VC10 was designed to fly from shorter African-based runways. As such the VC10 possessed exceptional hot and high flight characteristics, enabling it to take full advantage of the climate. All of this came at a cost, and the VC10 was an expensive aeroplane to operate due to its flight characteristics.

Its performance was impressive; it was the fastest passenger aircraft to enter service after Concorde. As a result, it soon found itself on transatlantic duties in the longer Super VC10 version. By the end of the 1960s, BOAC was well placed and ready to receive the next generation of airliner, personified by the large, wide-bodied Boeing 747.

Working alongside BOAC was British European Airways (BEA). This short- and medium-haul airline served British aviation needs in Europe and the Middle East/North Africa (MENA) and provided domestic flights. Initially formed as a division of BOAC on 1 January 1946, BEA became a separate state-owned enterprise on 1 August 1946. It initially operated a mix of aircraft, including war-surplus DC-3s, which Vickers Vikings gradually replaced. Initially working from Croydon and Northolt, BEA eventually moved its central hub to Heathrow, then known as London Airport, in 1954. BEA also operated from Liverpool's Speke Airport, providing flights across the north of the United Kingdom as well as international flights. By 1947, a few independent airlines, such as the Railway Air Services (RAS) and Scottish Airways, had been absorbed into BEA. It initially used Ju 52s for its services to Northern Ireland until DC-3s replaced them. BEA also started utilizing the DC-3's 6,000lb (2,725kg) load capacity on its first international scheduled cargo service to Brussels in 1947.

Initially, BEA operated at a loss, which led to the dismissal of its first chair, Gerard d'Erlanger, who had commanded the Air Transport Auxiliary (ATA) during the Second World War. D'Erlanger was replaced in 1950 by William Sholto Douglas, a formidable organizer. He had rebuilt the Royal Air Force's Fighter Command after the Battle of Britain. He then turned the it against the Luftwaffe as an offensive asset. Sholto Douglas was joined by a new chief executive, Peter Gordon Masefield, who had a background in engineering and, despite having a pilot's licence, had been turned down for war service by the RAF for poor eyesight. However, Masefield became a reporter during the war and, incredibly, was able to fly as a co-pilot and air gunner with the United States Army Air Forces. After penning a damming article into the workings of the Ministry of Aircraft Production, Lord Beaverbrook, who led the ministry, had Masefield appointed as his adviser. As a result of this appointment,

Masefield would go on to work as secretary of the Brabazon Committee, helping to shape Britain's post-war civil aviation recovery programme.

In 1952, the Airspeed Ambassador joined the growing fleet of aircraft. As a result of the experience brought by Sholto Douglas and Masefield, BEA celebrated carrying its one-millionth passenger. BEA continued to rationalize and develop its fleet, and in 1953 it became the first airline in the world to operate a turboprop aircraft, the Vickers medium-range aircraft, the Viscount, which worked alongside the Viking. BEA had been running an early prototype of Viscount, the V630, on its international London–Paris flights since 1950, using the experience to gain familiarity in working with this next generation of aircraft.

The following year the innovations continued, this time away from the flight line, with the introduction of Flightmaster, a mechanical Airline Reservation System (ARS). In addition, BEA could now sell seats as part of the wider Passenger Service System (PSS), which is linked to a Departure Control System (DCS), which helped manage the check-in process. This new system would transform how airlines were able to sell their service and process passengers.

In 1955, Masefield's hard work paid off, and BEA recorded its first profit. BEA continued to develop, adding the Vickers Vanguard to its fleet and then combined with a commercialization package which introduced new off-peak fares. These were exceptionally popular on the London–Paris route. The introduction of economy class flights and reconfiguring seating to enable aircraft to carry more passengers soon gave BEA a healthy balance sheet. Meanwhile, first-class passengers were treated to exclusive flights in the Ambassadors.

March 1958 saw BEA take the plunge into the jet age, despite Sholto Douglas's belief that turboprop aircraft were the way ahead for passenger carriage. The following year the first Comet 4B arrived, placing BEA on an equal commercial footing with Air France, who was now operating the Caravelle on their international services. 1959 also saw BEA sign a deal with de Havilland for its second-generation jet aircraft, the DH.121 Trident. Although the Trident was the first rear-engine, high-tail, tri-jet to enter service, which could initially carry just over 100 passengers, the ultimate version of the Trident would be able to carry 180 passengers. Through the 1960s BEA's profits and passenger numbers continued to grow.

Still, implementing the Civil Aviation (Licensing) Act (1960) opened the civil aviation market to independent operators. This Act removed the BOAC and BEA monopoly. It established the Air Transport Licensing Board (ATLB), which acted as judge on receiving the new scheduled route licences. On paper, the system looked fair but, in reality, both BOAC and BEA were able to challenge any threat to their operating routes by a presentation at licence hearings. Even if ATLB licences were granted to independent operators, such as British United Airways or British Eagle, they would struggle to gain traffic rights. Any such awards would see BOAC and BEA having to reduce their services accordingly to make room for the newcomers. Commercial aviation competition was a fearsome business, and neither BOAC nor BEA played nicely. They retained an almost complete monopoly, regardless of the new Act.

The sixties were the peak of BEA's influence with services throughout the United Kingdom and beyond. It was also a shareholder in many other airlines, including the Italian flag carrier Alitalia and Aer Lingus. In addition, the airline grew its operations outside of passenger travel by purchasing the Armstrong Whitworth Argosy cargo aircraft. Despite such a purchase, BEA began to feel that supporting the British aviation industry due to political pressure came at a financial and operating cost.

Regardless, the BEA jet fleet was coming of age, and the Tridents were providing sterling service, leading the way in various innovations. The most important was the world's first automatic landing via the Instrument Landing System (ILS), a radio navigation system, on a scheduled passenger service on 10 June 1965. This feat was followed in November 1966 by another Trident landing at London Heathrow in zero visibility, using the same system.

BEA continued to invest in the Trident as their second-generation jet airliner of choice, but overseas competitors were flocking to Boeing. The American manufacturer was now offering the new 727 and 737 series aircraft, which were proving cheaper to operate while generating a higher financial return for users. Spurred on by Air France's consideration of turning to Boeing, rather than Sud Aviation, for its second-generation jet airliner, BEA began trials with the 727-200, 737-200 and the Douglas DC-9-40. All 3 aircraft were tested to replace turboprop aircraft for BEA's short- and medium-haul routes. Despite favourable test results, the British government held firm on its stance that BEA should only use domestically produced airliners. As a result, the Board of Trade directed BEA to purchase the British Aircraft Corporation (BAC) One-Eleven short-range jet airliner. BAC had been formed in 1960 by the government-forced merge of English Electric Aviation Ltd, Vickers-Armstrongs Ltd, the Bristol Aeroplane Company and Hunting Aircraft. BEA placed the order for the One-Eleven in January 1967, a year after domestic rival British United Airways (BUA) had begun its scheduled daily flights as the United Kingdom's first all-jet operator.

As the decade drew to a close, BEA was under increased political pressure due to the publishing of a government-sponsored paper entitled 'British Air Transport in the Seventies', released on 2 May 1969. The paper revealed the findings of a committee led by Professor Sir Ronald Edwards, which recommended, among other things, the merging of BEA and BOAC, as well as the establishment of the Civil Aviation Authority (CAA). The paper also recommended further integration into the burgeoning package holiday market, a market fed by travel agencies displaying cutaway scale models of the latest aircraft and glamorous brochures of revelling holidaymakers. This, in turn, led to the formation of specific low-frills holiday carriers, and BEA Airtours.

Despite this, BEA continued to provide cargo carriage, and by 1969 it was moving 132,000 tonnes of freight, predominantly from tits new cargo hub at Heathrow that it shared with BOAC. With a profit of £6.5 million and a workforce of 25,000, BEA ushered in the 1970s with a bright future.

On 1 April 1972, as a result of the findings of Professor Edwards, the British Airways Board (BAB) was established with BOAC and BEA combining their operations, with each company forming a new division within BAB by September. By its final reorganization into what would become British Airways on 1 April 1974, BEA was the world's largest passenger-carrying airline outside the United States and the fifth biggest operator in the industry.

Air France

The post-war government of France was quick off the mark to begin the recovery process of its airline industry. One of the first tasks was the nationalization of the pre-war airline industry on 26 June 1945, with the privately owned Air France taken over by the state. By the end of the year, the airline was responsible for the entirety of the domestic air transport network and delivering the night mail across a network of 99,419 miles (160,000 kilometres). Then, on 1 July 1946, the first transatlantic crossing between the new Paris-Orly Airport and New York was made by a DC-4, with passengers enjoying flight-attendant service. This service would soon be renowned for its luxurious treatment of passengers with private cabins and à la carte hot food.

By 1948 Air France, like all airlines, was experiencing a considerable period of growth, utilizing all the positive aspects of air travel to the fullest and having a fleet of 130 aircraft. Alongside the now almost-customary DC-3s and DC-4, Air France also flew the 2,000-mile range, thirty-3-seat Bloch MB.161 Languedoc. The four-engine Languedoc made its first flight on 15 December 1939. Then, despite its pre-war heritage, it entered into serial production, with 100 units produced between 1945 and 1948.

By the early 1950s, the network covered by Air France had grown significantly, with Paris-Orly Airport becoming the airline's key hub, operating both long- and medium-haul flights. By 1953 the Air France network covered some 155,343 miles (250,000 kilometres). By 1954 Air France was making daily flights to North America with Lockheed Constellations and introducing the Vickers Viscount and the Bréguet Deux Ponts mid-wing, double-deck transport aeroplane to their fleet.

The same year Air France, along with national rail operator Société nationale des chemins de fer français (SNCF) and the public sector financial institution Caisse des dépôts et consignations established a separate domestic carrier – Air Inter. The goal of Air Inter was to rationalize the intercity air routes in France, and soon more investors were found to fund the project. A project of such scope was understandably slow to get going, but by 1960 scheduled passenger flights had started.

The end of the fifties also saw the airline introduce the Caravelle and 707 jet-powered airliners into service. These replaced the propeller-driven aircraft on all long-haul routes by 1966. The 707 reached destinations along the continental American coastlines and reduced the Paris–New York flight to just over eight hours. Another fundamental change was relocating the operations centre and maintenance teams from Le Bourget to Paris-Orly in 1960. The following year Air Inter began operating Vickers Viscounts while offering passengers saver cards.

The French government redistributed international travel rights to destinations in African and Pacific locations, including the West Coast of North America, in 1963. This route would be serviced by the recently formed private airline Union de Transports Aériens (UTA), created by the merger of Union Aéromaritime de Transport (UAT) and Transports Aériens Intercontinentaux (TAI).

As the decade progressed, so too did the competition. Despite the airline's international flights accounting for 90 percent of its flight traffic, the airline remained a small global operator. To help counter this commercial weakness, Air France invested in the new 727 medium-haul and the 747 wide-bodied aeroplanes, as well as building the new Orly Ouest Terminal.

The 1970s would be, for Air France and many airlines, a time of enormous change and one the company would grasp with both hands. In the United States, the post-war growth of civil aviation continued almost unabated. The Second World War introduced the motor industry production line to manufacturers. Post-war aeronautical companies, including Boeing and Beechcraft, were keen to continue using these streamlined manufacturing processes to their advantage. The shift to turbojet-powered aircraft saw the big manufacturers utilize the new techniques, outproducing smaller domestic and overseas competitors within a relatively short period.

European manufacturers faced the constraints of post-war austerity and rebuilding industry and infrastructure damaged by wartime and industrial stresses. However, in the United States, the desire for light aircraft remained, and the domestic market, along with an infrastructure unaffected by warfare, was ready to take commercial advantage. Cessna, founded in 1927, was one such company to offer cheap and reliable flying using its wartime design and build experiences to produce new models from 1946. The high-wing models

120 and 140 were the first of the company's post-war aircraft, including the 172, which first flew in 1956. The 172 was a massive success for Cessna and remains in production today, with a staggering 44,000 units built. Cessna would go on to produce a wide variety of propeller, turboprop and executive jet aircraft, including short-haul and executive types.

There was a post-war boom for the airlines, with the Big Four taking advantage of pre-trained ground staff, including engineers and hugely experienced flight crew. The flight crews, in particular, were essential to making the growth of the post-war airlines possible with their long-range flying experience, echoing what was happening in Europe and the Soviet Union. Air transport was about to become big business. Unlike the pre-war years' dominance, the Big Four United States airlines faced competition from many smaller airlines keen to establish their businesses. The competition between the established airlines and the new starters on domestic and overseas routes was promoted by the Civil Aeronautics Board (CAB)

One such airline was American Overseas Airlines (AOA), originally founded as American Export Airlines (AEA) in April 1937. AOA quickly capitalized on the increased range established by aircraft during the war and began to operate a transatlantic service on 5 July 1945 using Sikorsky VS-44 flying boats. The routes to Northern Europe soon expanded, with flights to the capital cities of Scandinavian countries, including Iceland, being offered. Transatlantic landplane flights soon followed, and the company's first aircraft, C-54/DC-4s, were replaced by Constellations and Stratocruisers. AOA, along with its routes and fleet, would be bought by Pan Am in 1950, helping to counter the new starter airlines. Pan Am also began something unique, a scheduled round-the-world flight. Despite passengers transferring aircraft, the service was a great success. In 1960, the route was taken away from the venerable DC-4s and Constellations by the new 707 which could complete the route in 56 hours, stopping at cities such as New York, Honolulu, Manila and Calcutta.

Pan Am slowly rebuilt their transatlantic and transpacific portfolios. By the end of the 1950s, the transpacific flights were serviced by 'sleeper'-equipped Stratocruisers. This would see the aeroplane's double-decker fuselage configured to sleeping berths on the top deck and a lounge on the upper. Meanwhile, the adoption of the 'Super' Constellation, with its improved range, gave the airline a non-stop transatlantic crossing capability. The Constellation was replaced by the DC-7C, the last significant propeller-powered aircraft made by Douglas. The DC-7C could carry up to 105 passengers over 5,635 miles (9,069 kilometres). The transatlantic route was soon adopted by the 707, into which Pan Am placed Douglas's first jet, the DC-8, to provide a New York–Paris–Le Bourget service. While the 707 would take a refuelling stop at Gander, Canada, the DC-8 would be able to make the transatlantic crossing non-stop. These first-generation jet aircraft would be replaced by the next generation of larger airliners, known as wide-bodies, led by the Boeing 747 in 1971.

TWA was quick to restart their transatlantic crossings, using the Constellations they had helped design with Howard Hughes and Lockheed during the Second World War. In February 1946, the first New York–Paris flight was followed in April by the first New York–Cairo flight, which echoed the wartime routes flown by TWA Stratoliners. Despite these promising early signs of a resurgent airline, TWA faced almost catastrophic financial issues. By 1947, after plummeting stock values and employee disputes, Jack Frye, whom Hughes believed to be responsible for the airline's economic woes, resigned as president. There followed a period of management reshuffles, with Hughes, supported by the eleven-member board and his cash injections, making the changes he saw fit to build a stronger airline. In truth, many of the financial woes were the result of technological advances, with many, including the turbojet engine, clearly going to be the way ahead but also expensive in these early post-war days. The government had previously subsidized research and development in the post-war world of civil aviation, a cost that would now ultimately be met by the consumer. This was a new fact of life facing the industry as a whole that Ralph Damon, the new president, fully recognized.

Damon's appointment by Hughes was inspired, and soon his ideas were making the airline money and becoming the gold standard for the industry. The critical change was the introduction of first and economy classes. Damon continued to grow the fleet, and the flight line soon boasted short- and medium-haul Martin 2-02-2s and 4-0-4s, which flew from various TWA hubs scattered throughout the continental United States. On 17 May 1950, TWA changed its name from Transcontinental & Western Air to Trans World Airlines. Despite operating a sizeable fleet of Constellations in various guises, TWA made its first order of Boeing 707s in 1956. The new aircraft would serve domestic and international routes, and orders for the Convair 880 soon followed. Regardless, the 880 was a commercial flop and, combined with Hughes not fully recognizing the growing influence of the jet engine on the airline, led to his removal from overall control of TWA.

In 1957, Damon moved on and was replaced by Carter Burgess. However, his tenure was short, and after clashes with Hughes, Burgess moved on to a fitness-machine manufacturer in 1958. Burgess was replaced by the recent Secretary of the Navy, Charles Sparks Thomas. Thomas had arrived at an essential time for TWA, overseeing the launch of its 707 fleet and the employment of the first African American flight attendant by a major airline, Margaret Grant.

1960 saw a further change, with Carter Thomas replaced by Charles Tillinghast Jr and Hughes relinquishing control of the airline. However, this was followed by a lengthy court battle which eventually saw Hughes sell his stock in the company for nearly $550 million. 1960 also saw the beginning of large-scale involvement of the banks in civil aviation, with Dillon, Reed & Co. loaning TWA $165 million for forty-five jet-powered airliners.

By the mid-1960s TWA, under Tillinghast Jr's guidance, had become a sizeable organization, covering everything from restaurants to hotels. As a result, the share price rocketed, making the airline an all-American success story. In 1964, TWA established its MarketAir service that provided specialist support to the commercial users with various products covering air freight movement, customs support and personnel air passage. The service was open to global and domestic customers. On 7 April 1967, the last of TWA's Constellations were retired, and the airline maintained a jet-powered-only fleet, becoming one of the first all-jet airlines.

The end of the decade saw TWA operating as the third-largest airline with flights as far as China, carrying the most transatlantic passengers, eclipsing Pan Am's route dominance. TWA's aggressive marketing and operating business models also saw it adopt a transpacific route via Hawaii. This came after a successful lawsuit, known as the Transpacific Route Case, against the Pan Am and Northwest Airlines (NWA) monopoly of operating passenger services over the route. With its corporate eye ever on the horizon, TWA was also among the first to place orders for the 747, the arrival of which would further expand its influence on air transportation.

Eastern (EAL) continued to maintain its strong presence along the East Coast of the continental United States. Propelled by owner Eddie Rickenbacker, EAL became the only airline to operate in the post-war market without subsidy and, for a short time, with a profit. During this early post-war period, EAL purchased New York-based Colonial Airlines, which opened up routes into Canada. However, Rickenbacker, like Hughes, failed to appreciate the impact the jet would make on the airline industry as well as the ever-growing influence of subsidized airlines. By the end of the decade, Malcolm MacIntyre had replaced Rickenbacker. Despite some wartime experience in aviation, the latter lacked the insight to run an airline. Regardless of his inexperience, EAL introduced the DC-8 on the New York–Miami routes in 1960. This was followed in 1963 with the arrival of the 720. Another innovation was the start of a New York–Boston shuttle, known as the Eastern Air Lines Shuttle. MacIntyre, perhaps keen to capitalize on the growing jet-powered aircraft market, had been wise enough to involve EAL in Boeing's development of the 727 as part of a commercial triumvirate including American Airlines and United Airlines. This meant that EAL was the first airline to fly the 727, on 1 February 1964.

In December 1963, MacIntyre was replaced by aviator Floyd Hall with Rickenbacker stepping down entirely at the end of the year. Hall had served with the United States Army Air Forces during the war and had flown for TWA as a pilot. His experience and insight into long-haul flights were now welcomed. Within three years, EAL started branching further into the Caribbean market. By the close of the decade, EAL looked at purchasing the 747. However, the four aircraft allotted were sold to TWA, leaving EAL struggling to find a replacement. Thankfully, a leasing agreement with Pan Am led to EAL receiving two 747-100s, operating the aircraft between New York and the capital of Puerto Rico, San Juan. Such was EAL's influence in the Florida region that Walt Disney World declared EAL its official airline, a title it maintained until Disney changed to Delta Airlines (DAL) in 1989.

Where Pan Am had focused on route growth and corporate influence, American Airlines (AAL) had been very much about developing the tools of their trade. Before the onset of the Second World War, AAL had been instrumental in developing the DC-3 series of aeroplanes. This set a pattern of cooperation between AAL and Douglas which would later produce the DC-10 wide-bodied, tri-jet aeroplane. In the immediate post-war years, AAL was quick to carry on building its route portfolio and began operating its first international flights as well as establishing an engineering hub in Tulsa, Oklahoma. In the 1950s, AAL was the first to operate non-stop transcontinental flights with their DC-7 fleet, supported by staff trained at the world's first specialist flight-attendant training facility in Fort Worth, Dallas. By the end of the decade, AAL had introduced the Lockheed L-188 Electra into their fleet, the first domestically designed and built turboprop aircraft.

For AAL, the 1960s provided the same technological and procedural challenges and changes as other airlines had been presented with, including the adoption of jet-powered aircraft. However, there were also social changes. Like many airlines, AAL led the way in changing attitudes and, in 1964, appointed David E. Harris as their first African American pilot. Harris, who would later become the president of the Organization of Black Airline Pilots (OBAP) and a member of Negro Airmen International (NAI), had learned to fly with the United States Air Force (USAF). His career with the USAF gave him an excellent background in flying large, multi-engine, jet-powered aircraft with tours on the B-47 Stratojets and B-52 Stratofortress. Leaving the USAF after six years of service due to racial discrimination, Harris started with AAL almost immediately, and the USAF's loss would be AAL's gain. After three years, Harris was rated a captain, the first African American. What followed was thirty years of flying with AAL, which saw Harris fly a host of aircraft, including the 747, the Airbus A300 and the BAC One-Eleven. By the decade's end, AAL had expanded its flights to the Caribbean by purchasing Trans Caribbean Airways, whose hub was located in San Juan, Puerto Rico.

United Airlines (UAL) had come a long way from its mail delivery roots under Walter Varney, and in the post-war years, like the other Big Four airlines.

Airmail, whose subsidy ended in 1952, was replaced by the need to generate revenue through passenger travel and take advantage of the potential financial boon. UAL, like AAL, continued to work closely with Douglas and, in 1955, alongside DAL, it introduced the DC-8 into its fleet. The DC-8 was soon joined by Boeing 720s and 727s and the Sud Aviation Caravelle. By the end of the 1960s, UAL had established a series of nationwide and overseas hubs, with Chicago-O'Hare becoming the primary one ready to receive wide-bodied aircraft such as the 747.

By the dawn of the 1970s, civil aviation had stabilized its business model. It was taking advantage of new technology that seemed to be arriving as rapidly as the scheduled flights the many operators were offering. Infrastructure was vastly improved, too, aided by solid legislation, improved knowledge and equipment, and a burgeoning holiday and cargo carriage market. The 1970s would see the airlines settle into a period of organizational stability that would help launch wide-bodied aircraft such as the 747 and the arrival of the airliner whose legacy would outshine all others: Concorde.

Chapter 5

Living the Jet Dream: Bigger, Farther, Faster

The 1970s saw an industry settling down. The jet had now established itself as the preeminent form of transport for scheduled and charter civil aviation. The airline industry was now very much leading the way in civil aviation. The 1970s would be the dawn of the wide-bodied and superfast airliners, offering the owner and user aeronautical technology, power and performance the pilots of war-surplus DC-3s could only dream of. This was backed up by various international agreements and legislation that helped spur the industry's growth at a low-operating cost. The sky had literally become the limit. All the while, designers and engineers continued to hone their craft with slide rules and French curves, creating flying works of art that defined an age.

The seeds for the growth of civil aviation and the eventual dominance of the large airlines were sown in the pioneering years immediately after the First World War. This period was one of regulation and experimentation, which sought to characterize and embrace a new era of peaceful development. The period also desired to rid the fledgling industry of repurposed, small-capacity military aircraft. The new age would see these aircraft replaced with purpose-build aircraft. The next generation of aircraft reflected the growing aeronautical knowledge base of what worked mechanically and aesthetically, as well as the societal and commercial changes. This gave rise to some beautiful designs and interpretations of what air travel should be and how it should appear. There were also economic factors: increasingly, new aircraft could often only be built by bank finance, and the banks often remained cautious in their patronage, so practical elements also had to be considered. After all, paying passengers and the weight of cargo that could be carried were worth their weight in gold for operators. Despite the numerous subsidies available, which focused primarily on airmail, the mid-1930s was a period where the presence of a tourist passenger on a flight was the ultimate expression of luxury, available to only the select few.

The glamour of the long-haul transcontinental flights for these lucky few was captured by images of sleek silver-bodied aircraft shimmering in the sun or of flying boats landing at sun-soaked tropical locations. The onset of war delayed the opportunity for the airline industry to develop a paying tourist passenger base for almost a decade. However, it allowed the sector to reap the benefits of well-trained aviators, engineers and support staff in the post-war years. Post-war austerity saw the sale of surplus military aircraft to civilian operators. Many were new start-ups with their own brands and ways of doing things, including marketing. The new airlines were now operating in an ever-growing media-hungry age, building on the pre-war marketing processes and sharing their stories and successes across the latest media sources, from cheap colour print to television. Public relations directors across the globe recognized civil aviation had an ever-changing face and were keen to share it, to sell the dream of flight and all that it encompassed.

Despite its exciting glitz and glamour, the jet age saw aircraft design, airliners in particular, remain relatively orthodox. Moreover, designs were influenced by the practicalities of utilizing the new power source to its fullest and keeping large-scale serial production costs to a minimum. Yet despite this, new icons such as Concorde were born, becoming the acme of aeronautical technological expression.

By the start of the twenty-first century, the aviation world had already embraced the opportunities made possible by social media. This new and powerful medium was often used as much for behind-the-scenes storytelling as it was for the passage of information. It also saw the age of the wide-bodied airliner extend across five decades and two centuries, proving that, for civil aviation at least, bigger was better.

Of Aircraft

The inescapable issue with anything mechanical is that it is designed to fulfil a specific task first and foremost, and aircraft, regardless of purpose, were no different. Early designs such as the Blériot XI or the Curtiss Reims Racer were utilitarian, initially equipped with the bare minimum, a simple wooden or wicker seat for the daring pilot. In comparison, the interior of the early dirigibles was refined, stylish, and bore more than a passing resemblance to the first-class accommodation of the finest ocean liners operating on the high seas.

In those early post-First World War years, the aeroplane had been refined. While remaining utilitarian, designers realized that more could be done to improve comfort, especially for passengers. Early aircraft, such as the DH.4A, were merely a conversion of existing and surplus wartime aeroplanes. So, conversions were somewhat rough and ready. Yet the concept was solid enough to warrant further investment in time, money and development. Designers like Otto Reuter quickly realized that the aeroplane had the edge over the dirigibles of Zeppelin. The Junkers F 13 introduced remarkable passenger comfort that was ahead of its time. The four-seat cabin featured a luxurious leather- or cloth-finished rear bench and two wicker bucket seats that would become the norm for early airliners.

Zeppelin was not to be outdone, and designer Dr Adolph Rohrbach created the Staaken E-4/20 monoplane, which made its maiden flight on 21 September 1921. This incredible four-engine high-wing aircraft featured a double-deck arrangement with the three flight crew sitting above the passengers. The Staaken could carry up to eighteen passengers and their luggage and featured a toilet and washing facilities. Rohrbach's all-metal design was revolutionary and could be classed as the first proper airliner, complete with wallpaper and carpet.

The post-war Inter-Allied Control Commission felt that the Staaken could be converted into a military aircraft. So, despite very promising early trials, the project was cancelled, and the Staaken was scrapped on 21 November 1922. Yet Rohrbach's design influence lived on, influencing aeronautical designs for another century, and innovations such as the hinged nose would be utilized by countless aircraft.

Passenger comfort and luxury came to the forefront with the advent of the large flying boat. These wonderfully evocative aircraft, which had been in development as long as landplanes, were the ideal canvas upon which designers could flex their creative talents. The DC-3 and Electra 10 could be configured according to use or remain utilitarian. A necessary feature which allowed aircraft to be readily converted into sleeper aeroplanes, especially in the case of the DC-2/3, with the minimum amount of fuss and often while airborne. While there would be individual examples of more refined interior design and decoration, especially for private use, it was the flying boat with its wider fuselage that automatically lent itself to a more sophisticated and comfortable interior. The Dornier Do X could carry almost 100 passengers in unrivalled luxury, with wood-panelled interiors with the finest carpets, chairs and occasional tables. It was the largest flying boat and heaviest aircraft to be built between the wars, and Dornier and Deutsche Luft Hansa A.G. (DHL) used the Do X's bulk to its full advantage. The elegant Do X caught the eye wherever it went, and early passengers were left with no doubt that they would be treated to an experience of a lifetime.

The basic layout would be copied by other flying-boat designers, builders and airlines to produce a flying experience that was simply second to none. Boeing, in particular, made full use of their 314 Clipper's size to offer an experience that would become the hallmark of 1930s passenger flying. Airlines would combine the Clipper's luxurious interiors with a spirit of adventure by flying to exotic destinations, particularly on the transpacific routes.

The onset of the Second World War prevented the flying boat from fully exploiting its design features, and the post-war world, extended ranges and the jet engine's introduction saw the flying boat's glamour consigned to history. Pan Am and Imperial Airways, in particular, had shown the levels of passenger comfort that could be achieved with an aircraft. The post-war airline operators initially concentrated on rebuilding or establishing pre-war routes and air fleets; this gave the numerous aeronautical companies time to adjust their output accordingly. One essential addition to aircraft in the post-war world was the galley. One of the first galleys was created by Delta Chief Engineer J. F. Nycum and added to Delta's DC-4s. The success of Nycum's galley was such that Douglas began adding the feature to all of its new DC-4 aircraft.

Not only did the addition of a galley add a touch of luxury hitherto unknown to many passengers and crew on longer flights, there would soon be à la carte menus provided for passengers and the provision of a duty-free trolley as aircraft became larger. Meals, in-flight entertainment, refreshments and comfort, generally reserved for first-class passengers, were now available to many. The race was now on to provide the best customer service, and airlines did not hold back, with reputations made or broken by the standard of service.

By the 1970s, most airlines had almost wholly relegated their propeller-driven fleets to short-haul and cargo-only flights. Jet aircraft now carried passengers, who were increasingly holidaymakers, to their destinations. Still, the aircraft of the time remained restricted by how many passengers they could carry. Pan Am in particular recognized that a larger aircraft would give better financial return to passengers and operators by a reduction in seat cost, as well as reducing holding numbers at increasingly busy airport terminals. So, in 1965, a 50,000-strong project team at Boeing, led by Malcolm Stamper, was assembled. Stamper then appointed engineer Joseph Sutter as head of the design team for the new aircraft.

The development costs were, for the time, astronomical: Boeing invested $2 billion in the project. Thankfully, this was mitigated by an advance order of twenty-five aircraft by Pan Am, who would work closely on the project with Boeing. In April 1966, Boeing was able to add $500 million to their development programme. Everything about the new aircraft was enormous; a specialist assembly plant was built in 1967 at the Everett Plant, Washington, which became the largest building by volume, ever to be constructed.

The new aircraft would feature two decks, with the flight deck in a distinctive hump above the nose, which could be hinged for handling cargo. Another critical aspect of the new aircraft was how to operate it from existing airports. As a result, the wing design became a key focus and the design team, racing against the clock, introduced innovative high-lift devices. This included three-part slotted flaps, which increased lift by 90 percent when deployed. Another vital issue to be overcome was designing an effective evacuation system for the 560 occupants that met stringent FAA regulations and could be achieved in 90 seconds.

Despite issues with the engines, which would be supplied by Pratt & Whitney, Rolls-Royce and General Electric, the new aircraft was unveiled on 30 September 1968 and introduced as the Boeing 747-100. The world of aviation changed at that moment. The first flight of the 747 took place on 9 February 1969 at the hands of test pilots Jack Waddell and Brien Wygle.

After further technical tests, the 747 was flown with 400 passengers, to debut at the 1969 Paris Air Show at Le Bourget Airport after a non-stop flight from Seattle, Washington. Unfortunately, it was joined by the still-in-development Aérospatiale/BAC Concorde, which slightly stole the new aircraft's thunder. Despite this, the 747 would receive its FAA airworthiness certificate in December 1969, which allowed it to be operated commercially.

The 747 would usher in the era of the wide-bodied aircraft and proved to be an excellent aircraft over its fifty-one-year production run. Boeing built over 1,500 units, including two VC-25s or *Air Force One*. For Boeing and operating airlines, the 747, or Jumbo Jet, continuously upgraded throughout its lifespan, was an icon. Its unique design was unmistakeable, gaining an almost cult following among enthusiasts and aviators. However, on 26 October 2020, British Airways announced the retirement of its thirty-one 747s, thus ending a classic era in civil aviation.

The team behind the 747, later dubbed The Incredibles, inspired a host of other wide-bodied aircraft from competitors, including the McDonnell Douglas DC-10, Lockheed L-1011 TriStar and the Airbus A300. The design elements of the wide-bodied aircraft remained constant, with two aisles being the standard, larger cabin space and more underfloor cargo space. They made long-haul travel more manageable, cheaper and accessible.

Airbus would have the final say regarding the ultimate expression of the wide-bodied aircraft design when they unveiled the A380, which could carry an astonishing 853 passengers. Airbus Commercial Aircraft was founded in 1970 by several European manufacturers, including Messerschmitt-Bölkow-Blohm (MBB) and Aérospatiale, with Hawker Siddeley joining in 1979. The first Airbus aeroplane was the A300 utilizing a fuel-efficient design which helped sales in the post-1973 oil crisis. Airbus's management quickly established its reputation as a design and build company of high quality led by Bernard Lathière. Lathières's energy in generating sales saw Airbus push aside its rivals to become the second largest aircraft company in the world after Boeing by 1980. In 1985 Airbus launched the A320, which featured a revolutionary control system that did away with the traditional yoke, using a side-mounted joystick instead. The pilot could also use the new Automatic Flight-Control System (AFCS), or fly-by-wire controls. This made the A320 groundbreaking and, even before the first flight, some 400 orders had been placed by airlines. The A320 had made history as the quickest-selling airliner and the vanguard of the next generation of aircraft.

Airbus continued to expand commercially, and by 1990, it had a manufacturing backlog of 900 aircraft; despite this, Airbus introduced the

A330 and A340 types. These modular designs shared standard fuselage, wing and cockpit specifications, reducing construction time. Airbus continued its growth into territory usually held by American manufacturers, with some companies cancelling orders from Boeing in favour of Airbus products. Meanwhile, Lockheed and McDonnell Douglas slowly withdrew from the airliner market to focus on other aspects of the aeronautical industry, including space, before merging with other manufacturers.

In the mid-1990s, Airbus began to look at developing a sizeable wide-bodied aeroplane that would challenge Boeing's 747 mastery of the sky, the A380. This project was announced on 19 December 2000, with Airbus unveiling an aircraft forty percent bigger than the 747-400 and backed by an initial investment of $10.7 billion. On 18 January 2005, the prototype was unveiled at the Jean Luc Lagardère Airbus A380 final assembly plant at Toulouse, France. This was followed on 27 April by the first test flight, with the type certificate from the European Aviation Safety Agency (EASA) and the FAA received on 12 December 2006. The first A380 was delivered to Singapore Airlines (SIA) on 15 October 2007, beginning total serial production. By the end of the Airbus production run in 2021, over 250 A380s had been built, with Emirates (UAE) receiving 123 in total, with most A380s used for long-haul flights. The A380 remains remarkably economical and efficient, capable of carrying large numbers of passengers over considerbale distances.

A380s also became renowned for their interior spaces, often well appointed with a host of features, including LCD screens placed into headrests and easily convertible sleeping pods. There were also some special adaptions, including bars, separate bedrooms, and a 'flying palace' for a member of the Saudi Royal Family.

As marketing for aircraft manufacturers remains as essential now as ever, Airbus decided upon something more eye-catching: the A300-600ST Outsize cargo freight aircraft. The A300-600ST was designed to replace the company's Aero Spacelines Super Guppy, transporting sections of completed aircraft between sites. The excessive loads are carried in a large cargo bay built onto the lower fuselage of a conventional A300, accessed by a large upward-swinging hinged door. The unique appearance of the A300-600ST soon earned it the nickname Beluga, and it proved to be an instrumental aircraft. On 9 January 2020, its replacement, the A330-743L Beluga XL, entered service, complete with a Beluga whale face painted on to the fuselage. In addition, the cargo bay was enlarged from 1,500m^3 to 2,209m^3, and a significantly improved range of 2,672 miles (4,300 kilometres) compared to the original Beluga's 1,056 miles (1,700 kilometres).

Despite these remarkable aircraft, which changed the face of air travel in the emerging age of the holidaymaker, the real game changer for the airline industry was the Aérospatiale/BAC Concorde. It was the marketer's dream, the enthusiast's delight and the pilot's desire. It evoked emotions rarely associated with a modern passenger airliner. Concorde was the epitome of high-speed, high-end travel. Used by pop stars and diplomats, this epoch-breaking aeroplane was the culmination of the aeronautical arts. It was as sleek as it was beautiful. Regardless of the Tupolev Tu-144 beating its western counterpart into the air by months, Concorde seized the imagination.

Its gestation started in 1954 as a series of studies which formed part of the Supersonic Transport (SST) concept, led by Welsh aeronautical engineer Morien Morgan. This focused on the wing design, which was deemed the central piece of the SST puzzle. A period of intense research culminated in a paper by aerodynamicists Johanna Weber and Dietrich Küchemann from the Royal Aircraft Establishment (RAE). The report identified that a thin delta wing with a high angle of attack would be the design most suited to the task. Morgan was then tasked by the Ministry of Supply (UK) to create a new working group and, on 1 October 1956, he formed the Supersonic Transport Aircraft Committee (STAC).

Over the next three years, STAC commissioned several studies and flying testbeds, such as the Handley Page HP.115, gathering data as the project progressed. By 1959, Hawker Siddeley and Bristol were awarded two study contracts for building an aircraft with the now recognizable delta wing mated to a long slender body. A series of partnerships ensued between the newly established British Aircraft Corporation (BAC) and various countries, companies and agencies, including the National Aeronautics and Space Administration (NASA). All partners worked together, seeking to perfect the physical form of the aircraft essential to perform supersonic flight. The culmination of this work and diplomacy was the signing of a development project legally presented as an International Treaty between France and the United Kingdom, on 29 November 1962. The treaty between the two nations symbolized a great deal, including a practical extension of the Entente Cordiale, which was alluded to in a letter written by Tony Benn, the British Minister of Technology. This led to the name Concorde being adopted for the project by British Prime Minister Harold Macmillan.

Concorde's first flight took place on 2 March 1969 with aircraft 001 taking off from the Aérospatiale plant at Toulouse, which was followed on 9 April by the BAC Concorde, 002, built at Filton, Bristol. Both aircraft were displayed at that year's Paris Air Show; Concorde 001 would break the sound barrier later that year on 1 October. A series of global sales tours followed, but the world had changed by the time of the Tu-144 crash at the Paris Le Bourget air show, which startled many potential airline customers and passengers. Combined with a growing awareness of ecological issues, the project began losing commercial ground. Airlines that had registered an interest in Concorde in the early 1960s shifted their focus onto more efficient aircraft like the Boeing

747 as the economic effects of the 1973 oil crisis took hold. The remaining customers were the two airlines, Air France (AFR) and BOAC.

Finally, on 21 January 1976, the first scheduled flight by British Airways (BA) took place, though due to concerns around noise pollution, not to the United States. These would finally be allowed after a court battle around the environmental issues was settled by comparing noise with the then *Air Force One*, a Boeing VC-137. The first transatlantic flights occurred on 22 November, between Paris, London and John F. Kennedy, New York.

Initially, Concorde, which was still government-owned, operated at a loss. It was not until the 1980s after BA and AFR purchased their Concordes from their respective governments and set their own ticket prices that they could operate at a profit. After that, Concorde soared, free to be used as a commercial asset. Soon, transatlantic flights were supplemented by weekly flights to Barbados. Travel companies were also keen to cash in on the Concorde mystic, and exclusive charter flights to a range of European destinations soon filled the flight planning boards. Each flight was filled with passengers keen to experience something only Concorde could supply.

The fatal Concorde crash at Charles de Gaulle Airport on 25 July 2000 was the beginning of the process which would see the graceful aircraft grounded three years later. Other influencing factors had been the effects on air travel caused by the 9/11 terrorist attacks, rising costs and Airbus's decision to no longer supply replacement parts. As a result, AFR's final Concorde flight took place on 27 June 2003, with BA following suit later that year on 24 October.

No civilian airliner inspired such emotion as Concorde on those final goodbye flights. That is a fine testimony to the men and women who helped shape the ultimate expression of elegance in an age of increasing design conformity.

Establishing the Brand: The Corporate Face

The act of flying has always captured and sparked human imagination. From the first dreamers to aeronauts pioneering controlled flight, flying was one of those few activities that barely needed to promote itself. After the First World War, aviation attracted men and women from all backgrounds, with flying meets and air races growing in popularity again. These were accompanied by wonderfully colourful posters celebrating the latest achievements of famous aviators who became household names. These new pioneers would attract a following that would bring cities to a standstill as the public thronged to catch even the most fleeting of glimpses of their heroes of the skies.

The newspapers, allied industries and industrialists soon picked up on the ever-growing popularity of flying. Eager to see that popularity turned into tangible developments for aviation, they began to support the greatest of adventures with publicity and money. To these adventures flocked the dreamers, the serious and the daring, all seeking the laurels that would come with being the first to be faster, further and higher.

The vast output created by the aviation industry public relations and marketing machines of the 1970s and beyond was built upon the sound foundation established in the early twentieth century. Early posters shared a great deal with those used by railway companies, often sharing the same colour palette, especially the early Zeppelin excursion posters. In many cases, the first aviation meet posters owed more to artistic licence than reality, showcasing the fantasy of flight rather than the actual practicality, literally playing on expectations. Nonetheless, the use of bright colours, and aircraft piloted by gallant aviators depicting the romance of flight, drew the public in. This ideal bore more than a striking resemblance to the pioneer Louis Blériot, complete with reversed flat cap. Pilots would often be depicted scattering flowers from the cockpit, or enchanting spectators who watched from the tops of spires. Aircraft were shown in all manner of visual aspects to best show off the drama of flight, with exaggerated swooping or altitude used to portray drama, excitement and freedom. The public lapped it up, and events were attended by tens of thousands of keen spectators.

The First World War would see a step change in how aviation was visualized. The romantic motifs and colourful palettes were replaced by industrial harshness and technical accuracy. Dreamy landscapes and moments of carefree joy were replaced by dogfights over battlefields. The once-adorning ladies were extolled to cast aside such feelings and join the production lines. While flying played a key role, these new posters were hard-edged and filled with a visual urgency not felt before.

Keen to put the war behind it and embrace the new world of free access, civil aviation began selling the dream again, albeit shaped now by war and matured by new advertising processes. Printing was increasingly in colour with marketing literature filled with multi-layered visuals, showing off routes across exotic locations as well as a timetable of flights. These would then be overlaid with dark silhouettes of modern aircraft punctured by squares of amber to indicating the internal lighting used on overnight passages. Photo montages, very much the cutting edge in contemporary graphic design, were also appearing, with striking shots of an aeroplane mated to dynamic designs and wonderfully vivid backgrounds. Photography would be put to excellent use as the decade progressed. Selling the flying dream, and it worked particularly well to show what this season's well-dressed aviators would be attired in, which in the main seemed to be jodhpurs and riding boots, a look championed by the actress Dorothy Sebastian and adopted by Louise Thaden as she competed in the Powder Puff Derby.

The posters continued to follow contemporary themes laid down by the rail companies, and soon the aircraft themselves returned to prominence over the locations. Imperial Airways in particular seized on the moment, offering posters as works of art that could be proudly pinned to bedroom walls, much to parents' alarm. As well as wonderfully evocative external images, a new series of views started to appear: interior depictions. Here colourful renditions of well-dressed passengers attended to by an equally well-appointed steward showed that the flight was all about relaxed glamour.

That's not to say that action-filled images were out of the picture. As designers became more confident with new technologies, mainly the airbrush, aircraft were depicted once more as translations of speed. The flying aircraft characterized by a mesmerizingly drawn spinning propeller. In the United Kingdom and United States, the use of contrasting coloured streak lines on sleek metallic fuselage sides increased and photos of these aircraft were embraced enthusiastically by marketing departments. However, the airlines, in particular, were becoming consumer aware. These action-packed images appealed to an excitable younger audience. At the same time, the more sophisticated and sedate consumer was treated to simple logo-led designs. These would be presented with a limited colour pallete and font, which were very much of their time. These designs echoed a popular artistic sophistication promoted by Walter Gropius, founder of the Bauhaus movement.

Smaller airlines also joined in, with influences ranging from art deco simplicity to complex landscapes. Aircraft such as the DH.86, the Ford Trimotor and the Lockheed Electra were being shown off flying over relaxing holidaymakers sunning themselves or crossing high mountains with ease. However, while these images sold a service, they also lacked drama. This gap in the visual market was picked up by the promoters of many of the air races that were taking place at the time. The Schneider Trophy posters, for example, mixed images of aircraft, complete with streaks of colour signifying speed, flying over white-peaked waves. Manufacturers soon followed, with Gloster and Supermarine producing awe-inspiring posters, as did ancillary goods manufacturers, from Shell to Lewis Leathers. Periodicals also picked up the theme, and across the globe, planes in action continued to captivate readers of aviation and engineering magazines with exciting portrayals of aircraft at speed.

By the mid-1930s, the excitement around flying had calmed as aeronautical developments now allowed expanses to be crossed with relative ease, and air travel became more readily accessed by the paying public. It also saw what was once extraordinary become routine. In doing so, the mysterious and exciting aspects of flight diminished, albeit a little, which reflected the adventurous spirit of the public at large. Although the new and numerous air routes and flights had been predictable for many, perception would be changed with the onset of the Second World War. In the meantime, the aeronautical industry had to become as creative as possible, especially the airlines. Fired imaginations spurred individuals to want to become pilots, technicians and ground staff. While many visual themes remained, there was one innovation that would be a winner: the cutaway.

These complex illustrations were the acme of the artist's skill. They placed the subject outside the blueprint in a recognizable environment populated by crews and passengers. Brightly coloured and featuring a mass of visual features, these images were extremely popular, making their way into comics and engineering magazines. This new genre of visual representation found its niche in portraying the larger aircraft of the time. By the end of the decade, they depicted aircraft from the United States and the United Kingdom, with the flying boat doing very well.

Flying boats were the marketers' dream: Deutsche Luft Hansa (DHL) sold their Dornier Do X as the ultimate expression of luxury. The Do X, then the largest aircraft in the world, was promoted by images of it basking in the sunny waters off Rio de Janeiro. Other airlines were quick to follow suit, and Pan Am saw its international flight portfolio as a great source of public relations material. The fleet, which now took to the skies, was given evocative names using the Boeing 314 Clipper, itself a great marketing tactic, as inspiration.

The Pan Am Clippers were luxurious seaplanes that helped establish routes throughout continental America and beyond. The term 'Clipper' was a very clever marketing ploy as it invoked the age of the speedy merchant vessels competing for the best time and speed between harbours. The name spoke of adventure, of crossing the trade routes laden with luxuries. *Clippe*r found itself used as a title followed by names such as *Pacific*, although the *Honolulu Clipper* was an exception. The Clipper name would continue to be used by Pan Am to name numerous aircraft until the airline's demise on 4 December 1991.

Flying boats also lent themselves to the artist's brush, where they could be shown in exotic locations featuring deep blue seas and lush, colourful shores. Imperial Airways was quick to follow, with aircraft pictured off the shores of Hong Kong or crossing Africa from Australia. Soon the flying boat became synonymous with a luxurious flight to tropical locations, an image carefully cultivated over a remarkably short period. They sold the dream of flight so well that the artists and designers turned the portrayal into an art form.

The Second World War saw many civil operations cease or wind down. However, the big companies remained eager to sell their achievements to the wartime public. So, they continued to push on with their advertising. Boeing, Avro, Goodyear, North American Aviation and Mercedes-Benz used their pre-war campaigns to significant effect during the period of hostilities.

At the war's end, the main changes for civil aviation were technological: radio navigation had become more sophisticated and weather prediction was now refined by radar. However, a significant difference was the advent of the turbojet engine. This engine would now relaunch the aeroplane as a tool of almost science-fiction wonder. Accompanied by a distinctive howl, the new world of civil aviation would become fantastically eye-catching and graceful. Silver-skinned machines would trace thin lines of vapour as they passed through the sky, flown by the new heroes of the jet age.

Alongside these new technological marvels flew the next generation of propeller-driven aircraft. These were flown by veterans who had made long journeys in cold and dangerous skies, in some cases only months before, in aircraft very similar to those they now flew civilian travellers in. For the smaller operators, war-surplus aircraft would provide sterling service, helping to establish routes, reputations and, more importantly, names. These aircraft would take passengers on flights which traced the routes carved out by the brave pioneers only a decade before. This was a different dream for the newcomers to the airline business. It was slow, steady and reliable, a proper salute for the repurposed masters of the air which had brought war to far shores but now reached out to connect communities once more.

For the marketing offices of the many hundreds of companies now involved in the growing market of flight, this new era would open imaginations. The corporate identity would lead to the development of increasingly colourful liveries, steward uniforms, logos and market visuals. Posters remained eye-catching but were increasingly influenced by global graphic-design trends. Visual communications also developed by adopting a more consumer-based approach. Photographs, often mixing posed shots with the documentary style and included as part of the now-familiar montage, began appearing on the pages of in-house magazines.

As safety became more rigorous and enforced by the principles identified in the 1944 Convention on International Civil Aviation, the need to share more complex information with passengers became commonplace. New innovations, such as individual seatbelts, evacuation procedures and how to wear and use lifejackets, would see the need for clear instructions led by well-trained staff. This information would also be backed up by a well-designed pamphlet, sized to fit in pockets fitted into the rear of seats or on bulkhead walls.

By the 1950s and 1960s, companies began to steer away from distant and often faceless and cold metallic flying shots of their latest airliner. Instead, the focus was increasingly upon the people that made their fleet fly and details of the latest technology. It was an opportunity to share new uniforms, interior information, depictions of the friendly customer and employee interactions and, most importantly, a chance to sell the brand. Pan Am was the undisputed master of this art, taking self-promotion beyond the blacktop. Their lead was soon picked up by TWA, KLM and Delta.

The application of logos and livery on aircraft has become more important to airlines and air service operators. In an increasingly crowded field, everyone needed to be seen. Initially, logos played a very muted role in the liveries of the numerous operators due to the small size of the aircraft they were flying. Aircraft like the de Havilland DH.83 Fox Moth and Lockheed Electra sometimes sported a single fuselage stripe. The fuselage would be finished on smaller aircraft in a single-colour paint scheme. Even on larger aircraft such as the H.P.42, the airline's name would only be painted upon the fuselage sides. It was not until the arrival of the DC-3 that things really began to change. Driven by the sheer number of operators flying the type and realizing that commercial success depended on getting a name out there, a new approach was needed to sell the brand. Operators began experimenting with different liveries, placing their corporate logo predominantly on the aircraft. Initially, the liveries, such as that used by Northwest Airlines, featured a shortened version of the company name, using a Fat Face-type style. This was placed in a prominent position on the fuselage on top of a contrasting coloured cheatline to help highlight the name. The round logo was almost military in its placement, sitting close to the elevators. The tailfin, at this point, remained mostly logo-free.

The wartime practice of adding markings to bomber aircraft showed that anything on the tailfin could be easily read. Its sizeable vertical surface area provided enough space for the corporate logo to be displayed. As larger, single-finned aircraft appeared post-war, the tailfin became the default position to place logos. By the 1950s, the graphic-design teams used by airlines were refining designs, many of which would become iconic. Pan Am used graphic designer Ivan Chermayeff to redesign their globe logo in 1955, while others, such as DHL, continued to use a simplified version of their original logo as designed by Otto Firle. Delta would slowly develop its logo, which often reflected the popular graphic-design influences of the time. Its 1950s designs looked like they would not be out of place atop a gas station fuel pump. It would not be until 1959 that the now famous 'Widget' logo, designed by Delta's advertising agency Burke Dowling Adams and inspired by Delta senior vice-president Richard Maurer, would appear. Since 1924 Delta has made sixteen changes to its logo, which may seem somewhat astonishing. Yet when placed against the need to remain relevant as a leading carrier, it is reasonable and shows a company keen to keep up with the times. The evolution of the logo continues for all airlines, designs reflecting aesthetic, cultural and social influences. Since its inception in 1984, economy airline Ryanair has had no fewer than four logo changes, while airline Aer Lingus has had eight since 1936.

Allied to the logo were the numerous livery changes. Design teams initiated complex identity programmes that covered everything from colour palettes to logo placements and uniform designs. From the 1950s, finishes were gradually refined into a series of styles that could be applied to any aircraft by a specialist finishing team. The finishing progressed from the pre-war cheatline, which had served civil aviation well, to the hockey-stick, which saw the cheatline carried upward towards the top of the tailfin. This style of finish was first used by Eastern in 1964 on their new jet liveries before being adopted by other operators. This was followed by the return of the all-over finish the following year as part of Braniff International Airways' (BIA's) massive rebrand, which saw airframes finished in one of several base colours. However, the changes, led by Alexander 'Sandro' Girard, did not stop there. Stewards were treated to a new uniform, which featured a space-age-style clear plastic helmet known as a 'rain dome' intended to protect the wearer's hair from the elements. Designed by Emilio Pucci, the new uniform, known as the Air Strip, was designed to be worn as a layered outfit which, unbelievably, would be slowly stripped off as the journey progressed. Such was the enthusiasm given to this new uniform by BIA that it even produced a striptease-style television advert to promote the new uniforms. The staff wearing this new uniform must have been exceptionally uncomfortable, in more ways than one, and thankfully it was short-lived. However, the Pucci influence remained. Staff were treated to uniforms finished in the more conventional fashions but adorned with Pucci's high-colour geometric patterns. Thankfully, his proposed catsuit, with matching hat and boots, also failed to pass muster, despite its colourful appearance.

Aside from the contemporary fashion, the BIA rebrand also introduced the Jelly Bean style of finish to the industry, which used the tailfin as a canvas for contrasting designs. By the 1970s, the finish of aircraft in civil use had developed into what is now called Euro white. This overall white finish could easily be adapted for leased aircraft without too much cost. This finish was highly flexible, and soon, the white background was treated like a clean sheet of paper. The cheatline was abandoned, and the airlines applied their names along the entire side of the fuselage, vertically and horizontally, thus giving this form of livery the term Billboard.

These simple approaches served the airlines well in the long run, they were cheap to achieve for the maximum visual impact. However, the industry still sought to push the visible boundaries, with two distinct types of specialized liveries beginning to appear. The first was the heritage livery applied to modern aircraft, with airlines such as Qantas, KLM and BA all using this type of finish. The second was the commemorative finish, used to celebrate organizational milestones or events of national importance, such as Air New Zealand's *Lord of the Rings* and *Hobbit* series of aircraft. These one-off colourful designs eschew the everyday concerns around costs as they are seen to be investments not only in raising awareness of the airline but also of the events they are celebrating.

Despite all this attention to detail in finishing the aircraft, it was only part of the branding story. Holidaymakers became the primary users of civil aviation in the post-war world and the need to provide meaningful engagement by airlines became increasingly important. Everything from monogrammed flight bags, signature meals and children's toys was on offer, as well as the all-important models of fleet aircraft. The flying experience was no longer reward enough, and the increasingly packed long-haul flights afforded by larger aircraft meant the airlines had to be on top of their game. For most, the adventure was just starting, and the age of the wide-bodied airliners was epitomized by Boeing's massive 747, which was waiting patiently in the wings.

The 1970s saw corporate identity become the cornerstone of a successful business, with colour television helping to spread the word. Marketing grew ever slicker featuring named staff as part of the story, personally welcoming passengers, all to the sound of contemporary musical soundtracks. Adverts highlighted the technical advances on the ground and in the air, with large monitored computer screens populated with information and crews addressing the camera directly. The airlines, especially those operating in and out of the United States, strove to convey that you were part of their family. These numerous adverts declared corporate standards and the desire to give you, the passenger, the best service; after all, the customer was now very much the interpretation of the business. The results were, as expected, positive and impressive. Package holidaymakers swarmed to locations all over the globe with locations, such as Benidorm in Spain, changing from quiet fishing villages to centres of tourism almost overnight.

However, the 1970s had also been the start of the dream going sour for some operators. It began a period of aggressive mergers and closures, which continued well into the twenty-first century. For those companies which folded, the banks, who had continued to fund new aircraft acquisitions, tried to recoup their losses by selling fleet assets to other airlines who had finance with them or selling older assets for scrap. Those lucky enough to merge would experience a gradual integration period. The old corporate identity would be replaced by the new owner's branding.

The first major issue to hit the industry would send ripples far beyond the boarding gate: the 1973 oil crisis. The Organization of Arab Petroleum Exporting Countries (OAPEC) declared an oil embargo on eight nations – including the United Kingdom, the United States and Japan – that had supported Israel during the Yom Kippur War. The effect of the six-month embargo was electric, with a barrel of oil trading at three dollars at the start of the embargo to twelve dollars by the end of it. Although the automotive industry was hardest hit, the airline industry also had to adapt. The first fundamental change was to reduce the use of auxiliary power units (APUs) and undertake movement on fewer engines or at lower power. Routes were also reviewed, with some being temporarily suspended.

The effect of the embargo had shown oil could be used as an economic and political weapon but also began the movement toward fuel-efficient design across aviation. However, it also showed how fragile many companies' operating models were and how vulnerable to rising cost they had become. The embargo also affected passenger numbers, with some companies recording low passenger figures a year after the embargo.

Further fuel crises, sparked by the Iranian Revolution, took place in 1979 and saw the price of fuel climb to almost $40 a barrel. However, this would steadily drop over the next twenty years. This time the airline industry was prepared, and the slight four percent loss in global output was picked up by other oil-producing nations, including the United Kingdom and the Soviet Union. By the decade's end, passenger numbers continued to climb slowly, fed by cheap all-inclusive holidays, larger aircraft and more efficient handling procedures. The 1970s also witnessed the appearance of the first major competitor to the American aircraft industry since the Second World War: Airbus.

Another key event to occur was the 1978 Airline Deregulation Act. This prohibited all states from regulating an air carrier's price, route or service to keep national commercial air travel competitive. In doing so, the Act also dissolved the notion of a flag carrier. Another aspect of the Act saw the powers of the Civil Aeronautics Board to regulate flight phased out, with safety-regulation management passed over to the Federal Aviation Administration (FAA). For airlines that had become reliant on overseas fuel, the shock increase in millions of dollars of costs, combined with the price war that followed the Airline Deregulation Act, was the start of a long decline towards bankruptcy.

As the 1970s slipped into the 1980s, travellers found they wanted more excitement and adventure and more choice. The era of increased disposable income meant holidays in overcrowded resorts were not to everyone's liking. Soon specialist tour companies began to appear, offering the traveller exclusive life experiences that a mere thirty years before had been strictly in the realm of the well-heeled. Safaris and themed expeditions soon became popular.

In 1984, two new airlines emerged that would change the passenger experience in very different ways: economy carrier Ryanair (RYR) and Virgin Atlantic (VIR). RYR soon established itself as a low-frills, low-cost short- and medium-haul airline that attracted a fair amount of controversy regarding how it operated and costed its services. VIR, led by the enigmatic Richard Branson, immediately set itself as a long-haul airline. VIR was keen to utilize the largest aircraft available at the time, the 747, and establish itself as a competitor to BA. Using the wider Virgin business empire to fund his latest endeavour, Branson soon created a commercial giant. Operating alongside Virgin Holidays, passengers could take a luxury holiday in several exotic locations, from the American continent to India. These two new airlines now met consumer needs brilliantly. Within a decade, both had become synonymous with the style of flying they offered. RYR's business model would soon become the template for many carriers eager to increase their profits. However, not everyone would not implement all aspects of the company's model.

The 1980s continued to see airline operator numbers fall, with few mergers occurring to save struggling operators, especially in the United States. Airlines that continued to operate did so amid stalling passenger numbers despite the best efforts of public relations and marketing departments. Price wars continued, especially on domestic flights, affecting many companies' bottom lines, including investment. The upshot would be fewer orders for manufacturers, including Boeing, which sent further ripples of uncertainty around the industry. In the United States, an additional bump in the road was the wholesale dismissal of 13,000 air traffic control staff by the chief of the FAA, Lynn Helms. This left a mere 4,000 air traffic controllers in post, which almost led to the collapse of the entire system. Combined with a faltering global economy, the 1980s were a far cry from the glory days of the 1960s, and airlines sought to rapidly address the changes as they occurred.

With its ever-expanding range of wide-bodied aircraft based on the A310 and the narrower A320, Airbus made huge inroads into the short- and medium-haul market. The European manufacturer kept its advertising visually clean and concise, focusing on the product alone. This approach was the polar opposite of the likes of Boeing, who traded on their product lineage. Airbus was also not afraid to swipe at the old guard of airline manufacturing, making jibes at their competitors' expense. Airbus also began to incorporate its customers into its marketing material, displaying the different airline tailfin patterns that could be found on their products. This approach was copied by several other aeronautical businesses, including Rolls-Royce. It showed customers they were genuinely central to the success of the company.

The 1980s also saw a change in how the message was getting out there. The holidaymaker was now seen as key to many airlines' survival, which dictated their advertising approach. Gone were the stuffy adverts of business people lounging in seats served by smiling stewards; these were replaced by fun adverts reiterating that flying was all part of the adventure. Agencies like Saatchi & Saatchi pulled out all the creative stops, especially for their television adverts, telling the story that flying was a great way to connect with the Global Village. Not only were these visually stunning adverts, but they incorporated soundtracks by legendary figures such as punk doyen Malcolm McLaren. The visual cues came thick and fast; United was a great exponent of this technique, placing their name as a footer against a range of eye-catching photographs. Other airlines continued to put their fleet and staff front and centre of advertising, carrying on with the theme of selfless service allied to technological innovation. However, some airlines, such as Pakistan

International Airlines (PIA), avoided any visual reference to their business. Instead, they chose images of domestic settings alongside the catchy strap line 'With Pakistan Airlines, its just like coming home. There's nothing better'.

This approach continued into the 1990s as the industry slowly recovered its lost confidence of the 1970s. Television began to play an increasingly important role in selling the brand, and social changes and more disposable income saw the popularity of the all-inclusive package holiday increase. Numerous companies sold the brightly coloured dream of sun, sea and sangria, all in easy reach, especially if the flight operator was also the holiday provider. It also made the holiday choice all the more appealing, capitalizing on the fact that the consumer would want the least possible hassle. British operators such as Thomson Airways (TOM) and Flybe (BEE), which at one time was the largest independent airline in Europe, were typical of the airlines that offered these services. In addition, straightforward hub and spoke flights serviced quick turnarounds, with some operators making several short-haul flights a day to the same location. Despite the often-chaotic terminal scenes lampooned by film and television adverts, the system operated well, especially with local customs agreements, such as the 1968 Customs Union used by EU member states.

The 1990s also saw the closure of several big names, including Pan Am winding its operations up in 1991, and Delta taking over the profitable remains. Aircraft manufacturer Fokker closed its doors for the last time in 1996, and Boeing took over McDonnell Douglas the same year. Such losses were expected and the gaps left were soon filled, and by the end of the decade, passenger numbers were rising to an all-time high of 1.6 billion – a figure helped by the growing confidence of Pacific airline operators keen to show off their increasingly politically settled region's riches in exchange for tourist dollars. Hong Kong-based Cathay Pacific (CPA) and Singapore Airlines (SIA) were keen to exploit such opportunities.

Further helping the airlines, especially when faced with often stiff commercial opposition and rising costs, was the emergence of the airline alliance. Though not a new idea, the concept had been mainly ignored in the post-Second World War world, as money and facilities had been plentiful. However, the actual fragility of the global network was exposed in the 1973 oil crisis. The first of what would eventually be six airline alliances, Star Alliance, was formed in 1997 by Air Canada (ACA), Lufthansa (DLH), Scandinavian Airlines (SAS), United Airlines (UAL) and Thai Airways International (THA). The key benefits for members were access to an extended network and the opportunity to take advantage of codesharing. Codesharing is a practice where a flight is marketed and sold by one carrier and operated by another, giving passengers a wider choice of destinations. Another crucial part of an alliance is sharing facilities and ground staff, which lowers costs for operators and consumers.

Star Alliance was followed by the Oneworld Alliance in 1999 and Sky Team in 2000. There would be three further alliances formed in the twenty-first century: Vanilla, U-Fly and Value.

The twenty-first century did not start well for the airline industry, marred as it was by the loss of an Air France Concorde on 25 July 2000, followed a year later by the terrorist attacks of 9 September 2001. In the wake of, what has become known as 9/11, passenger numbers dropped, taking almost eighteen months to return to pre-9/11 levels. As a result, several airlines ceased to operate, including TWA, Swissair (SWR) and Belgian flag carrier Sabena (SAB).

Thankfully, new technologies afforded by the growth of the internet were a saviour for many airlines, aided by the development of social media. This allowed airlines and individuals working in the aviation industry to tell their day-to-day stories. It involved the viewer, regardless of whether or not they were passengers. For the first time in its eighty-year history, civil aviation was once more in the public eye. The information available soon grew from blogs written by student pilots to new corporate websites. Aided by the growth of social media, civil aviation was in a brand-making heyday once again. The dawn of targetted social media, either through bought adverts or hashtags, helped spread messages individuals and companies wanted to share. However, social media was a double-edged sword, with specially assigned marketers and public relations professionals needing to be ready to counter bad news.

Some companies, like Virgin Atlantic, took early and full advantage of the marketing potential afforded by social media, effortlessly engaging followers and visitors. Not surprisingly, most of the conversations around social media were not driven by the big companies but by their followers. Photos that might not necessarily make the communications department cut would be shared among enthusiasts, essentially providing free advertising. For events such as the Red Bull Air Race, this was vital in spreading the word. In addition, stills of aircraft on the many blacktops around the world helped raise awareness of new liveries among the enthusiast community and beyond.

As well as changes in how civil aviation promoted itself, the twenty-first century saw the power of Airbus grow, its aircraft slowly usurping Boeing in fleets and orders. One of the key aspects behind Airbus's growth was its environmental credentials and use of increasingly efficient engines and designs. Boeing remained cautious, continuing to develop its 767 and 777 models. With its new wide-bodied 787 Dreamliner, which could carry almost as many passengers as the original 747-100, Boeing had hoped to take the fight to the Airbus A330/340/350 range of aircraft. But build issues dogged the 787's development and early use; by 2020 Boeing had addressed these. By 2020 the 787 was performing well on routes usually serviced by 747s or A340s. On 14 March 2020, a 787 operated by Air Tahiti Nui (THT) completed an astonishing, record-breaking 9,765-mile (15,715-kilometre) flight from Tahiti to Paris in a single trip, mainly because of low passenger numbers due to the COVID-19 pandemic. This trip no doubt allowed Boeing to breathe a sigh of relief as it maintained its position as the world's leading aircraft producer.

The first 101 years of regulated civil flight ended not as grandly as it should have. The COVID-19 pandemic of 2020 saw the 2019 pre-pandemic passenger numbers of 4.6 billion plummet to 1.9 billion in March 2020. Across the globe, fleets lay idle as staff was furloughed. Those airlines that could absorb the financial cost of almost complete grounding for three months returned to a much weaker market, with consumer confidence remaining low. Not all airlines or operators survived.

Chapter 6

Women in Aviation: The Pioneers and the Pilots

The Pioneers of Flight

Aviation has always pushed boundaries. The excitement of flight has drawn adventurers from across the globe with promises of excitement and immortality. Before powered flight, women had been drawn to the excitement of hot-air balloon flight, proving every bit as daring as their male counterparts.

In France, several women aeronauts emerged, with Ernestine Henry and Jeanne-Geneviève Garnerin forming a team to demonstrate their skills in the late 1700s. Garnerin would later be the first woman to make a parachute jump. Clearly aviation was in the family as Garnerin's niece, Elisa, would follow her into hot-air ballooning and demonstrate parachuting at least forty times in conditions that were far from ideal.

The trend of daring-do continued when Louise Poitevin, a Parisian aeronaut with more than a taste for the dramatic, took to the skies over early nineteenth-century Europe. Ascending with various props, often livestock, Poitevin would land with her 'teammates' safe and sound, thus proving the safety of flight and that no harm had come to her equine friends. Her final two landings in 1874 saw her land astride her horse, proving her mastery of the craft of flight as much as her equitation skills.

The influence of the French aeronauts on subsequent generations of women was profound, with the exploits becoming more and more daring. With the rise of the aeroplane, it was a natural progression for women to take to powered flight. This also brought women together to form the first women-only aero club, Stella, in 1909. In 1910 the first pilot's licence was awarded to French aviator Élise Deroche.

Élise Deroche was known as the Flying Baroness on account of being gifted the title of baroness by Czar Nicholas II after he witnessed her flying in St. Petersburg. As well as being renowned for her elegance and poise, Deroche was indeed a pioneer who was the spirit of early flight personified. Like all early aviators, Deroche had been captivated by the spectacles of the flight made by Brazilian Alberto Santos-Dumont in his bis-14 biplane on 12 November 1906.

Besotted by the idea of flight, Deroche, a highly skilled artist at the time, approached Gabriel and Charles Voisin for flying lessons, a request to which the brothers kindly responded. Her lessons made it apparent that she was a natural aviator, and she flew her first solo on 22 October 1909. She was smitten with flying at the expense of everything else. Determined to master her craft, Deroche was now focused on gaining her pilot's licence. The Aéro-Club rewarded her drive and passion by awarding her licence, Nr. 36, on 8 March 1910.

There followed a string of adventures which gained Deroche an army of admirers, as she soared above them, demonstrating her skill as an aviator, the distances flown becoming ever greater. With her growing experience came the urge to push her abilities. On 8 July, Deroche experienced a severe flying accident that could have killed her. Despite numerous fractures and internal injuries that paralyzed her, Deroche was determined to fly again. Her recovery was boosted by a burgeoning relationship with Charles Voisin and the chance to fly again, which occurred in February 1911 as she flew a Henry Farman biplane. Fate delivered Deroche another cruel blow, this time the death of Charles in an automobile accident on 26 September 1912. Flying became a salve once more, and in April 1913, Deroche enrolled at the Farman School, where she would meet her future husband, Jacques Vial. In the same year, Deroche would win the prestigious Coupe Femina. This award was presented on 31 December to the female pilot who flew for the most-prolonged period and distance. With the Coupe Femina came a 2,000-franc prize.

The First World War saw Deroche's Henry Farman aeroplane requisitioned by the French Army, her husband fall in battle, and her son succumb to the Spanish influenza pandemic. Small comfort was the burgeoning post-war racing scene, into which Deroche threw her pain and passion into it. Equipped with a new aeroplane, a Caudron G.3, Deroche re-established herself and trained hard for the return of the Coupe Femina. On 18 July 1919 she took off in the G.3 as the passenger of a pilot known only as Monsieur Barrault. Unfortunately, the aircraft flipped over shortly after take-off, killing Élise Deroche instantly, with M. Barrault dying shortly after. Élise Deroche was an inspiration to many women and had shown, no matter how heavy-handed and cruel fate can be, what was possible. After all, as Deroche had once declared, 'The sky is my stage' and it was a stage she graced well.

The post-war world was very different from the one that existed in the high summer of 1914. The First World War had seen women take on various roles during wartime, including working on and around aircraft. Lady Mary Bailey was one such woman and had volunteered as an aircraft mechanic during the war. Allied with her love of speed, aviation was a further opportunity to pursue her passion for all things high-speed. Despite having five children from her marriage to South African magnate Sir Abraham Bailey and learning to fly secretly, Bailey gained her pilot's licence in 1927. Almost instantly, she took advantage of the freedom afforded to her by flying and fed her need for excitement and adventure by becoming the first woman to cross the Irish Sea. A string of exploits followed, including setting the Fédération Aéronautique Internationale (FAI) World Record for Altitude of 17,283 feet (5,268 metres), flying a de Havilland DH.60 Cirrus II Moth biplane.

For her next challenge, Lady Bailey flew from Croydon to Cape Town, setting off on 9 March and landing on 30 April 1928 in a modified DH.60 Cirrus Moth. To help make the 8,000-mile (13,000-kilometre) journey, Lady Bailey had the Moth fitted with extra fuel tanks. She made the return trip later that year, leaving Cape Town in September and arriving back in Croydon on 16 January 1929. These flights represented both the longest solo flight and longest flights completed by a woman at that time. This, in turn, saw her receive the Britannia Trophy, an award still given today by the Royal Aero Club to aviators who have achieved the most meritorious performance in aviation during the previous year. To add to this accolade, Lady Bailey was awarded the Harmon Trophy twice, in 1926/7 and 1928, as the world's most outstanding female aviator.

She was still not satisfied with her aviation achievements, and in 1929 took part in the Challenge International de Tourisme 1929, an FAI challenge split into two parts. The first was a series of technical trials of aircraft to test construction and promote the design of more advanced aircraft. The second was a rally over Europe. Lady Bailey completed the challenge and entered again the following year, finishing thirty-first out of sixty entrants.

In 1930, Lady Bailey was invested as Dame Commander of the Order of the British Empire, as well as taking her place at the Women's Engineering Society Council. The following year Dame Bailey achieved her Certificate for Blind Flying, the first woman in the United Kingdom to do so. However, her adventures did not stop there. In the same year, she worked with archaeologists Gertrude Caton Thompson and Elinor Wight Gardner, taking aerial photographs of archaeological sites in the Kharga Oasis, Egypt. Not only did this task save time, but Dame Bailey's work also revealed unknown points of interest at the site. She died on 29 August 1969 at her home in South Africa.

Another British pilot was Amy Johnson, whose exploits would lead to her becoming a household name. Johnson, an economics graduate, had travelled to work as a typist in London from her native Hull in Yorkshire, in 1927. However, her fortunes changed when she joined the London Aero Club learning to fly and understand flight's technical and mechanical aspects. By 1929 Johnson was flying solo and had purchased a de Havilland Gipsy Moth, known as *Jason*. She now began planning a solo flight to Australia the following April.

In those early days of civil aviation, flight activities were open to anyone who could afford it. In the post-First World War world, surplus aircraft were cheap, and there was no shortage of aspirant and established pilots to purchase these aircraft. As a result, many flying clubs sprang up, with opportunities to learn to fly being cheap and plentiful.

On 5 May 1930 Johnson took off in her heavily laden biplane from Croyden aiming to beat the previous flight time of fifteen days set by Bert Hinkler. Johnson arrived in Darwin after flying 11,000 miles (18,000 kilometres) on 24 May after a journey that was far from smooth. She had to deal with misogyny, testing customs agents, mechanical issues and poor weather that tried her diplomatic and aviation skills to the maximum. Nevertheless, her welcome was akin to that of a modern pop star. On landing, Johnson was mobbed, and her reputation was made. For her adventure she received a £10,000 cheque from the *Daily Mail* and a CBE from George V's 1930 Birthday Honours List.

Johnson was fully aware that now that she was being watched by the press and public alike, she had to continue her aviation adventures. In 1931, she flew from Britain to Tokyo in a little over ten days after. The following year she married Scottish aviator Jim Mollison. This was followed by an attempt at breaking the London-to-Cape Town record set by her husband, as Amelia Earhart's successful transatlantic crossing had made that particular challenge redundant. Flying a de Havilland DH.80 Puss Moth, Johnson set off on 14 November 1932. She flew the 6,300 miles (10,140 kilometres) between the two cities in just under four days and seven hours, often flying in moonlight, beating her husband's record by 10 hours and 28 minutes.

The following year Johnson and Mollison crossed the Atlantic in a de Havilland DH.84 Dragon, making the flight between Pendine Sands, Wales and Bridgeport, New York State, and landing some forty hours later. They landed their aeroplane with the wind, leading to the aircraft crashing into a ditch and pitching Mollison through the windscreen. In America, Johnson strengthened her friendship with fellow pilot Amelia Earhart, who provided support following the crash.

There followed a period where Johnson was dogged by bad luck. Her entry with Mollison into the 1934 MacRobertson Air Race was cut short by poor fuel damaging the engines of their de Havilland Comet, *Black Magic*. Further pressure came from Mollison's drinking and suspected extramarital affairs, later confirmed by the discovery of a drunken Mollison with another woman. However, the year would see Johnson promoted to president of the Women's Engineering Society, an organization Johnson remained active in until her death.

In April 1936, she proposed a new airline known as Air Cruises. To promote her idea, she again set out to fly solo from London to Cape Town, flying a Percival Gull monoplane. Unfortunately, feats that had once raised headlines now barely raised eyebrows, and there was little press interest in her venture. Nevertheless, she set forth; unfortunately, a crash at an airfield on the edge of the Sahara prevented the trek from going any further. Undaunted, Johnson tried again and, on 4 May, took off on what would be her last record attempt. She arrived in Cape Town three days and six hours later, shaving 11 hours from the previous record.

By the time of her divorce in 1938 Johnson had shelved plans of long-distance flying, partially spurred on by the loss of her friend Earhart, and flew as a pilot for the Isle of Wright Aviation Company. In 1940 the company closed as most airlines ceased flying due to hostilities, and Johnson found herself flying with the Air Transport Auxiliary (ATA). Johnson lost her life while ferrying an Airspeed Oxford from Prestwick to RAF Kidlington near Oxford on 5 January 1941.

Another early pioneer was the American Mary Anita 'Neta' Snook. Snook was born on Valentine's Day, 1896, in Mount Carroll, Illinois. From an early age, she showed an interest in all things mechanical, a curiosity supported wholeheartedly by her father. While studying engineering at Iowa State College, Snook became interested in aviation and, after a couple of false starts, enrolled in the Curtiss-Wright Aviation School in 1917. America's entry into the First World War curtailed her lessons as civilian flying was grounded for the duration. Snook used the time to expand her mechanical knowledge and, in 1918, bought an unserviceable Canadian Curtiss JN-4 Jenny.

Using her time and knowledge, Snook got the Jenny airborne and, in 1920, flew her first solo in the aeroplane. There followed a period of barnstorming. Finally, late in 1920, she dismantled the Jenny and headed west, away from the frigid air of Iowa, to California. Once there she approached the aircraft designer and constructor Winfield Kinner for a role as an instructor at his airfield, Kinner Field. Given Snook's wide-ranging portfolio and experience covering a range of aviation aspects, Kinner employed her. Soon, Snook was running Kinner Field, the first woman to run a commercial airfield. On 3 January 1921 she was approached by Amelia Earhart to teach her to fly, a task she readily agreed to, with the two women becoming firm friends. Earhart would gain her licence in 1923.

In February 1921 Snook entered the men's air race at the Los Angeles Speedway and finished fifth. The following year she married Bill Southern, became pregnant and sold her business. After Earhart's disappearance she wrote a biographical account about her entitled *I Taught Amelia to Fly*. Snook returned to the air in 1977 when she was offered the chance to fly a replica of Charles Lindbergh's *Spirit of St. Louis*. In 1981 she was acknowledged as the oldest female aviator in the United States.

Snook was very clearly a talented teacher. Her protégé, Amelia Earhart, also known as Lady Lindy on account of her facial similarity to Charles Lindbergh, would go on to create an unbeatable role model for women everywhere. Earhart's interest in aviation was ignited by a visit to a California flying meet in 1920 at Long Beach with her father. Before this, Earhart, born in Kansas on

24 July 1898, had been a pre-med student at New York's Columbia University. A flight with First World War veteran, Frank Hawks, ignited a passion which would burn brightly throughout her life.

Taking a series of jobs to help with her tuition fees, Earhart started her journey in aviation with Snook before learning to fly under the tutelage of military aviation veteran John Montijo. Montijo, who picked up Earhart's training after an unfortunate incident with a tree, would continue instructing her in the military-looking Kinner Airster two-seater biplane, an aeroplane that was more than suited for training the student flyer and, by all accounts, forgiving too. The following year Earhart flew solo and subsequently invested in a canary-yellow Airster of her own.

During 1922 Earhart continued to hone her skills as a pilot, and the start of her remarkable career took off on 22 October when she achieved the women's altitude record of 14,000 feet (4,260 metres). 1924 saw Earhart sell her aircraft to be with her mother after her parents divorced, and she returned to Columbia University to continue her pre-med studies. This was followed by a move to Boston and employment as a social worker. However, the skies retained their siren call. Earhart applied when Amy Guest sought a companion to accompany her as a passenger on board her Fokker Trimotor floatplane on a transatlantic flight. The role was to fulfil a condition placed upon Guest by concerned family members. Earhart was selected by Captain Hilton Railey on behalf of a committee initially formed by the publisher George Putnam.

While not a pilot herself, Mrs Guest was married to British politician, Captain Frederick Edward Guest, a former Secretary of State for Air, which perhaps fuelled her love of flying. The aeroplane, known as *Friendship*, was crucial for an adventure that would take the first woman to cross the Atlantic. For Earhart, the opportunity was too good to miss. She would not be the pilot despite over 500 hours of flying experience. That task was given to Wilmer Stultz. Instead, she would fulfil the role of the aircraft commander. This was not a name-only role: it placed considerable responsibility on Earhart's shoulders. She would be the sole and final arbiter for all aspects of the safe operation of the aircraft, a huge responsibility in itself. The Fokker took off on 17 June 1928 from Newfoundland and made a safe landing on the waters off Burry Port, Wales, twenty hours and forty minutes later.

Despite her misgivings about taking part, Putnam, whom Earhart would later marry, had wound up the public relations machine that had pushed Earhart to the forefront of the public eye and began pushing the name Lady Lindy. The following year Earhart released a book about her adventure, leading to more media exposure and more flying time. She had become hot property, and Putnam, now enamoured with her, arranged for her to participate in the infamous Powder Puff Derby. This added to Earhart's skills portfolio; on 5 July 1930 she broke the women's speed record and achieved over 180mph (291km/h).

On 7 February 1931, Earhart and Putnam married. The adventures continued, including a trans-America flight in a Pitcairn PCA-2 autogyro from Newark, New Jersey, to Oakland, California, between 29 May and 6 June. To prepare for the trip, Earhart had taken time to familiarize herself with the exotic aircraft. Not only the first woman to fly the aircraft, she also set a new altitude record of 18,420 feet (5,613 metres) with the PCA-2. The same year Earhart was elected president of the famous Ninety-Nine Club.

The following year, Putnam planned a significant coup: a solo flight across the Atlantic. Many held the ridiculous notion that a woman would be unable to handle the still-arduous flight. Unperturbed, Earhart took off on 20 May 1932 in her Lockheed Vega monoplane, landing in Culmore, Northern Ireland, fourteen hours and fifty-four minutes later. Not only had she beaten the critics, but Earhart made landfall five years after Lindbergh's famous flight in the *Spirit of St. Louis*. The coincidence was no accident. Putnam had arranged a public relations coup and Earhart had delivered with élan. The world now lay at Lady Lindy's feet.

While the fame, photo calls and accompanying meetings were all well and good, the need for the next big adventure beckoned. So, in December 1934, Earhart prepared her next flight: crossing the Pacific between Hawaii and continental America. She took off ffrom Wheeler Field, Honolulu on 11 January 1935, landing at Oakland again the following day after eighteen hours and fifteen minutes of flying, covering a distance of 2,408 miles (3,875 kilometres).

This flight was followed in April by what could now be considered by the ever-restless Earhart, a short hop to Mexico. She expected to fly the 3,000-kilometre (1,684-mile) distance with ease, though the flight was cut short by sixty miles (100 kilometres) due to an insect getting lodged in her eye. Once recovered, Earhart set out to return to the United States, this time back to Newark, New Jersey, a distance of 2,082 miles (3,350 kilometres) that May.

These long endurance flights were building up to something, and Earhart's magnum opus was revealed in 1936: a 29,200-mile (47,000-kilometre) journey around the world. Earhart sought support, and a twin-engine Lockheed 10E Electra monoplane was purchased courtesy of a grant from Purdue University, Indiana. The 10E had a range of between 4,100 and 4,500 miles (6,589 and 7,242 kilometres); it was the ideal aircraft for Earhart. So, on 17 March 1937, Earhart took off on her journey, accompanied by Captain Henry Manning, Fred Noonan and Paul Mantz. The flight plan took the team west to Hawaii from Oakland, but due to a change in circumstances, she and Noonan returned to Oakland on 21 May 1937.

This was followed by a transcontinental flight to Miami, then to the island of Puerto Rico, and then to Natal, Brazil. After that came an Atlantic crossing to Dakar, Senegal. This was followed by a trans-African flight to the Gulf of Aden. The pair then flew across to the Indian subcontinent before stopping at Lae in New Guinea via Singapore and Australia on 30 June. By now, they had covered an astonishing 22,370 miles (36,000 kilometres) in thirty days.

The next stage would be the most testing for Earhart and Noonan's skills as pilot and navigator. The team would have to fly from Lae to the small island of Howland, a one-square-mile (2.6-square-kilometres) dot in the Pacific Ocean. Finally, on 2 July 1937, they took off. This was the last time Earhart and Noonan were seen.

There's been much speculation about the cause of their disappearance. From issues with using the military-grade radios that had been fitted during the journey, to confusion over the time zones used by Earhart and the US Coast Guard support cutter, the USCGC *Itasca*. There were also rumours that the aircraft had been intercepted by the Japanese and forced down, as Earhart was using the flight to gather intelligence about Japanese bases appearing on small islands. Regardless, Earhart and Noonan were considered lost after an exhaustive search of some 250,000 square miles of ocean. Despite possible sightings via satellite imagery and several expeditions, no conclusive trace of Earhart, Noonan, or their Electra has ever been found.

Despite the many dangers, the urge to take flight continued to capture the imagination of many women, including Bessie Coleman, an African American who faced more than her fair share of obstacles. Still, the desire to fly was stronger than the segregation and financial barriers she faced. Born on 26 January 1892 in Atlanta, Georgia, Coleman's early life was a mix of family break-up, economic hardship and caring for her siblings while her mother worked as a housekeeper.

In 1915, Coleman accompanied her brother Walter to Chicago, where she worked as a beautician. While financially rewarding, Coleman felt suffocated, her ambitions extending beyond the nail bar. Her imagination was spurred by the tales of French flying women and her brothers, who had fought in France. Flying was now the spur, and Coleman was encouraged by fellow African American Robert S. Abbott, publisher of the *Chicago Defender*, a paper established to fight for the civil rights of the African American population. Coleman left her job and took over the running of a small restaurant which allowed her to save and learn French.

20 November 1920 saw her leave the United States and head for France to learn to fly, her savings further bolstered by a gift from Abbott. On 15 June 1921 she was awarded her pilot's licence after training at the École de pilotage Caudron du Crotoy. Coleman had made history as the first African American female pilot. This further supported her view that flying was genuinely egalitarian. On her return to the United States in September 1921 she announced that she planned to open a flying school for African Americans. However, her return to the United States was brief. At the end of February 1922 she returned to France to perfect her skills in a Nieuport. By autumn,

Coleman had returned to the United States, delivering an exciting flying demonstration at Long Island on 3 September to honour the all-black 369th Infantry Regiment. Coleman's skills were beyond extraordinary, wooing the crowd with her daring displays and earning her the title of Queen Bess.

It was clear Coleman was an extremely talented flyer, often performing hair-raising stunts, further cementing her reputation as a pilot. Despite a crash on 22 February 1923 in Los Angeles, she continued to fly, regardless of the criticisms of her flying and personality. The exhibition flying continued, though the displays were often marred by the segregation laws that remained in place in a great many states. Unperturbed, Coleman would often refuse to fly unless members of the public were allowed to mingle. She would simply refuse to engage with organizers or fly if they were not.

Queen Bess was killed in a flying accident on 30 April 1930 when her Curtiss JN-4 Jenny, flown by her mechanic Williams Wills, lost control at 3,280 feet (1,000 metres). The Jenny threw Coleman from the aeroplane, killing her and Wills. Her wish to start a flying school had inspired fellow African American William Powell to establish the Bessie Coleman Aero Club in 1929 to promote aviation among the black community.

Coleman's death showed how dangerous flying was and yet how exhilarating it could be in life. This exhilaration was demonstrated no more clearly than in the famous 1929 Women's Air Derby, nicknamed the Powder Puff Derby by comedian Will Rogers. The race brought together the world's most experienced and hard-flying women. This was a gathering of the elite.

To enter the race, the pilots had to have a minimum of 100 hours of solo flight time, including 25 hours of cross-country flying. The rules were the same as those applied to men competing in the National Air Races. The only difference was that the women were expected to fly aircraft of a horsepower 'appropriate for women'. However, this rule would still offer pilots the choice of flying their aircraft in the heavy class, with engines of 510–875 cubic inches or the light class, which had engines of 275–510 cubic inches in size.

The race would see competitors trace a route, starting on 18 August 1929 from Santa Monica, California, finishing at Cleveland, Ohio, with eleven overnight stops to allow for rest, repairs, refuelling and relaxation. Aircraft taking part would take off at one-minute intervals, with the lighter aircraft leading the pack. This would ensure safety at altitude and give competitors a chance, once airborne, to push themselves and their aircraft hard between the start and finish. The race attracted the crème de la crème with Earhart joined by trailblazing women such as Blanche Noyes, Florence 'Pancho' Barnes, Ruth Nichols and Gladys O'Donnell.

At the time of entry, Blanche Noyes was a well-known actress turned pilot. She gained her licence in June 1929 after her husband, pilot Dewey Noyes had given her lessons earlier in the year. Unfortunately, her flight was marred by an on-board fire, which forced her to land. After extinguishing the fire and repairing the damage, she took to the air again and finished fourth in the heavy class.

After the race, Noyes would fly for Standard Oil as a demonstration pilot in 1931. Then, in 1936, she teamed up with fellow competitor Louise Thaden to take part in the Bendix Trophy flying a Beech C17R biplane. The team would establish a new world record of fourteen hours and fifty-five minutes for their flight between New York City and Los Angeles. This was an astounding achievement and a fitting tribute to her husband, who had died in a flying accident the previous December.

Noyes would join the Women's Advisory Committee on Aeronautics and the Air Marking Group of the Bureau of Air Commerce. This role would leave her the only woman allowed to fly government aircraft for many years. The committee was funded by the Works Progress Administration, which was part of Roosevelt's New Deal programme that sought to find work for the unemployed during the Great Depression. After the Second World War, Noyes headed up the air marking division of the Civil Aeronautics Administration to restore the pre-war air markings. These had been painted throughout the continental United States onto prominent buildings to help direct pilots. Noyes would die peacefully on 6 October 1961.

Born Louise McPhetridge on 12 November 1905, Louise Thaden graduated from the University of Arkansas in 1925. After graduation Thaden started work with J.H.J. Turner Coal Co. There she spent a great deal of time visiting the Travel Air Factory before its amalgamation with the Curtiss-Wright Corporation in 1929. As a result of her work, Walter Beech, who would later go on to found the Beech Aircraft Company in 1932, offered Thaden a job with Travel Air. Thaden's new role would be based with the Pacific Coast distributor of Travel Air aircraft, which included the Type R Mystery Ship racing aeroplane. As part of the deal, Thaden would be taught to fly, an opportunity too good to miss and she accepted the role keenly.

Thaden took to flying enthusiastically, and by 1928, her skills had gained recognition as she became the first pilot to hold the women's altitude, endurance and speed records in light planes. These achievements were followed in 1929 when she won the inaugural Women's Air Derby in the heavy class and was awarded a transport pilot rating to her pilot's licence. The following year Thaden returned to the East Coast and moved into public relations as a director with Pittsburgh Aviation Industries, as well as becoming the director of the Women's Division of the Penn School of Aeronautics. On top of these high-profile roles, she also took on the treasurer and vice-president roles of the newly founded Ninety-Nines.

The adventures continued, and in 1932 she was joined by fellow pilot Frances Marsalis to break the world endurance record in a Curtiss Thrush J monoplane. For 196 hours and five minutes, the two women traced a circuit over Long Island, New York, receiving food and water lowered by a rope from another aeroplane. They also took seventy-eight in-flight refuellings to keep their aeroplane aloft, a process developed by Captain Lowell Smith barely a decade before. Marsalis was killed in a racing accident in 1934 during a race meeting at Dayton, Ohio, after being caught in the turbulence of fellow competitors.

Thaden continued to fly, and in 1935, she was asked by fellow pilot and Ninety-Nines member Phoebe Omlie to join the new National Air Marking Program as a field representative. This was followed by her adventure with Blanche Noyes in the Bendix Trophy in 1936. In 1937, Thaden became the National Secretary of the National Aeronautics Association of the United States.

Thaden, now a representative and demonstration pilot with Beech Aircraft Corporation, retired from competition flying in 1938 and used her time to concentrate on creating airfields across the continental United States. She would also lend her skill to the newly formed Civil Air Patrol, an auxiliary unit of the US Air Force. There she initially served as a squadron commander at Roanoke, Virginia, before becoming the Mid-East Region women's coordinator and chair of the CAP National Commander's Training Committee.

In 1959, Thaden was appointed by Secretary of Defense Neil H. McElroy to the US Air Force's Defense Advisory Committee of Women in the Service. This promotion saw her experience the thrills of the jet age when she experienced her first jet flight in a New Hampshire Air National Guard T-33A Shooting Star. Thaden continued to promote women in aviation, believing that women were 'innately better pilots than men'. She died of a heart attack on 9 November 1979.

Phoebe Omlie née Fairgrave was another early female pilot. She took every opportunity available to advance herself and her skills in the world of aviation. Born on 21 November 1902, Omlie would leave a mark on aviation akin to the achievements of Earhart, Johnson and Barnes. Her interest in aviation was sparked when she witnessed a flypast in honour of the visit of President Woodrow Wilson to Minneapolis. After approaching a local airport, Omlie was treated to a flight, which was deliberately flown as aggressively as possible to put her off flying. However, the aerobatics Omlie experienced did the polar opposite, and after four lessons, she invested in a Curtiss JN-4 biplane. What followed was a series of hair-raising adventures and achievements culminating in her establishing Fairgrave's Flying Circus. Omlie delivered a truly barnstorming experience to onlookers, which included aerobatics, stunts, including hanging from the aeroplane by her teeth, and parachute jumps. Omlie's parachuting would earn her the record of the highest parachute jump by a woman when she descended from a height of 15,200 feet (4,600 metres).

By 1922 Omlie had married pilot Vernon Omlie, who would be killed in a flying accident in 1936, having established their own flying school in Memphis, Tennessee. Here they not only offered lessons in flying but also provided mechanical services. In 1927 Omlie gained her aeroplane mechanics licence but would also become the first licensed transport pilot in the United States. She then began flying the Monocoupe 90, a wonderfully attractive high-wing monoplane made by the Mono Aircraft Company from Framing Field, Illinois, which would host Charles Lindbergh on his cross-country commercial aviation promotional tour, called the Guggenheim tour.

The following year Omlie broke another record, the world altitude record for women, when she reached 25,400 feet (7,700 metres). This was followed by completing the Edsel Ford Air Tour and a flight across the Rocky Mountains, the first by a woman. Omlie's achievements seemed boundless, including winning the light class of the 1929 Women's Air Derby. Her reputation was further cemented by recognition of her success by the Democratic National Committee when she was enlisted to fly as a representative of Roosevelt's 1932 presidential campaign. Clearly, Omlie had made a good impression on Roosevelt. After his election, he appointed her special adviser for Air Intelligence to the National Advisory Committee for Aeronautics, a federal position and the first in aviation awarded to a woman. After her husband's untimely death, Omlie resigned from her federal position and returned to Memphis.

Omlie returned to Washington D.C. in 1941 as the Senior Private Flying Specialist of the Civil Aeronautics Authority with the remit to establish a training programme for the pilots the US military would clearly need. Omlie wasted no time and established sixty-six flight schools, including that which would train the famous Tuskegee Airmen. She would also establish a programme to train women instructors, who would instruct male and female pilots, including the women of the Women Airforce Service Pilots (WASP).

In 1952 Omlie felt the Truman administration's approach to aviation was overly restrictive and resigned from the federal government, returning to Memphis. She fell into alcoholism and poverty, dying of cancer on 17 July 1975. She never remarried and was buried next to her husband.

Florence 'Pancho' Barnes was a true character, both on the ground and in the air. The granddaughter of aeronaut Professor Thaddeus Lowe, the father of US military aviation, Barnes was born in Pasadena, California, on 22 July 1901. Barnes's early life was marked with adventure, her boisterous nature suited to a level of devilment that her Civil War veteran grandfather encouraged. However, the almost constant need for excitement, often provided by ponies and horses and ballet, which she excelled at, was allied to a healthy joie de vivre.

In 1921, she married for the first time. There would be four husbands during her life. Her first, Rankin Barnes, was a well-respected member of the Episcopal Church and was soon installed as a bishop. With the new role came a great deal of travelling. Often left with a young son, Barnes felt the settled life of a bishop's wife lacked the glamour and excitement she craved. In 1924 her mother died, leaving her with a sizeable inheritance, and soon her independence became more pronounced.

A series of adventures followed, including gun-running, albeit inadvertently, to Mexican revolutionaries. It was from this adventure that Barnes picked up the nickname 'Pancho'. On her return to the United States in early 1928, she accompanied her cousin, Dean Banks, on one of his flying lessons. Barnes was a natural aviator and went solo after only six hours of instruction. This followed a period of barnstorming, where she perfected her art while sharing the experience with passengers. Her hard work culminated with entry into the Women's Air Derby.

1930 saw Barnes take delivery of her Travel Air Model R Mystery Ship, which she would fly to break Earhart's speed record. This was followed by more adventures, most notably stunt flying in Howard Hughes's First World War Epic *Hell's Angels*. While flying as a film stunt pilot, Barnes lobbied for better pay for pilots and co-founded the Associated Motion Picture Pilots' (AMPP) Union in January 1932. She would also be a founding member of the all-women Betsy Ross Air Corps (BRAC) in 1930. The idea was led by fellow Powder Puff Derby racer Opel Kunz, an avid fan of Betty Ross, an upholsterer credited with making the first American flag in 1777. However, the idea was short-lived, and in 1933 the BRAC, which had been intended to supplement the US Army Air Corps in national defence and in times of national emergency, was wound up.

The Great Depression saw many fall on hard times. Barnes, renowned for her generosity in friendship, became well-known for her philanthropy and poured her energy into helping those less fortunate than herself, using her inheritance to do so. By 1935 Barnes had resorted to selling her Hollywood apartment and Mystery Ship. Using the proceeds, she bought eighty acres of land in the Mojave Desert, next to what would become the famous Edwards Air Force Base and NASA's Armstrong Flight Research Center (AFRC). There she established a welcoming bar and restaurant for the numerous pilots called the Rancho Oro Verde Fly-Inn Dude Ranch, equipped with a landing strip. The site soon became a one-stop show for pilots, with a swimming pool, casino and stables, all led by Barnes. Many pilots, and future astronauts, would frequent the bar, which was soon filled with photographs of pilots, aircraft and groups. Post-war, Barnes would find new ways to help pilots burn off excess energy as they would be able to take one of her hirelings from the stables for a gallop. This earned the ranch a new name on account of a remark made by General Jimmy Doolittle after one such excursion: The Happy Bottom Club.

A protracted legal battle in 1952 with the United States Air Force (USAF) over the price of land on which her Fly Inn was situated would take its toll and her precious stables and horses would be lost in a mysterious fire. The site sold above market price, but only after a bitter battle, which Barnes called the Battle of the Mojave. Thankfully the axe was buried in 1961 when she was bestowed the title Mother of the Edwards Air Force Base, with the officers' mess renamed the Pancho Barnes Room.

She stopped flying in 1950. There followed a period of poor health, including breast cancer. Although she remained popular, she died alone on 29 March 1975. Her passion for aviation was unsurpassable; she supported the aviation community with the aplomb of the mother figure she had become to military and civilian pilots alike, to the very end.

While many of the early pioneers would be grounded as wartime curtailed civilian flying, there was a group who would be empowering women in aviation and leading the jet age. Jacqueline Cochran had been brought up in poverty in Florida. Cochran, born Bessie Lee Pittman on 11 May 1906, pushed herself from an early age, determined to escape poverty. Her first marriage to Robert Cochran ended after the death of their five-year-old son, leading to a series of jobs, including nursing. By 1932 Cochran was in New York. After a discussion with Floyd Odlum, whom she would marry in 1936, about selling beauty products, she learned to fly, taking her first lessons at Roosevelt Field, Long Island, before moving onto Ryan Field at San Diego to complete her training. This was followed by purchasing her first aircraft; by 1933 she had gained her commercial pilot's licence.

In 1934 Cochran took part in the MacRobertson Air Race, flying a Granville Gee Bee R-6H, which was withdrawn at Bucharest due to mechanical issues. This adventure was followed in 1937 with an entry into the Bendix Race, flying the civilian precursor to the P-47 Thunderbolt, the Seversky SEV-2S. By 1938 Cochran was considered the best female aviator in America; a series of record-breaking achievements followed. By the time of her death on 9 August 1980, Cochran had won the Harmon Trophy no fewer than five times. But for many, her real achievement lay in opening the skies for women in wartime. As a member of the Ninety-Nines and subsequent president (1941–3), she used her influence to promote women in aviation. She started and succeeded in pushing for acceptance of women into the Civil Air Patrol, which remains an auxiliary body of the US Air Force.

One key area Cochran was keen to promote was a women's flying division within the United States Army Air Corps (USAAC). She perceived the purpose of such an organization as filling in the ancillary flying roles, such as those undertaken by the Air Transport Auxiliary (ATA) in Great Britain, thus freeing up the male flyers to fight. She envisaged herself as the de facto organizational lead and, in September 1939, she fielded the idea to the USAAC. This led to the commander of the USAAC, General Henry 'Hap' Arnold, recommending that she look at how the British organized their auxiliary flying support.

Along with seventy-six other women, Cochran went to Canada to complete the training in January 1942. Two months later, twenty-five pilots, and Cochran, went to Britain to start flying with the ATA. As a member Cochran became the first woman to fly a bomber, a Lockheed Hudson V, over the Atlantic.

While Cochran was still in Britain, Arnold formed the Women's Auxiliary Ferrying Squadron (WAFS), led by former air racer and test pilot Nancy Love, in September 1942. Love also held a commercial pilot's licence and had been instrumental in designing the aircraft tricycle nose gear arrangement. On hearing this, Cochran returned to the United States and helped establish the Women's Flying Training Detachment (WFTD), feeling that women pilots were capable of much more than ferrying. In August 1943, the two organizations merged to form the Women Airforce Service Pilots (WASP), with Cochran as director, overseeing the training of hundreds of women pilots, and Love as head of the ferrying division. In 1945, Cochran was awarded the Distinguished Service Medal (DSM) for her wartime service.

After the war she worked as a journalist, entering Japan as the first non-Japanese woman, and attended the Nuremberg Trials. In 1948, she joined the newly formed United States Air Force Reserve, where she reached the rank of colonel before retiring in 1970. However, the jet age would offer new challenges to Cochran. In 1953, she flew a borrowed Canadair Sabre 3 on a series of circuits, culminating in breaking the speed of sound on 18 May, the first woman to do so. After that, the firsts piled up, including flying the Northrop T-38A-30-NO Talon supersonic trainer to a peak altitude of 56,072.8 feet (17,091 metres).

Cochran was also a firm believer in women's rights and sponsored the Mercury 13 programme, which sought to test the suitability of women as astronauts. The programme was not supported by NASA. As NASA required all their astronauts to graduate from military jet test-piloting programmes and have engineering degrees, no women could meet these requirements. This led to a hearing before the Special Subcommittee of the House Committee on Science and Astronautics, which would determine whether or not NASA's selection was discriminatory. However, the space race had now turned into a political game, and Cochran conceded that time was of the essence and that, for the time being, men had the upper hand. Nevertheless, Sally Ride broke NASA's glass ceiling on 18 June 1983 as the first American woman in space. Cochran died at her home on 8 August 1980, leaving a legacy in aviation that would see her achieve more distance and speed records than any pilot, living or dead, male or female.

Outside the buzz of the United States, which had become the de facto centre of global aviation, especially from the late 1920s, women elsewhere continued pushing the boundaries of aviation. One was Jacqueline Auriol, a French contemporary of Jackie Cochran, who had gained her pilot's licence in 1948 after an eventful war as a resistance fighter. She continued to hone her skills to earn an advanced pilot's licence in a French-built Grumman G-44 Widgeon while recovering from a horrific crash aboard a SCAN 30 flying boat. In 1950 Auriol was awarded a military pilot's licence and, as well as being a talented stunt and test pilot, added fast-jet flying to her repertoire. With this new licence, she followed Cochran and broke the sound barrier and would set no fewer than five speed records in the 1950s and 1960s. Her final speed record of 2,038.70km/h (1,266.79mph) was established on 14 June 1963 when she flew a Dassault Mirage IIIR over a 64-mile (100-kilometre) closed circuit. This beat a record that had been set by Cochran the month before. Almost a year later, Cochran regained her speed record, achieving an impressive 1,303.18mph (2,097.27km/h) in a Lockheed F104G Starfighter. Auriol's exploits would see her awarded the Harmon International Trophy no fewer than four times.

Another contemporary of Cochran's was Pauline Gower, a Briton whose achievements in the field of British civil aviation were similar to many women flying in the United States. Initially, Gower and her friend Dorothy Spicer, a talented pilot and the first woman to gain an advanced qualification in aeronautical engineering, started their own air-taxi service in 1931. Despite having over 100 hours of flight time, being able to fly solo at night and investing in a new de Havilland Gipsy Moth biplane, the business folded. Undaunted, the pair moved on to performing with the Crimson Fleet Air Circus throughout 1932 before appearing at the British Hospitals' air pageant in 1933. There they would fly alongside Charles Scott, who would go on to win the 1934 MacRobertson Trophy Air Race in the de Havilland DH.88 Comet, and James Fitzmaurice, who made the first east-to-west Atlantic crossing in 1928.

In 1935, Gower was appointed as the council member of the Women's Engineering Society. The following year she became the first woman to be awarded the Air Ministry's Second Class Navigator's Licence. She continued to work alongside Spicer, promoting engineering. In 1938, her expertise was called upon by the government as civil defence commissioner in London with the Civil Air Guard, an organization formed to encourage and subsidize pilot training. With her professional standing now recognized, Gower pushed the government to create the Air Transport Auxiliary (ATA) to help ease the flying burden on military pilots. By December 1939, Gower had eight women ready for the first ferry pool based at Hatfield, Hertfordshire, England.

Appointed as head of the women's branch of the ATA, Gower took her newly appointed pilots and began the testing and selection process, the same that Cochran was subjected to in Canada two years later. Amy Johnson and air racer and flight test engineer Eleanor Curtis would add to this pool of pilots. They, in turn, would become founding members of the British Women's Pilots'

Association. Gower pushed the Air Ministry into awarding her wards equal pay to their male counterparts, which they received in 1943 and fought for her pilots to be able to fly any type of aircraft in service. For her wartime service and leadership Gower was awarded the Most Excellent Order of the British Empire (MBE). She married Wing Commander Bill Fahie in 1945 though she died two years later while giving birth. In 1950 she was posthumously awarded the Harmon Trophy.

The urge to spread wings was not restricted to western women. Egyptian Lotfia Elnadi was the first African woman to earn her pilot's licence. Born to an influential Cairo family, Elnadi approached Kamal Elwi, the director of Egyptair, for flying lessons. However, her family were far from supportive, so the studies were carried out secretly. On 27 September 1933, Elnadi was awarded her pilot's licence. Elwi had seen the publicity potential in Elnadi's endeavours, and his foresight was rewarded when she became world famous. This fame was cemented when she took part in the Cairo-to-Alexandra air race in December 1933, finishing first. However, Elnadi did not win, as she'd missed a checkpoint; she was awarded a consolation prize and the congratulations of King Fuad. She also attracted the support of noted Egyptian feminist Huda Sha'arawi, who led a drive to enable her to buy her own aircraft. Elnadi would continue to fly regularly, becoming the secretary general of the Egyptian Aviation Club before an accident in 1938 damaged her spine, preventing further flying. Due to her influence there was a surge of Egyptian women learning to fly until the outbreak of war in 1939. Elnadi moved to Switzerland after her accident and did not return to Egypt until 1989, where she received the Order of Merit of the Egyptian Organization of Aerospace Education.

To the east, Chinese actress Lee Ya-Ching was the first woman to be awarded a Chinese flying licence in 1936 going on to establish the country's first flying school. Ya-Ching began training at the Swiss Contran École d'Aviation, where she was awarded her pilot's licence in 1933, before moving on to the Boeing School of Aviation in Oakland, California, to perfect her craft in 1935. Ya-Ching returned to Shanghai in 1935, keen to promote civil aviation, and while the Shanghai Flying Club welcomed her, the Chinese government was frosty and initially refused to give her a Chinese pilot's license. Finally, after intense lobbying, and following an assessment of her flying made by a member of the Chinese air force, she was given her pilot's licence. She was then commissioned by the Chinese government to survey safe air routes and aerodromes around the country, a task that would see her wrack up 30,000 miles.

After performing this arduous task, Ya-Ching helped establish China's first civilian flying school and became the school's sole female instructor. In honour of Chiang Kai-shek's fiftieth birthday celebrations in 1937 she performed an aerobatic routine in front of 150,000 spectators. This was followed by a period of flying for the Chinese-owned airline, the Southwest Aviation Corporation (SWAC). With the outbreak of the Second Sino-Japanese War in July 1937, Ya-Ching offered her services as a pilot both in combat and transportation. Her offer was turned down, and while grounded she established a hospital and refugee camps. Still not satisfied she was doing enough she secured the loan of a Stinson SR-9B high-winged monoplane and began a goodwill tour to raise money for refugee care. She was supported by contacts in the film and aviation industries, including Louise Thaden.

Throughout the war, Ya-Ching would fly a Beechcraft C17R Staggerwing biplane on behalf of China Relief and an Aeronca Super Chief 65 LB high-wing monoplane for Relief Wings, flying throughout the Americas. Relief Wings was a humanitarian air service for disaster relief headed by Ruth Rowland Nichols and established in 1939. Nichols was the only woman to simultaneously hold world records for a female pilot's speed, altitude, and distance. At the war's end, Ya-Ching initially returned to a much-changed Shanghai before settling in Hong Kong, where she endeavoured to keep flying. By the 1960s, Hong Kong's economic decline forced her to return to the United States, settling in San Francisco, where she passed her written and practical flying exams in 1966 to become re-certified at fifty-four. She continued to fly, her prowess in the air never waning. She died in Oakland, California, at eighty-six and was buried in a large plot that allowed her the space in death she so loved in life.

Jean Ross Howard's ties with aviation were as remarkable and groundbreaking as any of the early pioneers. Ross Howard got her first taste of aviation through the Civilian Pilot Training Program, where she was awarded her pilot's licence in 1941. This was followed by roles within aviation, culminating in taking the post as pilot-secretary in 1942 with Piper, Taylorcraft, and Aeronca Aircraft companies, for whom she flew demonstration flights of their light aircraft for the US military.

Inspired by a speech by Jackie Cochran, Ross Howard signed up for the WASP programme, starting her course in January 1943 at Avenger Field, Texas. Although she failed the course, Cochran saw her potential and enlisteed her help in the training, where she remained until July. She then joined the Civil Air Patrol, where she worked as a programme director for the American Red Cross, flying out of the island of Capri until 1945. This was followed by flying with Civil Air Patrol's National Capital Wing, achieving the rank of major and starting a long career with the Aircraft Industries Association (AIA), which later became the Aerospace Industries Association. There she worked as a staff assistant for the AIA's Personal Aircraft Council, and by 1950 she had become a staff assistant to the AIA's public relations director. Ross Howard continued to fly, participating in the 1951 and 1952 All Women's Transcontinental Air Race, the restart of which was sponsored by Cochran in 1948. These races were designed to 'provide stimulation as a refresher course in cross-country flying for women whose services as pilots might once again be needed by their country'.

In 1954, Ross Howard convinced Larry Bell of Bell Aircraft to allow her to train for her helicopter certification at Bell's Helicopter School in Fort Worth, Texas. Bell agreed and Ross Howard achieved her accreditation after eighteen days of training, becoming the eighth American woman to do so. She became a keen advocate of the helicopter, recognizing its uses. She included it in her 1955 master's degree in history thesis from the American University titled 'Selected Economic Problems in the Operation of Common Carrier Helicopters'. The same year she founded the Whirly-Girls, an international organization for woman helicopter pilots supported by the enigmatic Howard Hughes. For the next decade she focused on growing the organization. By 1969 other officers had been appointed and she had become the Whirly-Girls' first president, a role she would fulfil until 1975.

Ross Howard's advocacy for establishing emergency services heliports, especially for medical helicopters, saw her become the first recipient of the Whirly-Girls Livingston Award in 1988. This award was instituted by fellow helicopter pilot Nancy Livingston and is awarded to a woman who has achieved distinction for her contributions, on behalf of women, in helicopter aviation. Other awards included the 1963 Washington Air Derby Association Trophy and the Lady Hay Drummond Award from the Women's International Association of Aeronautics in 1969. As well as the Whirly-Girls, Ross Howard served with several aviation bodies, including secretary and vice-president of the Ninety-Nines, secretary of the American Helicopter Society and Federal Aviation Agency's Women Advisory Council. She passed away on 29 January 2004 aged eighty-seven.

Born in 1926, Betty Miller's aviation story is as important as any. After graduating from high school, Miller worked as an aircraft communicator with the Civil Aeronautics Administration, now the Federal Aviation Administration (FAA), where she met her husband, Chuck. Later, they set up a flying school called the Santa Monica Flyers. Miller learned to fly in 1952, and would also complete her helicopter accreditation.

She honed her rotary- and fixed-wing flying skills, culminating in her 1963 solo flight across the Pacific, becoming the first woman to do so. The flight was organized by the Millers and their friend, light-aircraft test-pilot Max Konrad, a record-breaking long-distance pilot who had made a non-stop journey of 7,668 miles (12,340 kilometres) between Casablanca, Morocco and Los Angeles in June 1959. Using Konrad's experience and working with William Piper of Piper Aircraft, the group planned to deliver a twin-engine Piper PA-23 Apache from Oakland, California, to Brisbane, Australia. The journey would see Miller navigating the Pacific via Honolulu, Hawaii, Canton Island and New Caledonia. The Apache was fitted with special ferry tanks designed by

Konrad. Miller achieved the status of being the first woman to fly solo, non-stop from Oakland to Honolulu.

Miller took off on 25 April 1963, and after various stops, landed at Brisbane on 13 May 1963. For her achievement, She was awarded the Federal Aviation Administration's Gold Medal for Exceptional Service and the Harmon International Trophy for Aviatrix of the Year.

Like many of her peers, she was a member of the Ninety-Nines and the Whirly-Girls. Miller's physical standards were later used by NASA as the benchmark against which all female astronauts had to pass before being accepted onto the Astronaut Candidate Training Program. Miller died on 21 February 1981 at the age of 91.

Another long-distance pilot was Louise Sacchi, whose prodigious flight record included regularly crossing the Pacific and Atlantic as a ferry pilot. Learning to fly in 1939, Sacchi would soon build on her skills, gaining her navigator certificate in 1942 and becoming an advanced navigation instructor to the RAF at their flight school in Texas. In addition, she bolstered her skills by training as a mechanic; by 1955 she was a qualified airline transport pilot.

Sacchi continued to build on her experience and in 1965 became the first and only international ferry pilot, completing her 100th ocean crossing in 1968. This was followed in 1971 by setting the record for the fastest New York–London flight by a single-engine landplane, which remains unbroken. In 1976 Sacchi made her 300th ocean (Atlantic and Pacific) crossing.

Sacchi's skills as a pilot drew multiple awards and accolades, including being the only woman to receive the Award of Appreciation and Honorary Wings of the Spanish Air Force and the Godfrey Cabot Award for distinguished service to aviation. She was the first woman to be granted this award in recognition of her 340th ocean crossing, the most undertaken by any non-airline pilot. Sacchi passed away on 22 March 1997.

Geraldine 'Jerrie' Mock is perhaps the least known of the post-war female pilots. Yet, her achievements remain as awe-inspiring today as they were sixty years ago. Mock was born in Newark, Ohio, on 22 November 1922; her interest in aviation was sparked by a trip with her father in a Ford Trimotor at age five. She enrolled at Ohio State University to study aeronautical engineering, one of the first women to do so. Jerrie would marry pilot Russell Mock in 1945, together they would share a love of flying. At the age of thirty-two, Mock gained her pilot's licence, and as she and her husband flew around Midwest America, the urge to explore the confines of the United States took hold of the mother of three.

Keen for adventure, and at her husband's suggestion, she decided to trace the round-the-world route taken by the ill-fated Earhart. She now had 750 hours of flying time behind her and a 1953 Cessna 180 named *Spirit of Columbus* to carry out the endeavour. To prevent a repeat of Earhart's loss,

Poster for the 1910 International Air Meet at Dominguez Field, California, USA. (Unknown)

Persian Shah Kai Kawus in flight with his tethered eagles. (Princeton University Library)

Roger Bacon with an experiment, depicted by German physician Michael Maier (1568–1622).

The Montgolfier brothers' first manned hot-air balloon takes off from the garden of the Reveillon workshop, Paris, 19 October 1783. (Claude Louis Desrais)

Clémant Ader designs, including this Avions III, were as fantastic as they were imaginative. (Conservatoire National des Arts et Métiers)

A wonderful study of the Boeing Model 40 at Oshkosh Airshow. (Unknown)

De Havilland's DH.84 was everything an art deco-inspired aircraft should be and so much more. The fact that the type remains in service in the twenty-first century is testament to the longevity of its design. (Alan Wilson)

A wonderful study of a Boeing 247 at the Museum of Flight in Tukwila, Seattle. Note the rubber de-icing boots on the fin and horizontal stabilizers. (Aaron Hedley)

Perhaps the greatest piston-powered airliner ever built. Attendees visit a restored American Airlines flagship DC-3 at the Wings over Homestead Air and Space Show, Florida. (Defense Visual Information Distribution Service)

Lockheed were quick to sell the glamour of the Model 10 Electra using a Delta Airlines version alongside models posing with an Oldsmobile. (Robert Yarnall Richie Photograph Collection)

Experimental Aircraft Association (EAA) Ford 4-AT-E Trimotor. (Alexf)

A close-up of the top of the Schneider Trophy at its home in the Science Museum, London. (Morio)

The Short SC.1 was Britain's first VTOL aircraft and showed that innovation in aviation was far from muted. (Mike Freer)

Still flying, a Short SC.7 Skyvan, operated by Canadian Summit Air (SMM), taxis at Gloucestershire Airport. (James)

Above: De Havilland's Tiger Moth would rapidly take on legendary status because of its forgiving handling and popularity as a recreation and training aircraft. (Ossobe)

Centre: Supermarine had the S.6B, but de Havilland had the DH.88 which would influence the design of a wartime classic – the Mosquito. (RuthAS)

Bottom: The Reno National Championship Air Races attract aircraft of all types including this de Havilland Sea Vampire T.22. (Tatquax)

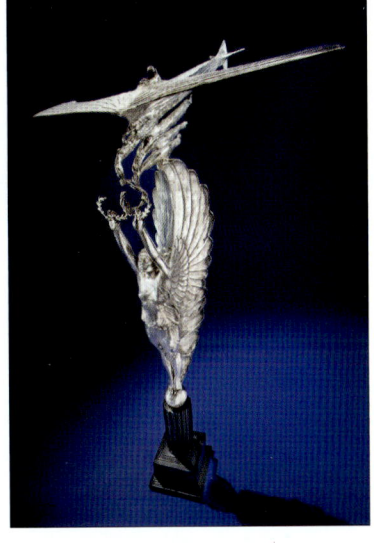

The beauty that is the Pulitzer Trophy. (NASM)

Above: Red Bull Air Race World Series pilot Kirby Chambliss performing a knife-edge manoeuvre in his Edge 540 in Perth, Western Australia in the 2006 series. (Hamish)

The Thompson Trophy. (Mj)

Left: A poster calling for citizens to buy stock in Dobrolyot. Ill: I.V. Simakov.

Below: Standard J-1, the backbone of the US airmail service in the early 1920s. (US Postal Service)

Left: The remains of one of the Air Mail Service Beacon System concrete arrows. Note the beacon mounting points in the central square section. (Dppowell)

Below: A Douglas M-2 in Western Air Express colours. (Gary Todd)

A Cessna 175 parked in from of a very high-frequency omni-directional range (VOR) at Frederick Municipal Airport, Maryland. (Acroterion)

A Lufttransport-Unternehmen (LTU) de Havilland DH.104 Dove 8. (Julian Herzog)

Douglas DC-6B of Balair (BHP) coming in to land at Basle Airport, Switzerland. (Eduard Marmet)

A British European Airways (BEA) de Havilland DH.4B approaching Berlin-Tempelhof Airport. (Altair78)

A fuselage fragment of a de Havilland Comet G-ALYP retrieved from the Mediterranean Sea; explosive decompression was found to be the cause of its crash on 10 January 1954. (Krelnik)

A Catair (Compagnie d'Affretements et de Transports Aeriens) Sud Aviation SE 210 Caravelle at rest. (Michel Gilliand)

An Aeroflot (AFL) Ilyushin Il-18 turboprop at Novosibirsk-Tolmachevo, Russia. (Gleb Osokin)

A Lockheed L-188 Electra in the livery of Brazilian airline Varig (Viação Aérea RIo-Grandense, Rio Grandean Airways). (Christian Volpati)

A British Caledonian (BCC) Airways Boeing 707-338C at Jomo Kenyatta International Airport, Nairobi, Kenya. (Steve Fitzgerald)

A Lufthansa (DLH) Boeing 727-230 coming in to land at Faro, Portugal. (Pedro Aragão)

Left: A Ceskoslovenske Aerolinie (CSA) Tupolev Tu-124V (Lars Soderstrom)

Below right: The game-changing Boeing 747. This early Pan Am 747-100 shows off its unique flap design. (Aldo Bidini)

Above left: A Vickers Super VC10 of East African Airways Corporation at London Heathrow. (Ruth AS)

A Rich International Airways (RIA) DC-8. The DC-8 was Douglas's first jet-engine airliner. (GBNZ)

Above: An Iran Air (IRA) Airbus A300-605R coming in to land. (Aero Icarus)

Right: A Pan International (DR) BAC One-Eleven. This aircraft would later crash attempting to land on an autobahn following the failure of both engines. (Lars Söderström)

A colourful ATA (Aerocondor Transportes Aéreos Lda.) Boeing 720-023B at Miami Airport, Florida. (RuthAS)

The acme of civilian aviation: Concorde. (Eduard Marmet)

Left: A wonderful cutaway poster from New Zealand Union Airways Ltd. enticing the viewer with the comforts of a modern airliner. (National Library of New Zealand)

Above: An advert for Aeroflot on the side of a Hong Kong bus. (Megabus13601)

The interior of a Siberian Airlines Airbus A320-214 showing its seating arrangements to gain maximum passenger carriage. (Gleb Osokin)

Above: A unique cutaway view of the Airbus A380's double-deck configuration. (Stahlkocher)

Left: Grande Semaine d'Aviation de la Champagne, Reims, celebrates the world's first air meet, the Great Week of Aviation. (Ernest Montaut)

An Airbus Beluga loads the European-built International Space Station Node 2 module for delivery to Kennedy Space Center, Florida. (Italian Space Agency)

Above: Often called Concordeski, the Tupolev Tu-144 worked predominantly within the interior of the USSR. (clipperarctic)

Below: An Air France Concorde rests at Charles de Gaulle Airport, Paris. (Michel Gilliand)

Above: Selling the dream. Pan Am's art department did not disappoint, drawing on the fashion illustration styles of the 1930s such as that practised by Leslie Saalberg.

Right: 1960 David Kleins poster for TWA captured the spirit of the emerging jet-age of travel, merging it with the vibrancy and excitement of New York.

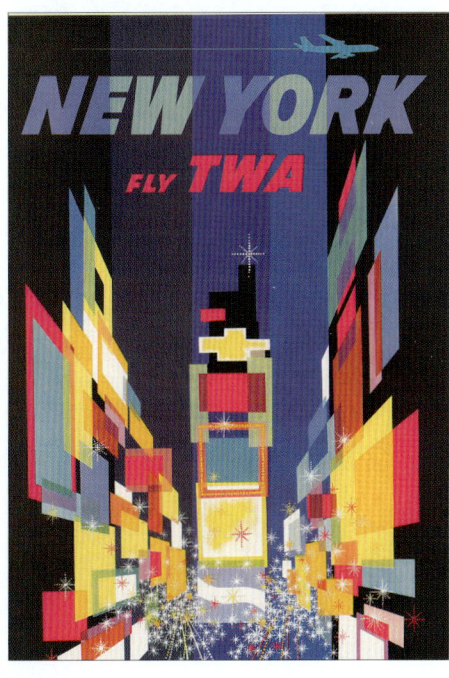

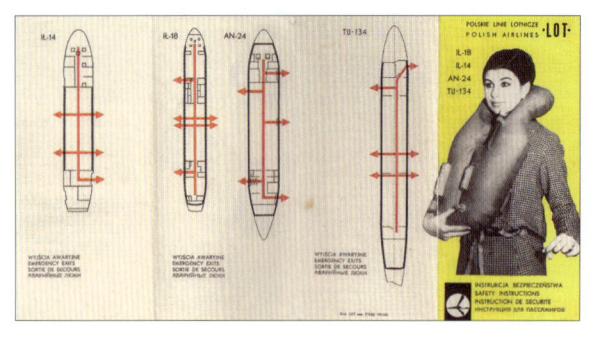

A 1968 LOT Polish Airlines safety instruction card for the Ilyushin Il-18, Ilyushin Il-14, Antonov An-24 and Tupolev Tu-134.

Above: A Ryanair Boeing B737-800 at Frankfurt Airport, Germany. Ryanair would soon establish a somewhat unique reputation in aviation. (Gameplayzz)

Left: The Airstrip uniform was, as can be expected, not all that popular among those having to wear Emilio Pucci's unique interpretation of a flight attendants' uniform. (Braniff International Airways)

Below: A Sichuan Airlines, Changhong livery, Airbus A330-343 shaking its tail feathers. Such eye-catching artwork soon replaced the plain expanse of exposed metal. (Md Shaifuzzaman Ayon)

A Virgin Atlantic (VIR) Airbus A330 with nose art at London Heathrow. Like its owner Richard Branson, VIR has always promoted itself with confidence. (Aeroprints.com)

Above left: Svetlana Savitskaya was second woman in space. On her 25 July 1984 Soyuz T-12 mission she became the first woman to fly to space twice, and the first woman to perform a spacewalk. (Unknown)

Above right: STS-93 Commander Eileen M. Collins shown wearing her orange launch and entry suit (LES). Collins became the first woman to command a Space Shuttle mission. (NASA Robert Markowitz)

Above: For Phoebe Omlie, née Fairgrave, fear was just a word and daring-do was the watchword of the day. Nothing was too dangerous, and the promotional card promised an eventful display. (Memphis Public Libraries)

Left: Flight continues to attract women of all ages. Here a young lady experiences her first taster session. (Ben Skipper)

Above: The unpaved runway at Lukla Airport complete with aircraft wreck, Nepal 1992. (Albert Backer)

Below: A wonderful view of the runway and its associated markings and taxiways at Cotswold Airport, England. (Adrian Pingstone)

Above: Automated foreign object debris (FOD) detection systems are one of the latest methods used by airports to detect FOD through the use of high-resolution camera and millimetre-wave radar as part of an intelligent runway management system. (Unknown)

Above: A visual approach slope indicator (VASI) at night showing its effectivity at guiding a pilot in poor visibility or low light. (Stephen Shields)

Left: A construction worker on a dowel drilling rig. The materials may have changed, but the methods have remained constant. (National Institute for Occupational Safety and Health)

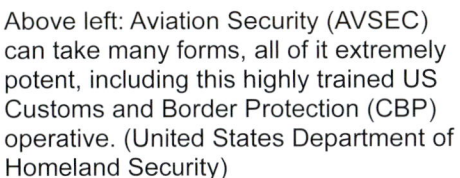

Above left: Aviation Security (AVSEC) can take many forms, all of it extremely potent, including this highly trained US Customs and Border Protection (CBP) operative. (United States Department of Homeland Security)

Above right: Light as a pilot aid is vital on this Lisbon runway 21; the lights aid a Low-Visibility Take-Off (LVTO) since the reported touchdown zone Runway Visual Range (RVR) is 200m. (Mathieu Neuforge)

Right: A Kenn Borek Air (KBA) Twin Otter takes off from a grass runway in Canada's Ivvavik National Park. (Daniel Case)

A wonderful snapshot of a Lufthansa Boeing 747-400 at Frankfurt Airport showing what is required to help make your flight a more pleasant one. (Igor Fedenko)

A Flybe (BEE) Saab 2000 turboprop passes in front of Sumburgh Airport ATC with its instantly recognizable Visual Control Room (VCR), Shetland, United Kingdom. (Ronnie Robertson)

Left: All airline crews need training to ensure the safety of passengers and one another. The most important of which is the in-flight safety brief. (Miguel Discart)

Above: Where nature becomes the influence; the amazing jewel of Changi Airport, Singapore, designed by Moshe Safdie. (Supanut Arunoprayote)

For some, financing doesn't necessarily mean a huge airliner. (Phuketian.S)

Wings of Alaska (SQH) mechanics work on the Pratt & Whitney R-985, of a de Havilland Canada DHC-2 Beaver floatplane. (Gillfoto)

A DHC-3 Otter supporting NASA's Operation IceBridge which surveys mountain glaciers in Alaska. (Chris Larsen/University of Alaska-Fairbanks)

Three civilian cargo planes line up at Cologne Bonn Airport in Germany, including a Fedex Boeing 777F and UPS Boeing 747-400F. (Raimond Spekking)

Opposite page, bottom right: Flight safety is everyone's concern and well demonstrated here at a simulated emergency landing rescue exercise conducted by the Pulkovo, Saint Petersburg's Airport safety services. (RIA Novosti)

A Boeing 747 Dreamlifter, capable of carrying three times the cargo of a conventional 747-400F, is seen here with its rear section open. (Eric Salard)

Technicians watch as space shuttle Endeavour is lowered onto the Shuttle Carrier Aircraft (SCA) at the Shuttle Landing Facility at NASA's Kennedy Space Centre in Florida. This manoeuvre is performed via the mate-demate device MDD, a specialized crane designed to lift a Space Shuttle orbiter onto and off the back of a SCA. (NASA)

The huge Antonov An-225 at Arlanda Airport, Sweden. (Larske)

The Airbus Beluga XL cargo with its familiar face takes off on its maiden flight on 19 July 2018. (Julien Jeany)

A United Nations Humanitarian Air Service 737-500 at Juba Airport, South Sudan. (UR-SDV)

A Beechcraft B200 Super King Air of the Royal Flying Doctor Service of Australia (SE Section) at Canberra. (BB-1812)

A Swiss rescue Eurocopter BK 117-C2 in action. (Matthias Zepper)

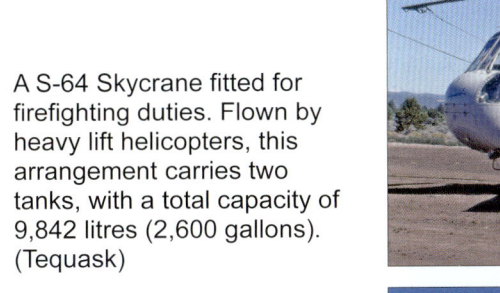

A S-64 Skycrane fitted for firefighting duties. Flown by heavy lift helicopters, this arrangement carries two tanks, with a total capacity of 9,842 litres (2,600 gallons). (Tequask)

A Mil Mi-26 on display at Lausanne Airport, Switzerland. (Audrix)

A Bristow-operated AgustaWestland AW189 search and rescue helicopter. (Adrian Pingstone)

A Beechcraft Model 17 Staggerwing at the Stampe Fly-In held at Antwerp International Airport, Netherlands 2019. (Ad Meskens)

An Embraer 200A Ipanema crop sprayer at Aeroclube de Ituiutaba, Brazil. (Evandro de B. Rocha Filho)

A Learjet 45 comes in to land at Oliver Tambo International Airport, Johannesburg, South Africa (Alan Lebeda)

Hot air balloons at Cappadocia, Turkey. The popularity of the sport has given rise to colourful and uniquely shaped balloons. (Sonse)

Pull up of glider ASK 13 at Degerfeld Airfield, Germany. (Olga Ernst)

The Extra Flugzeugbau EA300 flown by the Red Bull Air Race display team has built an almost cult-like following with their adrenaline-fuelled displays. (Contri)

On eagles' wings, Buzz Aldrin takes his first step onto the surface of the moon. (NASA)

Dr David Warren with the prototype flight data recorder. (Australian Government, Department of Defence/Defence Science and Technology Organization)

This unique self-built aeroplane, known as the Ward Gnome, was made by Mike Ward and first flew in 1967. It now resides at Newark Air Museum. (Ben Skipper)

Lettice Curtis, Jenny Broad, Wendy Sale Barker, Gabrielle Patterson and Pauline Gower of the Air Transport Auxiliary in front of an Airspeed Oxford, 1942. (IWM)

A 737 nacelle-mounted FOD deflector. (Airforcefe)

An image synonymous with aircraft bombings, the remains of Pan Am Flight 103. (UK Air Accident Investigation Branch)

FOD is removed from working areas in a variety of ways including this sweeper truck, fitted with a magnet. (Christine Cabalo)

British Airways Concorde 102 (G-BOAC) arrives at Manchester Airport after making its final flight on 31 October 2003. (Ken Fielding)

Dutch carrier KLM is implementing the semi-autonomous aircraft tow vehicles known as Taxibot at Schiphol Airport to help save fuel and prevent FOD damage. (Royal Schiphol Group)

The next stage in law enforcement: a remotely piloted aircraft system (RPAS), this one is a Queensland Police Service drone, 2018 (Queensland Police Service)

Hybrid air vehicle (HAV) Airlander 10 receiving it mission module fitting. (Philbobagshot)

Lockheed Martin's experimental X-59 QueSST holds its own as the next generation of SSTs are developed. (Lockheed Martin Corporation)

The next generation. A stunning night image of a Qatar Airways Airbus A350-1041 at an unknown airport. (Anna Zvereva)

The next generation in electrically powered aircraft, the Pipistrel Velis Electro, at Ajdovščina Airport in Slovenia after rollout from the assembly line. (Andrejcheck)

Wisk Aero's all-electric self-flying air taxi is shown in flight and will join Joby Aviation and Toyota in the Electronic Unmanned Aerial Vehicle (EUAV) and Urban Air Mobility race. (Wisk Aero LLC)

Qantas Boeing 474-400 City of Canberra lands at Ilawarra Regional Airport, Australia, on its retirement on 7 March 2015. This would be followed by hundreds more 747s over the next five years. (YSSYguy)

Air Malta Airbus A319-100 9h-AEM prior to its retirement on 13 October 2019, 100 years after the signing of the 1919 Paris Convention. (Keifer)

An Icelandair Boeing 737-8 MAX at Hamburg Airport, 14 June 2022. The 737 MAX had a bumpy start and only after various aviation authorities' clearances did the 737 return to service in 2022. (MarcelX42)

her husband with co-owner, Al Baumeister, fitted the single-engine, high-wing Cessna with a range of extras to ensure a safe flight. These included dual directional finders, short- and long-range radios, autopilot and three extra fuel tanks giving the aeroplane an impressive 3,500-mile range.

Planning her flight of 22,858.8 miles (36,788 kilometres) with the help of an air force friend, Mock was soon alerted that another woman, Joan Smith, was planning the same flight. Although Smiths' flight in a twin-engine Piper Apache would take her around the equator, Mock was quick to respond, getting her plans into the National Aviation Association while bringing her time of departure forward to 19 March 1964, two days after Smith's.

Flying east to west she enjoyed the flights over Africa and Asia, putting her flying experiences to good use, even when landing at an Egyptian military airfield. Instead of being detained, the bemused Egyptian airmen pointed her in the right direction, to Cairo's civilian airport. Mock was confident in her passage, skills and, most importantly, her equipment, vital to help her navigate the watery expanse below. On 14 April she departed Honolulu, Hawaii for Oakland, California, on what would be the longest leg of the journey. Safely making the crossing, she continued to her starting point at Columbus, arriving twenty-nine days after her departure on 17 April and importantly twenty-five days before Smith returned to the continental United States.

For her endeavour, Mock was awarded the Federal Aviation Administration's Exceptional Service Decoration and the National Aviation Trades Association Pilot-of-the-Year Award for 1964. Smith received the Harmon Trophy. Unfortunately, this would be awarded posthumously as she died in a flying accident on 17 February 1965. Mock was spurred on to more world record-breaking feats, including the 1969 speed-over-a-recognized course and being the first woman to cross both the Atlantic and Pacific and the first woman to fly over the Pacific in both directions. She died at home in Quincy, Florida, on 30 September 2014.

In the east, the Soviets had been busy cultivating home-grown talent, especially in the immediate post-war period of growth in its aviation industry. Valentina Grizodubova can be considered one of the leading pioneers in Soviet-era aviation. Born in Kharkiv, Ukraine, on 27 April 1909, she started her journey in aviation at a young age, performing her first solo in a glider at fourteen. In 1929, she graduated from flying as a member of the Volunteer Society for Cooperation with the Army, Aviation, and Navy, known as the OSOAVIAKhIM; this included flying from the Penza Flying Club and the Kharkiv Flight School. This was followed by graduation from the Tula Advanced Flying School in 1933 and subsequent appointment as a flight instructor. Between 1934 and 1938, Grizodubova performed duties with the Gorky Agit-Squadron, whose task was to deliver propaganda to Soviet citizens in various forms.

During this time, Grizodubova also established several world records in altitude, range and speed among women in a light aircraft in 1937, including the highest altitude reached by a woman in a two-seater seaplane, 10,718.5 feet (3,267 metres) on 15 October. These would be trumped by a famous flight from Moscow to the Far East the following year in a Tupolev ANT-37bis named *Rodina* (Motherland), an adventure Grizodubova was well prepared for, having accumulated some 5,000 flying hours. The trip was made with two other women, co-pilot Polina Osipenko and navigator Marina Raskova. Unfortunately the flight ended in a swamp after covering a distance of 4,008 miles (6,450 kilometres) in twenty-six hours and twenty-nine minutes due to navigational error because of severe weather. Nevertheless, for their efforts, the crew became the first women to be awarded the title of Heroes of the Soviet Union the following year.

With the outbreak of war, Grizodubova would command the 101st Long-Range Aviation Regiment, a specialist resupply unit which flew the Lisunov Li-2, a licence-build DC-3. The 101st was staffed by pilots conscripted from the Civil Air Fleet and would perform a range of dangerous missions under Grizodubova's command, primarily supplying partisans in Ukraine and flying supplies to besieged Leningrad. Internal struggles with senior commanders prevented her from becoming the first female General of Aviation; she completed her military career as a colonel.

With peace and the dawn of the jet age, Grizodubova would stay in the aviation field, working on electronic equipment for both civil and military aircraft at the Science Research Centre of Flight Test. Her work there would earn her the title of Hero of Socialist Labour and her influence would help boost the prospects of other femal pilots, She died on 28 April 1993.

Galina Korchuganova was born on 22 March 1935 in Barnaul, USSR, and benefitted from the Soviet era's sports club network, which introduced her to the sport of parachuting with the OSOAVIAKhIM. As a result, she was accepted into the Moscow Aviation Institute, where she studied aviation technology, graduating in May 1959. After this, she worked as an engineer at the Ramensk Avionics Construction Bureau. In 1962, she was among eighteen female space flight candidates selected for testing for inclusion in the Soviet space programme. During this phase she presumably went through flight training as, by 1965, she had set a world record flying a Yak-32 sporting jet in a 100-kilometre closed circuit. The following year she competed in the World Aerobatic Championship held in Moscow, where she would become the women's world aerobatic champion. This win saw her awarded permission to become a test pilot. After some initial obstructions, she was helped by Valentina Grizodubova, the head of the Science Research Centre of Flight Test, and in 1969 graduated as a test pilot from the Kirovograd Flight School.

Korchuganova would fly twenty different types of aircraft and accumulate a total of forty-two records. By her retirement in 1984, she had accrued over

4,000 flight hours, including 1,500 as a test pilot. Immediately after her retirement she worked at the Moscow Museum of Aviation and Astronautics. With the collapse of the USSR, she established an aviation club for fellow women and, in 1992, founded Aviatrissa, which quickly grew to 550 members. She died on 18 January 2004.

Svetlana Savitskaya was another key Soviet pilot who would start her career in parachuting with the OSOAVIAKhIM and become the first women to perform an extravehicular activity (EVA). By her eighteenth birthday Savitskaya had completed 450 parachute jumps, including two high-altitude jumps from 45,276 feet (13,800 metres) and 46,752 feet (14,250 metres). In 1966, Savitskaya enrolled at the Moscow Aviation Institute where she learned to fly and became a qualified flight instructor. After graduating in 1972 she trained at the Fedotov Test Pilot School, completing her course in 1976. By 1978 she was working as a test pilot with Yakolev where she became the first woman to reach 1,667mph (2,683km/h), in a MiG-25.

As well as test flying, Savitskaya was an accomplished aerobatics pilot specializing in flying light, piston-driven aeroplanes such as the Yak-18. Between 1969 and 1977, Savitskaya competed as a member of the Soviet national team for aerobatics and, in 1970, was part of the winning team at the FAI World Aerobatic Championships. In 1979, Savitskaya was selected for cosmonaut training and passed her exams in 1982, becoming the Soviet Union's fifty-third cosmonaut. She would go into space on two occasions, the first on 19 August 1982, returning on 25 July 1984. It was on her second mission, to the Salyut, where she performed her EVA to conduct welding experiments. From 1983 to 1994, Savitskaya was the deputy head of Russian space equipment manufacturer PAO S. P. Korolev Rocket and Space Corporation Energia before retiring from there and the Russian Air Force. Since 1996 she has served as an elected member of the Russian State Duma, and deputy chair of the Committee on Defence.

The achievements of women in aviation are tremendous, and there remains an almost endless list of their accomplishments, including Norwegian Turi Widerøe, who became the world's first female commercial air pilot for a major airline, Scandinavian Airlines, in 1969. In 1975, Yola Cain became the first Jamaican-born woman to gain a commercial and instructor's licence and fly for the Jamaica Defence Force. She was followed by Jill E. Brown, who in 1978 would become the first African American woman to fly with a major American airline, Texas International Airlines. Lynn Rippelmeyer would become the first woman to captain a Boeing 747 across an ocean in 1984. Jeana Yeager would co-pilot the Rutan Voyager, with Dick Rutan, on an around-the-world non-stop, non-refuelled flight, between 14–23 December 1986, flying some 24,986 miles (40,211 kilometres) in the process. Eileen Marie Collins would become the first female pilot and first female Space Shuttle commander, logging thirty-eight days, eight hours and twenty minutes in space. Indian pilot Aarohi Pandit would become the first and youngest pilot to cross the Atlantic and Pacific, solo, in a light sports aircraft.

No discussion about women in aviation would be complete without a special mention of the romance novelist Dame Barbara Cartland. A keen glider pilot, Cartland had been introduced to gliding in the 1920s. The primary method of getting airborne was being towed behind an aircraft or pulled by a winch. The towing cables would be released when the glider could sustain flight. Cartland wanted more than the short flights this method offered and a proposal was put forward to develop a long-distance airmail delivery glider that could be towed some 200 miles (360 kilometres). The first trials, held in 1931, were successful. The idea was later picked up by the airborne elements of the military for air assault. It would see towed gliders inserted with great accuracy behind enemy lines, landing in almost absolute silence and surprising defending troops.

Despite their superhuman exploits, that have captured the imagination and inspired women to follow in their footsteps, many of these pioneering women paid the ultimate price in pursuing their dreams of achieving aviation excellence. Their achievements and sacrifices should serve as a memorial to all women that the sky is not the limit: it is just the start of the adventure.

Chapter 7

The Infrastructure and Organization of Flight

Facilitating Comings and Goings

From tents arranged around an empty field to city-sized airports, the world of flight support has changed massively over the past century. The aviation industry infrastructure now encompasses a vast range of vital tasks and roles, from food to fuel supply, command and control, and personnel training and development. The many skills developed have kept up with and driven aeronautical developments. Allied to this is the role of banking, financing aviation growth.

By the time the Paris Convention was signed in 1919, numerous military airfields were rapidly being converted for civil use worldwide. Despite appearances, early aircraft were durable and capable of operating on grass runways easily, often using their irregular surface to aid take-offs.

The First World War spurred many innovations in flight, one of the key developments was that of the runway. In 1916 Michelin constructed the first concrete-paved runway to aid in the dispatch of aircraft they were producing for the war effort. After the war, Orville Wright reasoned that regardless of physical construction, there was a need for a recognizable uniformity in design and layout, regardless of cost.

The runway's surface refinement followed, including its construction from various materials, including ice, Sommerfeld tracking, and composite runways. Brick was also experimented with, but the combined effects of weather and the ever-increasing weight of aircraft led to the surface becoming damaged. In addition, the fragments provided a rich source of FOD, leading to the runways being ripped out or concreted over. For flying boat operations, the requirements are a little more involved and include providing a FOD-free waterway for landing and take-off, moorings and a ramp to necessitate any servicing or repairs. Aircraft such as rotary aircraft and airships require relatively small areas free of obstructions that could otherwise foul rotor blades or damage the aircraft's structure.

By the Second World War, the runway was the cornerstone of practical air travel; its construction went beyond its surface material. Runway dimensions had to be accurately calculated to handle all current and possible future aircraft, such as the Boeing 747 and Airbus A380. Location was also essential, with engineers initially carrying out a wind-rose survey to establish the direction of the prevailing winds. The runway would then be positioned so that aircraft could take full advantage of these prevailing winds.

Early airfields often featured three runways set out in a triangular pattern to take full advantage of any change in wind direction. While this was a great boon to the military and those operators who had the physical ground space to construct such a large set-up, many early airports were sited close to conurbations, which led to the standard strips. These were built parallel with another runway, as seen in larger airports such as London Heathrow or Los Angeles International. Such sites, with multiple runways, would identify their runways by the simple expedient of adding a letter suffix, such as 02L, the L indicating left. Airfields orientated to true north are given the suffix T. Further suffixes followed, including those for rotary aircraft (H) and Short Take-Off and Landing (STOL) aircraft (S).

To aid navigation airfields were given numbers representing the magnetic azimuth of the runway's heading in degrees. These numbers are added to the threshold in lettering large enough for pilots to read, with a reverse heading placed at the other end of the runway. For example, a runway's orientation will be marked as 02/20, with pilots directed to take off or land from the position most suited to their needs.

For some sites, especially waterways for seaplanes, traditional compass-point naming was employed. Though for some channels, the actual site might be termed an ALLWAY landing site, meaning the aircraft can land anywhere within a bay. By the time the jet airliner arrived, the runway had become more than a paved surface. It featured an array of markings and lighting to aid the pilot, with the addition of a Visual Approach Slope Indicator (VASI), Touchdown Zone Lights (TDZL), and exit taxiway lead-off lights.

At the end of the first century of regulated civil aviation, runway design and infrastructure had developed to such an extent that these key features would be the subject of an entire civil engineering discipline, meeting the International Civil Aviation Organization (ICAO) definition of a runway. In addition, designs were further influenced by the runway's purpose, e.g. whether it served a large international hub airport such as the King Fahd International Airport in Dammam, Saudi Arabia, the world's largest airport, or a local airfield used by microlight aircraft.

With the runway becoming the focal point of all flight operations, there was a need for the early operators and aviators to be provided with the most rudimentary forms of infrastructure and support. The most crucial aspect was shelter for their most significant investment – their aircraft. The hangar not only protected the aircraft from the worst of the weather but also proved a base to perform the necessary maintenance on the aircraft and safe and secure storage of essential equipment and parts. Hangars were the first large and permanent structures to be constructed. These contained a small office which doubled as a staff room, flight planning centre, customer waiting room and, on occasion, a customs office. There was also the need for dry, secure storage for mail that had been flown in and was awaiting pick up by the local postal service.

For those locations operating as airports, the multiple flights coming and going would necessitate separate accommodations for passengers. Some operators benefitted from taking over military airfields and inherited permanent and well-constructed shelters for passengers. Others, especially those in isolated locations, used tents as improvised shelters while wooden pavilion-type structures were made. By the 1930s, the industry had settled down, with two types of sites established for flight operations. Airfields serviced the needs of the private flyer, the flying school and the increasingly popular aero clubs. Meanwhile, airports became the main departure point for passenger flights. American Airlines set the trend for the formalization of passenger flight operations by establishing two different types of flight systems.

The hub-and-spoke system was based on the concept of the wheel. The hub would be the home airport, such as Hartsfield–Jackson Atlanta International Airport, for the operating airlines, in this case, Delta Air Lines. This system uses the hub to feed secondary locations with flights, their paths radiating outwards like spokes. This method proved exceptionally efficient for long-haul flights, with the capacity of an aircraft efficiently utilized. It also rationalized the use of different aircraft types. The smaller aircraft would serve these spoke routes, while larger aircraft would fulfil the hub-to-hub flights. The system also sees the flight broken down so that a London–Seattle flight could go via the hubs of Atlanta and Denver before arriving in Seattle. This could theoretically make it necessary for the passenger to make up to three flight changes, leading to missed connections due to changes in circumstances or long waits for connecting flights. There is also increasing awareness of the environmental impact of short-haul flights, which may not be as efficient as medium-spoke or hub-to-hub flights.

A second, more straightforward system used by airlines for regional passenger and cargo carriage is the point-to-point system. This type of system is more flexible than the hub-and-spoke system, opening the airspace to charter flights and providing a non-stop service between locations. Some scheduled flights use the point-to-point system; however, the demand for these flights can fluctuate, and as a result, such flights can be cancelled or routes closed. Another setback to this method is an increase in maintenance costs and an increased pressure on airspace, especially with charter flights, that can lead to delays across the board. However, flight planning is nothing if not flexible. By the advent of the Second World War, operators of airlines and private companies were using a hybrid mix of both systems to good effect. The increase in both systems also saw the rise of the Multi-Airport Regions (MARs). This system would see a geographical region or key city, such as London, served by several airports, each meeting the differing needs of commuters. While a boon for air travel consumers, the organization and coordination of the airspace by air traffic control and planning at all levels would become increasingly complex. Developments, leading to the solutions we know today, were aided by practices and technologies adopted during the Second World War.

The airline industry shaped how flights would be sustained at ground and air levels. In the 1930s, there was a boom in establishing new airports and airfields and rationalizing physical structures. Many smaller terminals, often operating from grass runways, slowly disappeared as training became more regulated and new monopolies began to appear. Airlines were now carrying more passengers as aircraft such as the DC-3 started to appear in airline fleets. The need for a formal passenger and aircraft management system saw the arrival of the first International Air Transport Association (IATA) standardized tickets, which featured multiple flight coupons. Flight planning became more professional as the various regulations affected minimum heights flown between routes, and the identification of those airports which featured the best facilities for aviators and passengers. Such moves were intended to feed professionalism and slowly eradicate some of the fly-by-night practices that had cropped up in the first decades of regulated civilian flight. This meant passengers were flying in an airworthy aircraft, flown by a competent crew, arriving at a well-serviced airport.

The airport soon became a veritable gateway to the skies, with architects invoking neo-classical and en vogue styles. The new buildings began to take on the two key familiar forms that today's traveller is so familiar with: the terminal building and control tower. Early terminals echoed the grand designs used by railroad companies, ensuring the traveller was left with a lasting impression, not only of the grandeur and scale of the journey they were about to embark upon but also impressed on them that they were part of something new and bold. Air travel captures the imagination like no other form of travel in many respects; it fulfils the personal dream of flying. As the new terminals acted as a sentinel to these dreams, they had to be as impressive as they were beautiful, echoing the latest designs of sleek, silver-skinned monoplanes by Boeing, Lockheed and Douglas. Some airports, such as the International Pan American Airport, catered for flying boat passengers, treating them to an entirely different experience as they boarded the waiting Boeing and Sikorsky flying boats via piers.

Whether hosting landplanes or flying boats, the terminals were decorated beautifully, with giant globes tracing airlines routes, chandeliers, and beautiful murals and tiling. They were also home to practical features that would be instantly recognizable to today's traveller; the check-in desk, weighting points, cafés and ticket offices, as well as home to basic security checks for Dangerous Air Cargo (DAC) and the all-important departure lounges, points, gates and piers.

Much like its railroad-based cousin, the terminal soon developed to meet the needs of all, while adapting to allow for the passage of more than one flight at a time. As a result, architects, flight planners and estate managers began to refine the simple terminal design into more sophisticated shapes. These took the forms of elliptic or concourse-type terminals capable of hosting several aircraft around a single point. The acme of these increasingly complex design arrangements was the multi-concourse with carrier lounges designed to meet the needs of the busiest hub airports.

The advent of the Second World War, the 1944 Chicago Convention and the arrival of the jet aircraft soon saw the introduction of more stringent safety practices. These new buildings, some located at old wartime airfields, and new safety standards, were incorporated into new designs. The initial post-war austerity soon gave way to a glamorous era of travel, with automation and the addition of labour-saving devices, such as loading and unloading carousels. Airports became more than just departure points; they soon featured a dazzling array of duty-free shops and viewing areas for enthusiasts of the latest jet aircraft. The arrival became just as important as the departure, with spaces designed to effectively disperse passengers. Security concerns become more apparent, and the need to separate those passengers cleared from customs was addressed. Departure lounges grew as check-in times became established and offered the passengers myriad ways to pass the time while waiting to get airborne. With the increasing availability of motorcars, rail and bus connections, the need for larger areas to build services became as important as the construction of the runways.

As aviation technology grew, so too did the airports and specialist firms of architects and designers, shaping the airport's future. The terminals and control systems separated as time progressed, with the control towers slowly moving away from terminal buildings to standalone structures. The first such structure was built in Cleveland in 1930. These ensured the integrity of the building and workforce and saw the control tower become home to the numerous staff needed to watch over radar systems for weather and aircraft movements. The increasingly busy skies saw the introduction of innovations to make travel more efficient. Among these was the introduction of the satellite terminal. This simple terminal first appeared at London Gatwick. It featured a building separate from the other terminals, allowing aircraft to park around its circumference. Passengers gained access via an underground walkway from the main terminals. The design was adopted by other airports to attract more aircraft, which were increasingly operating as standalone businesses, separate from state or airline patronage.

The increase in the size of terminals also saw the instruction of powered horizontal walkways, buggies for the disabled, and Automatic People Movers (APMs). These train-like vehicles ferry passengers and their luggage around the airport without interrupting flight line operations or creating more physical structures. As the twentieth century progressed, the terminal building became the embodiment of national engineering prowess and architectural prestige. Designs by architects such as Frenchman Paul Andreu spanned the globe and utilized modern building techniques while echoing local architectural styles, such as Soekarno-Hatta International Airport in Jakarta, Indonesia.

Other key developments in airport and airfield design saw the gradual installation of national aviation fuel storage sites and pipelines to help keep the new passenger airlines aloft. Many of these pipelines had been installed, especially in North America and Great Britain, during the war. These allowed airlines to access the vast quantities of fuel they needed without putting the supply chain and the public at risk. Refuelling is understandably highly regulated, with the safety and quality of fuel and its delivery to the aircraft monitored by trained ground staff and aircrew.

Alongside fuel viability, many airports and airfields retained and expanded their workshops and specialist staff, an investment in effort that would pay dividends in terms of fleet management costs. Such is the capacity of some flight line maintenance facilities that they can often perform the most complex overhauls outside of manufacturing as required by local and international laws.

The support goes further with airports being furnished with specially equipped and trained fire and rescue services, a far cry from a simple extinguisher placed a safe distance from the aircraft. Furthermore, to supplement the delivery of services, cargo and goods, the flight line motor pool has grown from the uncomplicated bicycle to supplying fire-fighting appliances, heavy-duty tractors, multiple bowsers and power appliances. Also included are the supply trucks for in-flight catering, high-lift loaders and all-important gangways, all considered vehicles. The delivery of these critical elements is known as cross-servicing, and their efficient delivery is essential in making air travel safe and successful.

While the numerous ground engineering and support undertakings ensured the aircraft were kept aloft, the catering departments' task was keeping passengers and crew fed and watered. The early meals were uncooked and often brought on by passengers as homemade or bought at an airport café. The arrival of the fully equipped galley, complete with oven to provide hot food, was a welcome addition to flights, with the first in-flight meals served on the the London-Paris route. Again, airlines played to their strengths, producing simple and visually appealing national dishes that captured the essence of the airlines and the nation to which they belonged. Pan Am and Air France, in particular, became renowned for their luxury, choice and quality of the food options. For short- and medium-haul passengers, fare remain limited, often at extra cost. Still, the standards are, for obvious reasons, among the highest in the travel industry.

The logistics behind the in-flight meal are staggering; airports are home to literal armies of caterers producing nutritionally balanced meals that will stand the rigours of flight. These meals, numbering in their millions, must include various options to cater to the contemporary diet and palate, from gluten-free to religiously observant options. Flight attendants are trained to dispense meals but also alcohol and hot and cold drinks at set times and by request. The cabin-generated waste, including food trays, is then collected for recycling, which remains the first priority, or responsible disposal.

By 2020 the airport terminal and all that made it work had become more than simple physical constructions. The design of the airport and all its structures has moved away from mid-twentieth-century functional brutalism, embracing its role as a gateway to something magical. The designs are interpretations of what makes air travel welcoming to the masses. Architects firms such as Safdie and Zaha Hadid entertain the visitors with glimpses of paradise at Changi Airport, Singapore, and science fiction dreams realized at Beijing Daxing.

Other means of identification were required with the proliferation of airfields and airports. Since some towns host several airports, identifying these became important for safety and ensuring pilots landed at the correct location and passengers went to the correct departure point. In 1947, the newly established ICAO and IATA began to issue specific sets of codes to airlines and airports to aid travel and navigation. The four-letter ICAO codes, also known as airport codes or location indicators, identify a specific airport and are used in the flight planning and Notice to Airmen (NOTAM) processes to ensure the safe passage of flights. ICAO codes also identify other sites of interest, including weather stations, international flight service stations or area control centres. In contrast, the IATA codes, more familiar to the traveller, use three-letter codes to identify a site on timetables, tickets and reservations, and all-important baggage tags.

ICAO and IATA codes were also applied to airline operators from 1947, with IATA two-letter codes used for the same purpose as their airport codes. The ICAO code was initially a two-letter code, the same as the IATA code; these were changed to a three-letter code in 1988. These codes not only identified the airline for flight planning but were also incorporated into call signs for telephony to help identify individual flights and aircraft. Another critical aspect, vital for clear communications, was the adoption of English as the language of aviation as part of the 1944 Chicago Convention. This was further refined to ensure clarity, especially between words and figures, and all instructions were clearly understood. A whole new language, disciplined and military in nature, of air traffic management, appeared to ensure the safety of all. While appearing cumbersome, its use has ensured safety since adoption, and its misuse or total abandonment has often led to catastrophic disasters.

Commanding and Controlling the Skies

Since its inception, Air Traffic Control (ATC) and regulation remain a vital part of safe air travel regardless of type and location. First introduced in 1920 at London's Croydon Airport, the initial purpose of the first ATC service was to provide specific flight information to aircrew, which included air traffic, weather and moon status, and location information. A further development, primarily in the United States for its burgeoning airmail service, included the development of the Airmail Radio Stations (AMRS). AMRS was a command-and-control system developed by the United States Army during the First World War. It proved highly useful to early aviators crossing the 2,897-mile-wide (4,662 kilometres) country. AMRS was followed by Flight Service Station (FSS).

The FSS provided was able to take full advantage of the increasing number of radio-equipped aircraft, providing aircrew with a range of flight-specific and relevant information. Initially, FSS was provided individually by nations, but as flights became increasingly international, and the airways increasingly congested, there was a need to amalgamate services, such as those offered by the United States and Canada. These new larger regional information hubs are further supported by Remote Communications Outlets (RCO), which help relay FSS information. They are joined by Remote Transmitter/Receivers (RTRs), the task of which is to relay information from terminal air traffic control facilities.

Typical information accessed and passed by an FSS includes pre-flight briefings, opening and closing flight plans, security information and monitoring Navigational Aids (NAVAID) for aircraft on the ground and in the air. They are also responsible for coordinating lost aircraft searches and enforcing Visual Flight Rules (VFR), which are informed by Visual Meteorological Conditions (VMC) and allow the FSS to make informed flight safety decisions. For example, should flying be considered unsafe, the FSS may make an exception for aircraft to fly using Instrument Flight Rules (IFR) for pilots who are qualified to do so and are airborne, or for flights that are considered essential through the implementation of Instrument Meteorological Conditions (IMC).

One incident of FSS implementing IMC controls was the volcanic eruptions at Eyjafjallajökull, Iceland, in April 2010, which lasted until mid-May, and saw huge plumes of volcanic dust expelled into the atmosphere. The resulting cloud covered the airspace of some twenty countries in Northern Europe, grounding flights for fear of FOD-related incidents and stranding around ten million passengers.

Another significant weather-related role of the FSS is monitoring the four polar and subtropical jet streams. The jet streams had been theorized for some time. Their presence was finally confirmed by Japanese meteorologist Wasaburo Oishi in the 1920s while he was using weather balloons to track upper-level winds near Mount Fuji. The term 'jet stream' was first coined by German meteorologist Heinrich Seilkopf in 1939. These meandering high-altitude winds run globally from west to east in the tropopause boundary between the troposphere and stratosphere. The winds are rarely a continuous loop around the globe and can change rapidly. Not fully understood until later in the Second World War, the jet stream can give transoceanic flights a welcome boost, saving time and fuel. Such is its nature that the jet stream is treated with extreme respect as it can create issues for even the most experienced pilot, including the incidence of Clear Air Turbulence (CAT) caused by vertical and horizontal wind shears encountered while traversing the jet stream. CAT events can cause traversing aircraft to lose height rapidly and have led to at least one incident which occurred on 28 December 1997 when a United Airlines (UAL) Boeing 747-100 on a Japan–Hawaii flight dropped 100 feet (thirty metres), injuring thirty passengers and three crew. Flights going east to west cannot utilize the jet stream and so make their flights along a great circular route, which follows the intersection of the sphere and the plane that passes through the centre point of the globe.

FSS in North America also works with Flight Information Centres (FICs), a Canadian concept initiated in the 1990s. The FICs employ flight service specialists, similar to those found in an FSS, whose responsibility is managing and disseminating flight safety-related information. FSS staff can also provide or pass on information should there be an absence of air traffic controllers or if a control tower is closed. However, FSS is not responsible for giving instructions or clearances: that reasonability belongs to ATC alone.

The role of ATC and its staff has developed over the past century, embracing advances in technology and utilizing these, especially radar, for the safe passage of aircraft in the air and on the ground. The United States led in the development of command and control due to its increasingly busy airspace in the late 1930s. In 1935 the first Air Route Traffic Control Centre (ARTCC) was opened in Newark, New York; this was followed by another twenty-one. The role of the ARTCC echoes that of ATC to safely direct the movement of aircraft between departure and destination. A mid-air collision in 1958 saw the United States Federal Aviation Administration (FAA) take complete control of the nation's ATC/ARTCC systems, with other nations' civil aviation authorities soon following. As a result, ATC has an influence on flight operations taking place up to eleven miles (eighteen kilometres) from its geographical location.

The focal point of the ATC remains the dominating tower, crowned with the now-familiar 360-degree visual control room (VCR). The VCR is manned by controllers tasked with providing Ground Movement Control (GMC) to all ground, landing and departing traffic from the runways and aprons. ATC systems were developed over time to prevent confusion by creating three distinct operational control areas: local control or air control, ground control and flight data/clearance delivery.

Other functions fulfilled by the ATC have been refined over time, including providing flight information, approach control and the local alerting service. ATC is also home to the various runway offices and control rooms which house the vital surveillance displays, which are fed information by the primary and Secondary Surveillance Radars (SSRs). Other systems managed by the controller include the Surface Movement Radar (SMR), Surface Movement Guidance and Control System (SMGCS) or Advanced Surface Movement Guidance and Control System (ASMGCS). The SMR, in particular, is an essential piece of equipment when guiding traffic at night or in adverse weather conditions, especially at larger airports.

When drawn together, these services are vital in maintaining flight safety, especially when around 9,000 commercial aircraft are in the air at any one time.

Training and Development

Well-designed flight training was key to the success of any aeronautical endeavour. The experience was first-hand and taught by aeronauts to aviators. As aircraft design became more refined and many different types of aircraft appeared, the skill set required to fly and operate these increasingly complex pieces of machinery burgeoned. The initial approach of doing little more than simply getting in, winding the engine up and hoping for the best was far from ideal; there was much more to flight than flying.

The early pioneers had to wear at least four hats: pilot, navigator, flight engineer and ground crew. The limitation of early designs helped ensure that the technological complications came at a steady pace, occasionally sped up by lessons learned or innovations from either disaster or warfare. As civilian aviation became more regulated and technical elements around flight more sophisticated, the need to recognize qualified pilots increased. From 1 January 1910, Aéro-Club de France began issuing aviators certificates after a suggestion by the secretary, Georges Besançon. Many of these certificates were issued retrospectively to established aviators, including the Wright and Farman brothers, Louis Blériot and Glenn Curtiss. The Royal Aero Club of Great Britain followed suit the same year, with the Aero Club of America issuing licences a year later. However, these were not compulsory until 1917. The arrival of licensing meant that flying was to be undertaken by trained pilots, thus starting the professionalization of the trade.

The onset of war in 1914 saw numerous men and women, including Russian Princess Sophie Alexandrovna Dolgorunaya, take to the air to defend their homelands. In the post-war world of the 1919 Paris Convention, article 12 cements the commitment to ensure competency by stating: 'The commanding officer, pilots, engineers and other members of the operating crew of every aircraft shall, in accordance with the conditions laid down in Annex E, be provided with certificates of competency and licences issued or rendered valid by the State whose nationality the aircraft possesses'.

Not only were pilots expected to be licensed but also engineers, which meant that many in those embryonic days one had to hold several licences to operate a passenger or cargo service or simply fly an aircraft as a leisure activity. This was intended to ensure the safety of all, which still lagged, with aircraft rarely fitted with that most rudimentary pilot safety device: a harness. The article kick-started an industry which grew from veterans of the First World War offering ad hoc lessons to centrally controlled national pilot training programmes. Programmes such as the American Civilian Pilot Training Program (CPTP) were established within twenty years, offering not only flying training but also engineering and navigation syllabuses and licensing.

This early period also saw the beginning of the development of the pilot licence, with categories being introduced over the coming decade, which reflected the evolution of flight technology. These categories included aeroplanes, rotorcraft, gliders and lighter-than-air aircraft. Over the coming decades, competency classes of aircraft which could be flown on an individual licence were gradually introduced, with single- or multi-engine, water, helicopter and free balloon among the possible options. The final element was the type rating which identified the aircraft type, such as turbojet, that can be flown by the pilot and is only awarded after the pilot can demonstrate safely that they have the skills required. Further specialist ratings and categories for activities covered night flying and sport pilot certification, introduced as a specialist category in 2004, and Unmanned Aircraft System (UAS(Drone)) licences. These ratings and classes give the pilot, commercial or private, goals to aim for, such as achieving a private pilot's licence with an instrument rating and helped design a viable and attractive career pathway. Aside from the practical elements of flying training, the student must also study and pass theory, including navigation, meteorology, communications, flight performance, planning and air law.

The Second World War saw scores of men and women taking to the skies with the skills and training they'd learned in those pre-war years. The need for pilots did not wane, and by the war's end, the United States Army could graduate basic pilots after eighteen weeks of training. This seemed exceptionally rapid compared to the Royal Air Force, whose pilots completed their elementary flight training in around thirty weeks. When viewed alongside the fact that the United States was carrying the burden of the air war in the vast Pacific theatre of operations, it made sense for the United States to design a programme which produced pilots and aircrew quickly. It also shaped and standardized how the pilots of the dawning jet age were trained.

The costs of such courses are always steep. For many, the qualifications are gained through military service in the ground and air trades or personal finance. However, many early aviation supporters recognized that there would be those keen to fly who, for whatever reason, could not meet the costs of training or joining the military. This led to the establishment of airline or operator cadet schemes, youth training schemes such as the Air Scouts, and the opportunity to apply for scholarships. Among the first scholarships were those offered by organizations such as the famous Ninety-Nines organization of women pilots and the United Kingdom-based Honourable Company of Air Pilots (UK).

The flight engineer was a vital role that sat alongside the pilot's, especially on the larger multi-engine aircraft throughout most of the twentieth century. The flight engineer's task was to manage the numerous systems that enabled the aircraft to fly. These members of the flight deck crew would invariably come from a military background where complex multi-engine aircraft were more likely to be encountered, with some occasionally trained by the operating company. However, as flight decks became home to increasingly sophisticated management software, the flight engineer's role has slowly become redundant, with the odd exception.

The Second World War also saw the proliferation of ground crew trades, from mechanics to survival equipment fitters. These trades would form the foundations of a resurgent post-war airline industry with ready-trained mechanics and technicians eager to continue working in the industry. Many further and higher education establishments saw the opportunities for training such staff, and soon so did the industry itself. Evening school classes were quickly replaced with formalized full-time courses which would award students with an academic qualification and certification on certain aircraft types. As aircraft have become more sophisticated, engineering training has reflected these changes, with specialist courses and continuous training offered by manufacturers and operators. This ensures that individual certification is correctly maintained and assures customers, aircrew and operators that the aircraft is maintained at peak performance.

One key innovation that would become essential for the success of commercial aviation in particular, was the proliferation of flight attendants. These vital members of the aircraft's crew followed in the footsteps of the first steward, Heinrich Kubis, appointed by Zeppelin in 1912. By the 1920s, cabin boys had appeared, serving passengers with light refreshments and attending to their in-flight needs in a role that was little changed from Kubis and his Zeppelin team. However, cabin boys would only serve on the more prestigious flights; their presence was often the exception rather than the rule.

It would not be until 1930 that a regular flight attendant service was included on aircraft, with the first appointment made by Boeing on their United Airlines service in 1930. Ellen Church was the first of Boeing's new 'Sky Girls' who met the stringent recruitment criteria of being not only a registered nurse but also 52kg (115lb) in weight. Initially, Church had applied to be a pilot with Boeing Air Transport (BAT) but was turned down by the San Francisco manager, Steve Stimpson. Unperturbed, she suggested to BAT that highly trained nurses could help calm those passengers who were afraid of flying. Taking her up on the idea, BAT hired her and seven others to be the first flight attendants, albeit for a three-month trial starting on 15 May 1930.

As well as supporting passenger safety and comfort, the appointment of the flight attendant also marked a new level of luxury in air travel and soon gained popularity. However, the idea was slow to be adopted overseas. In May 1934 Swissair became the first European airline to employ a flight attendant, Nelly Hedwig Diener. Diener was killed in July that year in what became known as the Tuttlingen Accident which saw the Curtiss AT-32C Condor II biplane carrying her and eleven others, shaken so violently in a thunderstorm that the right wing set fell off, leading to the deaths of all twelve aboard.

The role of the flight attendants would be refined over the next eighty years, with many of the original flight attendants using their nursing skills to significant effect in the Second World War as flight nurses. By the twenty-first century, the flight attendants' task had developed into a pivotal role providing an extra set of eyes and ears for the aircrew regarding in-flight safety and passenger care. A hierarchy has also been established, with senior flight attendants fulfilling the roles of chief purser, in charge of all flight attendants; a purser may also be found managing the staff of a section of a larger aircraft such as the Boeing 747 or Airbus A380.

Throughout the working life of anyone in aviation, there remains the requirement to be up to date with skills honed in the classroom by simulation or practical assessment. Staff must also be mentally and physically fit to ensure the safety of all when on the ground and in the air, with those involved in flight safety, such as firefighters and cabin crew, exercising their skills regularly.

Financing the Dream

A final key piece of the infrastructure of aviation remains its financing. Regardless of the heady days of plentiful cheap surplus military aircraft in the aftermath of the two world wars, the funding for the aviation industry remains challenging. In the immediate aftermath of the First World War, the aviation industry was still very much in its infancy. For some, money was no problem. Self-financing, often from inheritances and wealthy relatives, helped individuals, often veteran airmen, take to the sky. They were joined by dreamers and adventurers who saw the opportunities in the growing world of aviation. Many took multiple jobs to pay for their lessons and licences. Those aircraft companies based in the Allied countries developed a vast range of aircraft models, each an improvement on its predecessor. Such developments were funded either by money made during the war or government and private loans, often both. The defeated Central Powers, Germany in particular, struggled with the constraints of the Treaty of Versailles, war reparations and rising inflation. Yet backers were found, with Dornier and Junkers able to fund programmes that produced innovative aircraft.

As the dream of speed and distance became uppermost, the prizes offered by the media outlets, such as the British newspaper the *Daily Mail*, grew in number and amount. These were organized alongside famous challenges that promoted technical innovations as much as speed, e.g. the Schneider Trophy. These were joined by the numerous air races run by clubs and individuals, offering considerable sums to the winners. Besides, money came from pleasure flights and air shows featuring famous names and aircraft. Then there were the showmen and women of the many flying circuses whose famous barnstorming spectacles, often verging on suicidal, proliferated throughout 1920s America.

These immediate post-war years were financially successful for aviators, designers and manufacturers, with some happy to produce existing aircraft rather than develop ideas. This period saw fortunes made and, on occasion, lost. Still, the spirit of adventure remained, with events drawing huge crowds, good numbers of participants, and the all-important sponsorships and occasional understanding benefactors. Someone was always making money somewhere, but not often the people providing the entertainment. Stories of near-penniless barnstormers and air racers working their way across early-depression America in their Travel Air biplanes were plentiful.

By the decade's end, the industry was affected by the almost continuous search for capital to invest in new aircraft, more stringent safety requirements which cost money to implement, and training costs. The banking industry had remained a reasonably reliable source of income, but the Wall Street Crash of 1928 had either broken banks or left them wary of making investments.

The resulting Great Depression saw many of the early entrepreneurial fliers, in the United States, in particular, hang up their goggles or merge with like-minded fliers to form small operations, taking advantage of the money to be made in airmail. In addition, airlines such as Imperial and Qantas were pushed hard to deliver as most of their revenue came from servicing lucrative government contracts.

It was not until the mid-1930s that the paying passengers' money impacted aviation and allowed operators and manufacturers to service debts, develop new models and expand services. Boosted by slowly returning financial confidence in the banking and government lending sectors, flag carriers and the larger operators were able to invest in new routes, infrastructure and aircraft. For its part, the industry was now edging ever closer to stressed-skin aircraft, which featured a range of constantly changing and increasingly complex innovations. Boeing, for example, ran out of money for projects in the pre-war years of the 1930s and was forced to lay staff off at its Seattle plant while money was sought. Conversely, Douglas remained buoyant with sales of its DC-2/3 aircraft. The eve of war also witnessed colossal government investment in physical infrastructure, training and aircraft manufacturing and development. In addition, civil aviation provided training expertise by organizations such as the Curtiss-Wright Technical Institute, Glendale, California.

The post-war years again saw the return of the surplus aircraft, but this time it was Douglas's troop carriers and Boeing's bombers that were being snapped up. Buyers could essentially purchase the infrastructure of a working airfield at one of the many disposal sales throughout North America and Europe. They needed a suitable airfield location, a little capital to cover fuel and administration costs, plenty of vision, and they were literally in business.

Unlike the First World War, fighter aircraft were designed for a particular purpose, restricting any post-war conversions, so the chances of the return of the flying circus were virtually eliminated. Another critical factor was the professionalism brought to the industry by the returning airmen and women. Many of the airlines had been grounded or nationalized during the war. They were keen to get back in the air and take full advantage of what war material they could buy or commandeer. This approach helped all parties immensely as the post-war world was met with austerity. Regardless, somewhere in the psyche of those post-war directors, designers, fliers and ground crews lay the realization they had to carry on developing the commercial possibilities of aviation as anything was now possible. This invigorated outlook was fed by the dawn of the jet age and all it promised. For many, the old days of winging it and scratching a living in forgotten corners of the globe were consigned to history.

The banking industry had become more refined over its first thirty years of dealing with an industry whose post-war forte would lie in military, commercial and government aviation. As a result, specialist lenders such as the United Kingdom-based Lombard and J.P. Morgan in the United States began to appear.

The manufacturers had also been busy, investing their wartime profits and knowledge, especially from the defeated Axis powers, into developing new technologies. Ground infrastructure was now growing beyond the scope of being managed by airlines. Now governments and private investors saw the opportunity to develop existing sites or build new ones to facilitate the aircraft that took to the skies with new routes. Many airlines took advantage of new facilities left behind by the war, repairing and improving a new network of airports. Some countries utilized captured and abandoned aircraft and airfields to rebuild their national airlines and air routes. The development of the jet engine would see airlines clamouring to access the new technology for either turboprop or turbojet aircraft, while manufacturers were keen to use these technologies to power larger, faster aircraft with ever greater ranges.

As the manufacturing industry moved away from wartime twenty-four-hour mass production and an almost bottomless government chequebook, they needed to be innovative, especially in the face of technological change. The crossover of military technology, the radar in particular would have a huge and positive impact on safety. Allied with the arrival of the jet engine, it soon became the must-have piece of equipment, especially for those working within Air Route Traffic Control Centres (ARTCCs). New technology is never cheap, and the investments required were considerable. While the infrastructure had always been able to pay for itself through the levy of user charges and local taxes, there was still the need for investment in both airlines and manufacturers. The next generation of civil aircraft was going to be expensive. The need to invest in mass-produced airliners capable of operating safely at near-unheard-of heights, with an increased capacity for passengers and cargo, was at the forefront of development.

While there were industry tie-ins between airlines and manufacturers, these still had to be funded. The airlines adopted new, more innovative ways of measuring income while saving expenditure, achieved partly by developing responses to average load factors, which fluctuate over the years, peaking in the holiday season in particular. Other measures included recognizing that short- and medium-haul journeys were more profitable than long-haul flights. This enabled airlines to develop a business model where increasing the frequency of short- and medium-haul flights would balance out any loss made by long-haul flights. Profit margins were also identified and maximizing strategies grew, where the intended goal was reached by addressing the intended benefit against any strategic pitfalls. By doing so, airlines determined three distinct advantages identified in a particular order: stimulate demand, reduce operational costs and increase revenue. This approach, especially after the 1973 world oil crisis, allowed airlines to focus on the best economic survival and growth routes. It also helped prove to partners in the banking industry, who had developed a sophisticated system of aircraft finance, that operators understood the low financial margins the airline industry operated within. It allowed the airline to demonstrate that investment from the banks would always be rewarded, debts could be serviced, and business remained viable and open to future success.

The significant cost for airlines remained the aircraft, and the banks organized a series of lending options that would allow individuals and airlines the chance to purchase new aircraft. Private aircraft are by far the easiest to finance or purchase. The individual or group of aviators, each getting a share, will select and buy an aircraft between lender and purchaser, similar to a mortgage. Commercial aircraft purchases are understandably more complex and are purchased through three schemes: secured lending, operating leasing and finance leasing. In addition, of course, there are other buying methods, including cash buying, which has become increasingly rare, as well as operating leases and sale and tax leases. By the end of the twentieth century, forty-two percent of commercial aircraft were leased, which reflected the growing cost of aircraft and made leasing an increasingly attractive option for both large-sized aircraft purchases such as the Airbus A380 or large fleet purchases for short- and medium-haul operators like easyJet (EJU) based in the United Kingdom.

The airline industries 2019 Aviation Benefits Report identified that the cost of travelling had dropped remarkably since the 1990s while passenger numbers had continued to soar. Passenger and freight miles are expected to climb, guaranteeing the operators' financial health. The market for new aircraft continues to grow, especially among the Asia Pacific nations. These nations will account for the purchase of half of the projected 40,000 airliners that will be built by Boeing and Airbus over the next twenty years, proving that commercial civil aviation remains as vibrant now as it was a century ago.

Chapter 8
Civil Aviation Beyond the Holiday

Over the past century, the story of civilian aviation has been dominated by two distinct themes – the first to push the technological boundaries of aviation, the second to develop the airliner as the ultimate expression of civil aviation. In the years after the 1919 Paris Convention, the burgeoning interest in flight led to the establishment of a host of services that could utilize the aircraft for more than air sports and pleasure flying. Soon air ambulances, fire bombers, private and cargo aircraft were taking to the skies. These were joined by aircraft undertaking scientific research in arctic climes and those delivering aid to famine-affected areas. In all its guises, the aircraft has helped us in ways never thought possible. The range of aircraft developed and used over the years, fixed-wing and rotorcraft, have shown that civil aviation is greater than the sum of its parts.

Extra Commercial Activities

The role of cargo carriage was quickly established as the primary source of income for many airlines and commercial operators in the post-Paris Convention years. By the end of the peaceful inter-war period, three different aircraft types were in service to meet the operator's needs and maximize profit and air cargo. First, passenger aircraft retained space for baggage in the belly hold, with the opportunity for passengers to carry on hand cargo as an onboard courier. Another way of monetizing passenger aircraft is temporarily converting the cabin into a cargo hold by removing seating; this process is known as cargo-in-cabin.

The second type features dedicated cargo bays and a main cargo hold in the belly. These aircraft can be loaded in a variety of ways. Some, like the Bristol Type 170 Freighter and the Boeing 747-400 Large Cargo Freighter (LCF), can be loaded either through swing-up or clam-shelling opening noses, like the Boeing 747. Others like the Douglas DC-3, utilized fuselage side doors, while the Tupelov An-124 can be loaded via the nose or tail. For the operator, the nose/tail loading combination makes sound commercial sense as it allows for the rapid loading of large airline units or Unit Load Devices (ULDs). The ULDs, which can be manufactured in various shapes and sizes, are mounted on low-profile slave pallets that work in partnership with a specialized aircraft floor known as an air cargo rolling deck, allowing easy manoeuvring into a safe position by staff once loaded. These roller systems are often found in the terminal building's luggage and cargo handling areas to help speed up cargo processing. The final type of aircraft used for cargo is the combi-aircraft which carries the cargo load in a dedicated bay often fitted with an air cargo rolling deck, either in front of or behind passengers. Cargo is loaded via oversized fuselage side doors with aircraft such as the Boeing 737-100 and the Hawker Siddeley HS 748.

As the size of the loads and financial rewards increased in proportion to global aeronautical, social and economic development, air freight became an all-important element in the functioning of many airlines. The numbers increased year on year, so much so that by 2018, IATA estimated that the revenue raised for that year would be a staggering $59.2 billion. The Asia Pacific region, which was also managing thirty-nine percent of global air freight, also saw a sixty-six percent growth in e-commerce in the five years up to 2020. By 2020, before the global COVID lockdowns, the annual tonnage of goods carried had reached 62.4 million metric tons. By the end of the first 101 years of civil aviation, the primary method of purchasing physical goods had become e-commerce, with numerous online sellers trading across the globe, adding to freight figures.

With loads increasing, aircraft manufacturers responded by designing and building larger aircraft. While many cargo aircraft were little more than converted airliners, in some sectors of the aerospace and space industries there was a need for something a little larger. This led to the development of aircraft capable of carrying outsize cargo.

The ultimate expressions would be the Boeing Shuttle Carrier Aircraft (SCA), of which two 747-100s were subsequently converted, and the Antonov An-225 Mriya, which had been designed to fulfil a similar role with the ill-fated Buran-class orbiters. The SCAs were intended to return the space shuttles that had landed at Edwards Air Force Base to the Shuttle Landing Facility (SLF) at the Kennedy Space Center. In the USSR, the sole An-225 completed its tasks until the demise of the USSR halted the development of the Buran project. This followed a hiatus in all flight operations for eight years before refurbishment and entry of charter service with Antonov Airlines (ADB), which specialized in carrying outsized loads. The six-engine An-225 could take a massive 551,000lb (250 tonnes) in a load area of 38,846ft^3 (1,100m^3) of cargo.

In comparison, the SCAs carried their shuttle payloads on their backs, leading to extensive modification of the fuselage, which included strengthening and adding ballast. The 171,961lb (78,000kg) shuttle, complete with aerodynamic engine cowling, was attached and detached to the SCAs via a specially designed Mate-Demate Device (MDD), which consisted of two towers and a lifting beam. The SCAs were also used for mid-flight separations to enable flight testing of the shuttles. These unique aircraft provided ferry services for the National Aeronautics and Space Administration (NASA) between 31 January 1971 and 21 September 2012.

While impressive in size and capacity, the SCA and An-225 followed in the footsteps of a certainly different, if not striking looking aircraft, the Aero Spacelines Guppy. Aero Spacelines was founded in 1960 by John Conroy, a former B-17 pilot who had been captured during the Second World War. Conroy was every inch the innovator, starting several companies, including Aero Spacelines, and seeing the opportunities that space travel, in particular, had to offer the commercial industry. His many innovations include the Guppy range of heavy-lift aircraft and the development of a series of specialist land and seaplanes. Conroy's most striking design, a competitive bid against Boeing's SCA, was the colossal twin-boom Conroy Virtus. This was based on utilizing two Boeing B-52s as fuselages with the space shuttle slung between the two. Needless to say, the proposed design was considered too large to be of practical use. Despite the Virtus proving to be of otherwise sound design, NASA opted for the SCA.

It was with the Guppy range of heavy-lift aircraft that Conroy really struck gold. His designs were conversions of the Boeing's successful 377 Stratocruiser into specialist transport aircraft capable of moving outsized loads for NASA. The modification was extensive and complex, but the resultant aeroplane with its familiar bulbous upper deck would be eye catching, if nothing else. Initially, NASA was uncertain about the radical concept. Still, the aeroplane worked and was soon nicknamed the Pregnant Guppy on account of its appearance. On 19 September 1962, the Pregnant Guppy, with its unique teardrop-shaped fuselage, made its maiden flight, which proved to be a success, from Van Nuys, California. This was watched by uneasy airport management, who doubted the safety of the design, and the first responders of the local emergency services, with some trepidation. However, the success of Conroy's project was its entirely removable tail-section to load the oversized elements of its Gemini programme. This feature alone reassured NASA, who were soon using the aircraft. This included the delivery of the Douglas-made S-IV, the second stage of the Saturn I rocket from Santa Monica in California to Cape Canaveral, Florida, in a single flight. Such transits were soon saving NASA up to three weeks of overland transit time, speeding up the assembly process for the project.

The single Pregnant Guppy was followed in 1965 by the Super Guppy, which carried the third S-IVB stage of the Saturn V rockets used in the Apollo programme. The Super Guppy was converted from a former military C-97J Turbo Stratocruiser, powered by Pratt & Whitney turboprop engines. Unlike the Pregnant Guppy, the Super Guppy was loaded via a sideways-swinging nose section, which showed off the teardrop fuselage cross-section beautifully and also made loading a little easier. As well as serving NASA, the Super Guppy, with its much larger cargo bay, would also support Airbus in the early 1970s. In 1981, Union de Transports Aériens of France built two Super Guppys for Airbus, who had purchased the rights to produce the aeroplane. These joined the two other Super Guppys already in service with Airbus. They flew with Airbus until the introduction into service of the Airbus A300-600ST (Super Transporter) Beluga in September 1995.

The final aircraft of the Guppy range was the swing-tailed Mini which made its maiden flight in 1967. Again, this was used by NASA to transport critical parts of its ongoing activities, including the Pioneer 10 space probe. Other commercial loads included the famous Goodyear Europa Blimp, which uses inflation for its shape, making it easier to transport.

By 2020 only one of Conroy's Guppys remained in service, a Super Guppy operated by NASA. However, this veteran of the early space programme continues its work, now transporting the Orion multi-purpose crew vehicle (Orion MPCV) prototype from Kennedy Space Center in Florida to various sites in the continental Unites States for testing.

The final heavylift aircraft is the Boeing 747-400 Large Cargo Freighter (LCF), designed to transport Boeing 787 Dreamliner elements between sites in Europe, South West Asia and the United States. Also known as the Dreamlifter, the converted 747-400 passenger aeroplane was given a bulging fuselage and featured a swinging tail. The LCF made its maiden flight on 16 February 2007 and is used to ferry 787 fuselage and wing sections that are too big for container transportation between assembly sites. The volume of the main cargo compartment is 65,000ft^3 (1,840m^3) which is smaller than the Airbus Beluga XL cargo compartment of 78,000ft^3 (2,209m^3).

Despite their impressive size, the various outsize cargo-carrying aircraft will always remain few in number due to lack of fleet size, costs and requirement to land at airports and airfields large enough to cater for their unique needs. Despite this they play a vital role in the aerospace industry.

Working alongside their larger brethren is a range of cargo-carrying aircraft in all shapes and sizes, from the evergreen Antonov An-2 biplane to the Ilyushin Il-76 and the Lockheed L-300/C-130, all capable of landing on short or rough airfields. These aircraft are often used by governments or chartered by large organizations to deliver humanitarian aid in times of crisis. While unloading at an established airfield is always preferable, that may be impossible to achieve due to infrastructure damage or a disaster-struck location. In such instances, aircraft will often utilize military delivery styles, including gravity or extraction-type airdrops at low level or unloading into a helicopter to carry a load to its intended location.

The World Food Programme also manages its own fleet of chartered aircraft and helicopters as part of its United Nations Humanitarian Air Service

(UNHAS). The UNHAS operates predominantly in areas where United Nations' support is needed most. When undertaking operations, the UNHAS provides the delivery of aid in compliance with the United Nations Aviation Standards for Peacekeeping and Humanitarian Air Transport Operations (UNAVSTADS) and ICAO.

A Helping Hand in a Time of Need

The use of aircraft in the aeromedical evacuation role has its roots in the work of French pilot Marie Marvingt. Marvingt became the first female combat pilot and the world's first certified flight nurse. She also sought to establish an air ambulance service as early as 1910. Regardless of Marvingt's work, the first civilian-operated aeromedical service would be established far from France.

Reverend John Flynn established what would become the Royal Flying Doctor Service of Australia (RFDS). Flynn had been spurred into action upon reading about the unfortunate death of stockman Jimmy Darcy in 1917. At the time, a flying ambulance was suggested to Flynn by Lieutenant Clifford Peel. He suggested using aircraft to provide medical aid to those living in the Australian outback. Peel would be posted as missing on 19 September 1918 when his Royal Aircraft Factory R.E.8 disappeared in France during a mission.

A little over a decade later, with the help of a large bequest from agricultural industrialist Hugh McKay and support from Hudson Fysh of Qantas, the newly formed Aerial Medical Service (AMS) became operational. The service was soon put to use using a de Havilland DH.50, capable of carrying four passengers. To help summon assistance, engineer and inventor Alfred Traeger developed a pedal-powered radio. These simple radio transmitter and receiver sets initially used Morse code to summon assistance before being designed to use voice communication. These simple sets, which increased in number through the sparsely populated outback, were used for various purposes, including connecting people socially. Their success in providing early medical assistance was the making of the AMS in those early days, allowing the service to expand beyond Queensland.

In 1942 AMS became the Flying Doctor Service, and in 1955 it was awarded the prefix Royal. By the 1960s, the RFDS would move away from contracting aircraft. By 2020 it was operating seventy-seven aircraft from twenty-three airfields across Australia, each helping to save the lives of 40,000 Australians since the inception of the service almost a century earlier.

Other aeromedical services were established in the wake of AMS's foundation, including in 1934, the establishment of an air ambulance service in Morocco by Marvingt. This was followed in 1936 by establishing an air ambulance service as part of the Highlands and Islands Medical Service in Scotland.

What Marvingt and others proved was that aviation was an exceptionally flexible medium in which to provide humanitarian assistance. Aeromedical provision continued to expand after the Second World War, and the helicopter gave the service an edge with its introduction. Now injured individuals could be picked up from a remote mountainside as quickly as from a busy shipping lane.

The helicopter reduced the time from pick-up to arrival at an emergency care centre considerably. Remote areas would see the helicopter become the first responder, taking the patient to an airfield where a Medical and Emergency Response Team (MERT) would be waiting with a specially equipped aircraft. The MERT is staffed by specially trained flight medics and is led by a retrieval doctor. At the same time, the aircraft would provide a rapid medical evacuation (medevac) service. The staff and equipment on board would be used to stabilize and care for the patient during the flight. Another unique medical use of aircraft includes the flying hospital as the DC-10 Orbis Flying Eye Hospital. Orbis's first flight was in 1982 and was founded by ophthalmologist Dr David Paton to provide a flying ophthalmic hospital and teaching services around the globe.

Other uses of aircraft, both fixed and rotary wing, include law enforcement. While a common enough sight today, the first use of an aircraft to assist with law enforcement was the use of airship R33 to help police with traffic control at the 1921 Epsom horse race in the United Kingdom. This was followed fifteen years later by the use of a police autogyro at the Battle of Cable Street in the East End of London.

Like the air ambulance, the steady development of the helicopter from the 1950s led to its adoption by law enforcement agencies worldwide. Today the helicopter in law enforcement service has become a highly sophisticated piece of equipment. It is often fitted with an almost military suite of sensors through which aircrew can direct officers on the ground with precision.

Adopting the rotorcraft and the helicopter as a tool was only a matter of time. Since the early drawings by Leonardo da Vinci of his man-powered helicopter, the desire to operate an aircraft with similar mechanics to that of bird flight remained at the forefront of design. However, it was not until the early twentieth century that attention returned to creating an aircraft capable of fulfilling a vertical and/or short take-off and landing (V/STOL). The first successful design, a quadcopter, was made by the Bréguet brothers and Professor Charles Richet in 1907, the Bréguet-Richet Gyroplane. Although tethered, the Gyroplane, using a single engine driving four sets of propellers, two running clockwise and two counter-clockwise, managed to stay aloft for around a minute on 29 September. This was followed on 13 November by fellow French engineer Paul Cornu and his tandem-rotor helicopter. Cornu's flight lasted a mere 30 seconds, but his design, featuring two 'paddle wheel' rotors was the first such equipped aircraft to be crewed and controlled in flight.

After the First World War, the main effort for the growing aviation industry was the desire to produce easily manufactured fixed-wing aircraft; rotorcraft design remained very much on the fringes due to its complexity. In 1924, French engineer Étienne Edmond Œhmichen flew his variable-pitch multi-propeller Oehmichen No. 2 for 1,640 feet (500 metres). This proved that the concept of rotary aircraft was sound and worthy of attention regardless of the engineering difficulties. This early use of multiple propellers had little to do with lifting power but more with being able to counteract the inevitable torque reaction of the fuselage in flight.

The following year Juan de la Cierva, a London-based civil engineer, created the autogyro. The autogyro would become a testbed for Cierva's ideas, and many of his discoveries would be incorporated into later helicopter designs. One of these was autorotation; here, the primary rotor system of the autogyro was turned by the action of air moving up through the rotor rather than engine power driving the rotor. This allowed a freely rotating single rotor acting as a wing, with the forward-mounted propeller pulling the aircraft forward by exerting torque reaction. Cierva continued to develop his idea until his untimely death in 1936, when the DC-2 he was flying in crashed in poor weather shortly after taking off. However, Cierva's design did see service during the Second World War, with Avro producing the Model 671 Rota, based on the Cierva C.30 and used to help calibrate ground radar stations.

Post-war, the design and mechanics of the helicopter were refined by engineers like Igor Sikorsky. His R-4 two sweater entered service in 1942. One of the main changes to rotorcraft design was the main and tail rotor (MTR) design which became the established layout for helicopters. Bell and Boeing Vertol followed Sikorsky in developing the aircraft, with Bell creating their unique giro-stabilized two-bladed rotor. However, development and running costs for helicopters remained initially prohibitive, and most work went into producing helicopters for military use. In Europe, there had been several post-war attempts at producing helicopters. Still, it would be French concern Aerospatiale and British firm Westland who would initially steal the lead.

The wars of the 1950s and 1960s further refined designs. The introduction of the Boeing Vertol CH-47 Chinook with its tandem rotating rotors and Bell's famous UH-1A Iroquois took full advantage of the new turbine engines, such as the Lycoming series, which gave the aircraft immense power. These allowed helicopters to generate more airspeed and carry larger loads. They replaced many, often inefficient or complex, powering rotors, and used compressed air to power rotor blades. The 1960s also saw the advent of aerial crane helicopters such as Sikorsky's S-64 Skycrane based on the military C-54 Tarhe, capable of carrying various loads to inaccessible places such as high-rise buildings. The S-64 can also take a water cell for fire-fighting and be used alongside

CIVIL AVIATION BEYOND THE HOLIDAY

fixed-wing aircraft, providing the same service on land or sea.

By the late 1950s, the Soviets had realized the potential that helicopters presented, and two design bureaus took the lead in meeting the requirements of the state. Kamov produced helicopters with unique contra-rotating coaxial rotors, which did away with the need for a tail rotor. Mil would go on to build the Mi-10 flying crane and the Mi-26, which has become the largest and most powerful helicopter to have gone into serial production. While it may seem a little extreme to build helicopters capable of carrying up to twenty tonnes (44,000lb) of cargo, it made perfect sense given the areas in which they operate. The most unique load ever carried by the Mi-26 was that of the 25-tonne block of frozen soil encasing the 23,000-year-old Jarkov mammoth.

Despite the many thousands of helicopters produced since the Second World War, the cost of operating helicopters keeps their use restricted to specialist roles. Within the emergency services, coast guard, and search-and-rescue (SAR) roles, the helicopter has repeatedly proved its worth. However, its small size can act against it, especially in fire-fighting. As the proliferation of unmanned drones continues apace, survey and non-critical or sensitive 'flying eye' tasks, such as inner-city traffic reports, are slowly being replaced by aircraft no bigger than a seat pad. The added bonus of using drones is that they are cheap and easy to repair and replace. Drones also offer the trained operator and the customer the same benefits as a helicopter with a fraction of the concerns.

Private commercial companies still operate helicopters, including Bristow Helicopters Ltd, founded by Alan Bristow in 1955, which uses a wide range of helicopters for charter work within the engineering sector and for passenger carriage. Other companies offer niche services, such as Canadian Alpine Helicopters, founded in 1961, who operate a diverse portfolio of services, including tourism, heli-skiing and wildfire management. Helicopters can also be found in the luxury travel market. For example, the American company Paramount Business Jets offers charter services alongside personal flights.

As can be expected over the years, helicopters and rotorcraft have broken a range of records. These are managed through the Fédération Aéronautique Internationale (World Air Sports Federation) Rotorcraft Commission (Commission Internationale de gravitation (CIG)).

The agricultural world was one of the first to recognize the advantages of light aircraft, like helicopters, to provide rapid and widespread use of pesticides and fertilizers. Often called crop dusters, aircraft began this role in the mid-1920s, with many pilots using the money made from their agricultural work to invest in their next aeroplane. Early aircraft were often converted biplanes, such as the de Havilland Tiger Moth, fitted with wind-powered pumps. These pumps then passed the liquid pesticides or fertilizers for delivery through a spraying system attached to the wing's trailing edge.

As with other sectors in civil aviation, a specialist type of aircraft was soon developed. A unique range of aircraft with excellent STOL began appearing in the 1950s for use on larger farms across the globe. Such aircraft included the single-engine Commonwealth Aircraft CA-28 Ceres and the PZL-106 Kruk. A vital feature of these aircraft is the unique raised cockpit to allow the fitting of the spray tanks. When the area to be sprayed is larger than expected, or multiple fields require spraying, aircraft such as the An-2 Colt, the Lockheed Lodestar or the de Havilland Canada DHC-2 Beaver have been used. As agricultural spraying aircraft are fitted with tanks and a spraying system, they are often re-tasked for local fire-fighting tasks. Helicopters, equipped with external tanks and a flying boom, are occasionally used for spraying.

Captains of Industry and Leisure

The private ownership of aircraft remains one of the cornerstones of civil flight. Many privately owned aircraft are used for sports or leisure activities, with owners also utilizing the aircraft to experience the pleasure of flying. For some, the desire is to own a private aircraft as a tangible symbol of their success or status. This has led some companies to initially develop a range of small- to medium-sized, often multi-engine aircraft, targeted at those wanting something more exclusive from their flight experience.

The early years of regulated civil aviation had shown that not all aircraft needed to be bulky converted wartime bombers. The Junkers Ju 13, with its well-appointed interior that pointed to an age of refined comfort, more than proved this. The inter-war years were very much a period of design refinement, but several companies stood out over the years, including Cessna, founded by farmer Clyde Cessna in 1927, and Beechcraft, launched five years later by Walter and Olive Beech.

Cessna's early output included the high mono-wing SR-3 racer and the DC-6 four-seat tourers. Unfortunately, the Great Depression saw Cessna cease trading in 1932. Still, two years later, Cessna's nephews Dwane, an aeronautical engineer, and Dwight Wallace bought and reopened the company. The first of the new Wallace Cessna aircraft was the four-seat Cessna Airmaster range featuring various models offering different specifications. This series of aircraft featured construction methods that were no different to that of early aircraft. However, the welded fabric-covered fuselage and wooden wings and other features rapidly made the series expensive to build. After the hiatus in civilian aircraft production in the early 1930s, combined with the experience gathered from producing the AT-17 Bobcat/T-50 advanced trainer aircraft, Cessna now looked to the future.

Given that semi-stressed and stressed skin monocoque aircraft had become the norm, Cessna followed the market trend and, in 1945, flew the all-metal

Cessna 140 light utility aircraft for the first time. This was followed in 1946 by the Cessna 120, featuring a conventional fixed landing gear. Both aeroplanes were similar in appearance to the pre-war Airmaster range. Though both were built in far greater numbers due to the production skills gained during the war.

The Cessna 140 ceased production in 1951 and was quickly followed by the two-seat Cessna 150, another light utility aircraft. This new aircraft now featured a fixed tricycle undercarriage arrangement. With 23,839 units produced, including the 152 model, between 1958 and 1985, it became the fifth most produced aircraft up to 2020. The crown of the most produced aircraft goes to another, earlier Cessna though, the four-seat Cessna 172 Skyhawk. Despite a pause in production between 1989 and 1996, it remains in production, with over 44,000 units produced. The 172 would earn colossal praise and admiration in the aviation world, but perhaps its most memorable moment occurred on 28 May 1987. German aviator Mathias Rust set out to establish an air bridge in a rented 172P to help alleviate what he saw as the growing political tensions between East and West. Flying from Finland over Estonia, Rust luckily avoided being shot down by Soviet air defences and eventually landed in Moscow on Bolshoy Moskvoretsky Bridge by St. Basil's Cathedral. Unfortunately, the Soviet authorities were less than impressed by his unauthorized flight. A four-year prison sentence was awarded in late 1987 for hooliganism. Thankfully as part of the thawing of East-West relations, Rust was freed at the beginning of August 1988.

Cessna would produce an enviable range of aircraft as it gained experience. The post-war years saw Cessna build an aircraft portfolio in various configurations. This peaked with the introduction of the twin turbojet-powered Citation in 1969, designed to break into the light business jet market.

Like Cessna, Beechcraft were early producers of utility aircraft and, in 1932, unveiled their Beechcraft Model 17 Staggerwing biplane. Walter and Olive Beech had formed the Beech Aircraft Company in 1932 due to dissatisfaction with his role at Curtiss-Wright, where he worked as president of the airplane division and vice president of sales. Beech was not short of ideas, and in 1937 the Model 18 all-metal low-wing twin-engine monoplane was unveiled. Built between 1937 and 1969, the Model 18 was an extremely popular aeroplane capable of fulfilling various roles, including business aircraft. The post-war years also saw Beech, like Cessna, increase its product range to include medium-sized, multi-engine aircraft, including the pressurized ten-seat Model 80 Queen Air series. Introduced in 1960, the Queen Air would serve as a foundation design for the turboprop-powered Model 90 King Air, which followed in 1963. Beechcraft was a little late to the business jet market, and in 1998 the Model 390 Premier I took to the air as a direct competitor to Cessna's Citation. Where Cessna will be remembered for the antics of an idealistic teenage pilot, a Beechcraft aeroplane will be associated with the death of singer Buddy Holly. The distinctive V-tailed Bonanza, which has been in continuous production since its first flight in 1945, making it the longest aircraft production run, was chartered by Holly on 3 February 1959. The plan had been to avoid an uncomfortable journey in a poorly heated tour bus. Aboard were Holly, Richie Valens, the Big Bopper Jiles Richardson Jr and pilot Roger Peterson. Peterson was an inexperienced pilot who was not cleared for the instrument-only flight required for the poor weather conditions. The aircraft took off shortly before one in the morning and, within five minutes, crashed, with the loss of all four men. A subsequent investigation would highlight failings in the meteorological briefing, inadequate pilot knowledge and poor instrument-flying knowledge.

Both companies would eventually become part of the Textron industrial conglomerate, with Cessna joining in 1992 and Beechcraft following in 2013 after an earlier merger in 1994 with Hawker Beechcraft Corporation (HBC). The same year Beechcraft had ceased production of its Premier range of aircraft due to wider economic uncertainties.

The final of what was to be known as the 'Big Three' of American general aviation aircraft production was Piper Aircraft Inc. Initially founded as the Taylor Brothers Aircraft Manufacturing Company in September 1927 by Clarence and Gordon Taylor, the first decade was far from the success they hoped for. The Great Depression hit Taylor Brothers hard, and in December 1935, William Piper, an engineer, bought the assets of the company when its bankruptcy sale was held. Piper retained Clarence Taylor as the president until 1937, when, after a series of disagreements, Piper bought out Taylor's share in the company. The latter went on to form Taylorcraft Aviation, while Piper renamed the business Piper Aircraft Corporation.

At the same time, Piper started producing its high-wing Cub series of two-seat light aircraft, producing over 35,000 units of all variants, including the PA-18 Super Cub, between 1936 and 1994. The Cub was followed by the equally famous two-engine low-wing PA-28 Cherokee which was launched in 1960 and was still in production sixty years later. As well as designing a range of light aircraft that suited many needs, Piper also designed and produced agricultural and pressurized light aircraft, often powered by turboprop engines. However, unlike Cessna and Beechcraft, Piper did not make their own turbojet-powered aircraft; despite testing the concept with the single-engine Piper PA-47 PiperJet, Piper partnered with Honda for their HA-420 HondaJet light business jet in 2006.

The first significant change for private flying and certainly, for the executive and luxury market, occurred on 7 October 1963 when inventor William Lear unveiled the Swiss American Aircraft Corporation Learjet 23. Although the Learjet was not the first private jet – that crown belonged to Lockheed, who first flew their four-engine ten-seat L-329 JetStar in September 1957 – it was the first to set out to be a luxury aircraft.

Based on the Swiss Flug- und Fahrzeugwerke Altenrhein FFA P-16 single-engine ground-attack aircraft project, the Learjet was designed to be the last word in business travel. To prove its commercial worth, an early wingtip fuel tank-equipped Learjet made the return journey between Los Angeles and New York in less than eleven hours in 1965. Undoubtedly, this staggering turnaround was helped by the Learjet's impressive cruise speed of 834km/h (518mph), which was achieved at 40,000 feet (12,000 metres).

The Learjet could seat up to eight comfortably and set the physical appearance for a generation of small business jet aircraft with its rear fuselage-mounted turbojet engines. Within two years, Lears had released two further models and renamed the company Lear Jet Industries Inc. By the end of the decade, Lear Jet Industries had merged with Gates Aviation Corporation to form the Gates Learjet Corporation. Lear was also fortunate, as besides the JetStar, his nearest competitor was the larger Grumman Gulfstream II, which first flew in 1966 and could carry nineteen passengers.

Lear had hit a magic formula with his new jet aircraft. The mid-1970s would give the Learjet two key milestones. The first was the production of the 500th aircraft, and the second was the completion of one million flight hours. This was proof that the personal luxury aircraft was here to stay. The performance of the Learjet was also an essential factor for the positional customer to consider. The original service ceiling of 45,000 feet (14,000 metres) was extended to a staggering 51,000 feet (15,500 metres) in 1977, giving the Learjet 28 model gave the operator an edge over commercial air transport providers. The 1980s saw Gates Learjet Corporation grow as a company. It ended the decade as the Learjet Corporation after an integrated acquisition with a portfolio of three different models. Learjet owners now included Elvis Presley and Marlon Brando.

In 1990 Learjet became part of the Canadian Bombardier Aerospace family, which saw the brand's continued growth and development, with the launch of the Learjet 60 and 45 models, followed by the Learjet 40 and 70/75 models in the 1990s. For Learjet and Bombardier, the ultimate expression of the Learjet philosophy of style, speed and luxury was the composite Learjet 85 model. On 30 October 2007, Bombardier sought to slide the new high-speed aeroplane into the mid-size and super-mid-size markets. Despite the early interest, there were not enough potential buyers to justify the development costs. On 29 October 2015, the project was cancelled in favour of Bombardier's Global 7500 and Global 8000 ultra-long-range business aircraft. The end of the first century of civil aviation saw Bombardier make a difficult choice. After the production of some 3,000 Learjets, the company ceased production.

Regardless of Learjet's demise, there remains a strong appetite for luxury aircraft, with numerous companies and brokers offering tailor-made aircraft options and travel solutions. For example, Brazilian aerospace company Embraer provides a range of light long-range business aircraft that can be customized to the owner's taste. In contrast, American firm Cirrus Aircraft offer a range of smaller short-range aircraft, including the Collier Trophy-winning single-engine V-tailed Vision Jet. While only able to carry a maximum of five passengers, the Vision Jet showcases the unique safety feature of the Cirrus Airframe Parachute System (CAPS). Another critical innovation is the passenger-operated Garmin Safe Return emergency autoland system. Such elements reflect how far flight safety has developed over the past century, a world away from the hostile environments in which the early pioneers flew.

Flight for Pleasure and Adventure

Flying Clubs
The formation and growth of flying, or aero, clubs, grew with the popularity of powered flight. They provide the membership with lessons, flying experiences, flight planning support, and the opportunity to rent, share or fully own an aircraft. The clubs can be found globally at old military airfields, grass strips or purpose-built locations. Here members and visitors can access fuel for their aircraft, up-to-date air traffic control information, hangar use and refreshments. As well as practical support, many clubs provide a complete social diary, including visits to other airfields and hosting air navigation competitions. Other activities include local air days or air shows, where a selection of rarely seen, often vintage aircraft can be viewed on static display or in the air.

Adventure and Sports Flying
While for many, the simple pleasures that come with flying are reward enough, the Fédération Aéronautique Internationale (FAI) or World Air Sports Federation provides more than enough challenges for those more adventurous spirits. Established in 1905, the FAI has remained independent of any external influence throughout its history and now has members from more than 110 nations. The FAI rapidly became the vanguard of early aeronautical endeavours with its vital role in ratifying over 18,000 aeronautical records since 1905 and coordinating over 700 international air sports events a year. These critical elements of the FAI's work are managed by the FAI General Aviation Commission (GAC) and General Air Sport Commission (CASI). CASI's responsibilities include the production of definitions and terminologies, measuring standards and methods, and awarding sporting licences. As well as CASI, the FAI covers several air sport activities, each with its governing commissions, from aeromodelling to space travel, ensuring aeronautical

excellence. To help categorize the records achieved over the years, the FAI has developed a series of over twenty-five separate classes and sub-classes which cover everything from ballooning to uncrewed aerial vehicles.

Aerobatics

The FAI CIVA (Commission Internationale de Voltige Aérienne) is responsible for promoting and regulating international aerobatics championships for powered and glider aircraft. Aerobatics combines and tests the key principles of airmanship: judgement, control, discipline and knowledge, mixing all four to deliver compelling examples of flight for both pilot and spectator. The early pioneers made it up as they went along, rectifying errors and mastering the techniques used to induce jaw-dropping manoeuvres. These early manoeuvres were joined by tactics pilots had learned in the skies during the First World War, such as the famous Immelman Turn, where pilots carry out a half-turn at the top of a loop. These new manoeuvres produced breathtaking aerial displays that drew in huge crowds, both on the flight line and on celluloid in films such as Howard Hughes's First World War epic *Hell's Angels*. Hughes used some of the best pilots of the time, including Florence 'Pancho' Barnes, who had honed her craft on the barnstorming circuit of the 1920s.

The early aerobatic pioneers used aircraft that were often trainers, obsolete types or light enough to sustain the pushes and pulls of the g-forces that pilots exerted on the airframes. This led to numerous aircraft accidents and incidents, often with deadly results, leading to designers strengthening aircraft frames, developing particular wing forms, increasing the aircraft's power-to-weight ratio while perfecting fuel and oil systems that operated when inverted through injectors and high-pressure pumps. These changes, combined with experiences gained during the Second World War and the subsequent expansion of awareness around aeronautics and access to more robust materials, saw the development of specialist aerobatic aircraft. Whatever its configuration, the aerobatic aircraft was a series of design and technical compromises to get the best performance for the pilot.

The foundation of an excellent aerobatic aircraft was its handling capabilities and response to the pilot's intention. This meant developing powerful and sensitive controls coupled to a light aircraft that was strong enough to handle all that was required. This knowledge manifested in the first flight of the Pitts Special S1 light aerobatic biplane in September 1944. Designed by American Curtis Pitts, the Pitts Special, which remains in production and includes a kit version, the S-2SE, soon established itself as the foundation upon which all subsequent aerobatic aircraft would be based. It became the favourite of several well-known air show performers, including Betty Skelton. She earned her Civil Aviation Authority private pilot's licence at age sixteen. Skelton would continue to fly while working for Eastern (EAL), becoming graded on various aircraft, including multi-engine types. In 1945, after tutelage by Clem Whitteneck, a famous aerobatic pilot, Skelton took part in an amateur air show. Unable to fly as a pilot with either the military or airlines due to her gender, Skelton pursued a career in aerobatics, using air shows to sate her appetite for flight. Initially purchasing a Great Lakes 2T-1A Sport Trainer biplane, which excelled at aerobatics, it became a championship-winning aeroplane in Skelton's hands. In 1948 Skelton bought a Pitts Special, known as *Little Stinker*. Along with her Chihuahua, Little Tinker, who wore a custom-made parachute, Skelton would become the US Female Aerobatic Champion in 1948, 1949 and 1950.

After her last win, Skelton stepped back from competitive aerobatics, exhausted by the non-stop work required. Nevertheless, she would excel in the sky and on the ground, gaining seventeen world records, working as a test pilot and being an active member of the Civilian Air Patrol. The Pitts Special became an extremely popular aeroplane among pilots, offering the right balance of size and agility afforded by the four ailerons of twenty feet (six metres) in the biplane wingspan.

The first formalized competitions began to take place in the United States, with the Freddie Lund Trophy becoming the sought-after prize. Named in memory of the Waco test and stunt pilot Freddie Lund, the trophy ran for ten years between 1931 and 1941. Air show pilot Mike Murphy won the trophy three times. He would later be the lead in organizing international post-war FAI aerobatic competitions. The first major post-war aerobatic competition would occur in the United Kingdom between 1955 and 1965. Named the Lockheed Trophy, the freestyle aerobatic competition laid the foundations for the first FAI World Aerobatic Championship.

The first FAI CIVA World Aerobatic Championships (WAC) took place in August 1960 in Bratislava, Czechoslovakia, spurring the development of new designs across the northern hemisphere. New designs continued to develop, and the monoplane type of aircraft became more popular. Soon designs were appearing that could withstand more than the 5gs pilots often pulled during their sequences. By the turn of the twenty-first century, new aircraft were taking to the skies, utilizing experience with new technologies to give pilots even lighter aircraft, such as the Slick 360. In 1995, FAI introduced an advanced-level World Aerobatic Championship. Held every other year, it features aircraft with restricted power and performance to challenge the skills of all taking part. As well as aerobatics championships for powered aircraft, CIVA also organize similar events that cater for acrobatically rated gliders.

The challenges for pilots continued to grow alongside the technical development of aircraft. Due to the nature of the competitions, a formulaic approach was taken, one which would measure the four principles of airmanship. Not only did this help pilots prepare, much like show jumpers pacing a course, it also aided those scoring the competitions. However, the manoeuvres were often complex, and as competitions became more challenging, pilots needed total concentration. Reading from a long-worded list of manoeuvres was simply impracticable. So, in 1960, Spanish pilot José Luis de Aresti developed the Aresti Notation of recoding aerobatic moves. Using a series of lines, symbols and colours, Aresti condensed the most complex instructions into a readable format. All competitve flying is carried out in a 1,000-metre-wide aerobatic box split into X and Y axes for ease of judging.

There are also national organizations which run aerobatic activities for domestic competitions. For example, the International Aerobatic Club (IAC) is the National Aeronautic Association body responsible for aerobatics in the United States. In the United Kingdom the Royal Aero Club has designated the British Aerobatic Association (BAMA) to fill this role.

Aeromodelling

Aeromodelling was one of the first forms of flight, with examples of model aircraft dating back to the reign of the Egyptian pharaoh Ptolemy IV Philopator. It continued as an essential stage of developing aeronautical ideas, as well as being a leisure activity, with pioneers such as Blériot, Farman and Messerschmitt using model aircraft to test the operation of the innovations they were in the process of developing. The earliest aeromodelling contests began in 1905 and were intended to be as entertaining and accessible as the powered aeroplanes they depicted.

The 1930s, in particular, saw aeromodelling swiftly develop with models either free-flying or handled via a control line that allowed the operator to pitch the axis of a powered aircraft. In addition, this period saw the first international aeromodelling competitions organized in England and Germany, giving modellers a chance to show off their skills.

The 1950s saw the arrival of early radio-controlled models and the introduction of three official FAI International Commission (Comité International d'Aéromodelism, (CIAM)) aeromodelling classes: the Nordic A-2 Glider, Wakefield rubber power, and internal combustion engine power 2.5cc. 1952 saw the first world championship for all three classes at three separate locations throughout the summer months. The following year the topic of allowing the inclusion of radio-controlled (RC) aircraft was considered. The final decision was made in 1957 that radio-controlled aircraft could be involved in future FAI aeromodelling world championships and radio-controlled aircraft joined the control line as another class now added. In 1955 the first unofficial combined class world championships took place at the French Air Force base Mainz/Finthen near Wiesbaden in Germany; this class would be formailzed, offically recognised five years later.

The 1960s saw a refinement in involvement, and the first World Championships for Radio Control Aerobatics were held in Dübendorf, Switzerland. This event heralded a new dawn for aeromodelling, with the range of aircraft on display and the standard of flying captivating audiences. Contemporary reportage showed that the world championships engaged the crowds as much as the full-sized counterparts, with American Ed Kasmirski taking first place. By the decade's end, the CIAM had not only developed a pattern to organize a world championship but had also introduced safety helmets for team race pilots. It had also held its first international judges' course for radio-controlled aerobatics.

The 1970s saw continued growth, and in 1972 the first World Championships for Space Models took place, beginning of the discussion around allowing helicopter models into the championships. This happened in 1985, which also saw the CIAM adapt to technological and practical changes in the world of aeromodelling. There was also a programme of FAI recognition for other aeromodelling achievements established, and RC Electric Motor Gliders gained world championship status.

The 1990s saw the publication of the first CIAM newsletter, the *CIAM Flyer*, which helped spread the word of the commission's work, which would be involved in a great deal of networking throughout the decade. Due to various political upheavals, several championships were either cancelled or delayed, which led the CIAM to establish ties further afield, especially in South East Asia.

The new millennium saw the CIAM develop into a more inclusive media-friendly organization. In 2003, RC balloons were accepted as a new category, and the term 'aero model' was replaced by 'model aircraft'. In 2010, the first CIAM scholarship was awarded to Briton Oliver Witt, the purpose of which is to fund the development of aeronautical knowledge of pilots aged sixteen to twenty. The highlight of the 2019 CIAM plenary meeting was the adoption of drone soccer as one of the new classes, showing that the CIAM has maintained a firm understanding of the ever-changing world of aeromodelling, which is reflected in the categories it now oversees, including solar-powered aircraft. With over one million enthusiasts enjoying aeromodelling globally, it is fair to say that the CIAM is the most significant FAI discipline regarding the number of active flyers.

Ballooning

The activities of the aeronauts and early aviators of aerostats and gliders, which were still very much in their infancy, remained popular after the advent of powered flight. The First World War had shown that, despite the advances of powered flight, powered and unpowered human-crewed aerostatic aircraft remained viable methods for commercial and private enterprises.

Following the end of the First World War, the balloon initially found itself relegated from military and commercial roles as the allure of powered aircraft gripped the masses. However, the scientific community was keen to understand more about the meteorological workings of the atmosphere, so the balloon remained a useful tool. Instruments, and occasionally scientists, would ascend in specially designed capsules suspended beneath the balloon. New records were set with the Swiss physicist Auguste Piccard and Belgian colleague Max Cosyns ascending 53,153 feet (16,201 metres) on 18 August 1932. This set a challenge which continues to attract the brave with the current record altitude set by Alan Eustace at 135,889 feet (41,419 metres) on 24 October 2014.

The following year German aeronaut Alexander Dahl took the first picture of the Earth's curvature in an open hydrogen gas balloon on 31 August. However, advances in balloon technology and understanding the dynamics involved remained sketchy, and a dangerous activity. In January 1930 the *Osoaviakhim-1* fell from the sky after the Soviet-made high-altitude hydrogen balloon had reached an altitude of 72,000 feet (22,000 metres). The cause of the disaster was the loss of buoyancy by the balloon causing it to fall uncontrolled with the loss of the crew, Pavel Fedoseenko, Andrey Vasenko and Ilya Usyskin. Nevertheless, the Soviet approach to ballooning remained focused, seeking to achieve the next record or make new scientific discoveries. These were often met with disaster but proved that lessons learned would go on to save aeronauts' lives worldwide.

By the Second World War, the balloon remained little developed, with the plain fabric envelope still inflated by one of three specific types of gas. The traditional hot air type developed by the Montgolfier brothers in the eighteenth century remains popular today with commercial and hobby aeronauts. The multi-gas balloon uses the lower molecular-weight gas from the surrounding atmosphere to generate lift. Initially, this type used highly flammable hydrogen to fill dirigibles. As a result of the numerous disasters, hydrogen was replaced first by cheaper, though not as efficient, coal gas and finally by helium.

The final type of balloon is via the combined Rozière type, which utilizes heated and unheated lifting gases stored in separate gasbags to assist with the lift. This type was used most famously in the *Breitling Orbiter 3* for its non-stop circumnavigation of the world in spring 1999, the first by a balloon. Crewed by Briton Brian Jones, and Swiss explorer Bertrand Piccard, the *Breitling Orbiter 3* completed its journey nineteen days, twenty-one hours and fifty-five minutes after taking off on 1 March. Further development of inflation are the balloons inflated by solar power, an idea realized by British innovator and aeronaut Julian Nott. Nott proved the efficiency of the solar-powered balloon on 1 November 1981 by crossing the English Channel in a balloon utilizing his ideas. Nott was the first person to pilot a Super-Pressure Balloon (SPB), also known as a pumpkin, on account of its shape. The SPB is typically used for upper atmospheric scientific studies and, due to its design, maintains a steady altitude over a long distance, remaining airborne for up to 150 days. SPBs were also used in the 1985 Soviet-led Vega programme which saw SPBs float at heights of thirty-one miles (fifty kilometres) over the surface of Venus. As well as being known as pumpkins, SPBs are also known as Ultra-Long-Distance Balloons (ULDBs). Nott died from injuries sustained after the gondola he was travelling in broke free of its balloon attachments in March 2019.

After the Second World War the balloon experienced a renaissance of experimentation and adventure, helped by the development of fabric design technology and a greater understanding of weather patterns. American Ed Yost, who completely redesigned the hot air balloon concept, gave the balloon's new lease of life. Yost designed a lightweight set of propane burners attached to fuel bottles carried by the balloon. Not only did this allow the balloon to self-inflate anywhere but, more importantly, it could re-inflate in flight, extending ranges and aiding control. Other innovations by Yost include redesigning the shape of the balloon into its now familiar teardrop and adding controllable vents for manoeuvring. Yost also invented the deflating parachute system, which sits at the top of the balloon, covering a vent until drawn aside by the pilot to vent air from the balloon. Yost proved the efficacy of his design on 22 October 1960 when he took to the sky for an untethered flight lasting one hour thirty-five minutes. This proved that his newly designed burners and balloon were safe, and his innovations worked.

He went public with his new inventions and refined designs. Raven Industries, which he'd formed in 1956, sold its first civilian hot air balloon in November 1961. This was followed by a series of record-breaking adventures for Yost, including the amount of time spent aloft in a hot air balloon achieved during his attempted Atlantic crossing, which failed. This was followed two years later by the transatlantic crossing of the *Double Eagle II*, a helium balloon designed and built by Yost. This was the first west-to-east transatlantic crossing carrying pilots Ben Abruzzo, Maxie Anderson and Larry Newman. The trio took off from Presque Isle, Maine, on 17 August 1978. After a flight lasting 137 hours and six minutes, they landed safely at Misery, Normandy, France. This flight proved that the balloon and ballooning were as capable as any aircraft of achieving marvellous feats and opened the door to a host of land and water crossings, culminating in the *Breitling Orbiter 3* adventure.

Yost's work led to ballooning offering a flight experience accessible to most. Associations were established to promote ballooning as a sport, including the Balloon Federation of America (BFA) and the European Ballooning Federation. Commercially balloon rides remain an exceptionally popular activity, as is balloon racing, where competitors complete various tasks to compete. These races are fun, with competitors completing multiple activities, such as dropping a sandbag marker on a ground target and navigating. Ballooning activities are regulated through the FAI Ballooning Commission (Commission Internationale de l'Aérostation (CIA)).

Gliding
Another early form of flight was that of gliding with sailplanes. Before powered flight, there had been a series of successful experiments by aviators such as pioneer Sir George Cayley (27 December 1773–15 December 1857). Cayley had established the early principles of aerodynamics and 'downhill' flight, or gliding. The dawn of powered flight initially saw the sailplane replaced as a popular form of flight. However, many control mechanisms were applied to early aircraft, such as wing-warping and weight-shifting controls. These types of control remain in use today within the paraglider, hang glider and ultralight types of aircraft.

The 1920s witnessed a resurgence in the growth of gliding, especially in Germany, which became a world leader, developing gilding into a sport. This was partly driven by restrictions placed upon it by the Treaty of Versailles. In the United Kingdom, the *Daily Mail* offered £1,000 to the aviator who could stay aloft for three hours at a gilding event held at the Itford Hills, Sussex. The immediate post-war period saw the development of gliding as a rival sport to powered flight. The 1920s was a period of development, crowned by the invention of the variometer by gliding pioneers Augustine Haller and Wolfram Hirth, which allowed the sailplane pilot to judge whether they were flying in rising or sinking air. This allowed pilots to avoid flying around hills and mountains and utilize easier-to-access airfields. Hirth would also discover the standing wave phenomenon, which helps the sailplane pilot ascend into the stratosphere. However, to fly at this height, the pilot is required to wear breathing apparatus, but most pilots only partially use the lift supplied by the standing wave to clear obstacles such as hills and mountains.

Gliding continued to teach pilots and observers of the nature of weather and its effects on the flight. By the mid-1930s, pilots were using thermals to soar ever higher and further, with pioneers such as British pilot Philip Wills setting records, including long-distance cross-country flights up to 55 miles (88.5 kilometres). Although this was dwarfed by the thirty-six hours forty-seven-minute endurance flight of the Texaco Eaglet, which featured a custom-made fifty-foot wingspan, piloted by Captain Frank Hawks on 6 April 1930.

As gliding progressed, so did the controls and, more importantly, the instrumentation, with sailplanes increasingly well equipped to facilitate blind flying. Other developments included the refinement of wing and fuselage designs, which allowed skilled pilots to fly hundreds of miles in reasonable comfort and safety. The acme of 1930s glider design and technology was the German Minimoa with its futuristic gull-wing and enclosed cockpit. The

Minimoa featured a glide ratio performance of descent of 0.3 metres to 7.92 metres flown, the best of the time. The Second World War saw gliding take on a new role beyond sport, which gave birth to various innovations, including waterproof glues and wing design refinements.

In the immediate post-war years, the British EoN Olympia, a development of a pre-war design by German sailplane and gilding pioneer Hans Jacobs, was designed and built by furniture company Elliotts of Newbury. The design was joined by the Slingsby Sky, another British sailplane. Like its contemporary, it was an extremely successful glider and gained a fine reputation as a competition sailplane. The 1950s saw gliding develop as a sport and sailplanes develop as aircraft, with the first glass-fibre-bodied sailplane, the German Akaflieg Stuttgart fs24, or Phoenix, produced from 1957. The fibreglass body gave the sailplane its now recognizably smooth fuselage and wing surface, which helped achieve more incredible speeds, and better handling characteristics.

The next stage in the sailplane development was started by the Polish firm Szybowcowy Zakład Doświadczalny: the semi-reclined pilot's position. This new position helped reduce the cockpit size and forward shape of the sailplane and was first seen in the aircraft, the SZD-19 Zefir, which was in the air in 1958. The Zefir was followed in 1960 by the SZD-24 Foka, a high-performance aerobatic sailplane, which further refined the sailplane's design. These innovations were followed in 1964 by the Akaflieg Darmstadt D-36 Circe, which led the way towards contemporary sailplane design, including a retractable landing wheel, the addition of the now familiar T-tail and an improved glide ratio. Another vital innovation, especially for the hobby sailplane pilot, was that aircraft were being designed to be easily broken down into component parts, allowing easy transportation via a special road trailer.

The sailplane has continued to keep up with technology. The pilot of the single-instrument sailplane of the 1930s would be astounded by the often-complex instrumentation found in the cockpit of the twenty-first-century sports sailplane. However, there would also be some reassuring similarities with sailplanes launched by towed light aircraft, known as an aerotow, which can tow several gliders if the home airfield has the room or by a ground-mounted and operator-controlled drum. Once airborne and they are satisfied, they can fly safely; the pilot releases the cable and can enjoy the flight experience. Two other methods of launching were using a bungee catapult and auto-towing, which saw the sailplane pulled aloft by a wheeled vehicle. This method involved running the cable around a ground-mounted pulley and the vehicle driving towards the sailplane, using a steady speed to get the wing flying and the sailplane airborne. This particular method is now virtually unheard of, and for good reason.

Should any of these methods be unavailable, or the pilot simply wishes to have a little more power, the motor glider is a great solution. The engine is considered a Means of Propulsion (MoP) by the Fédération Aéronautique Internationale (FAI) International Gliding Commission (IGC) that gives the motor glider sustained soaring flight. The first motor glider, a Carden-Baynes Auxiliary, took off in 1935 powered by a 250cc Villiers single-cylinder, air-cooled, retractable engine placed in the rear of an Abbott-Baynes Scud 3 sailplane. The idea was devised by vehicle designer Sir John Carden, with the engine pushing the sailplane from a position by joining the fuselage and the wing's trailing edge.

Post-war development of the concept saw various types of motor gliders appear. Touring Motor Gliders (TMGs) are fitted with either full feathering or fixed propellers on a front-mounted engine that enables the aircraft to take off like a conventional aircraft. TMGs are not as efficient as full sailplanes. Still, they use the best aspects of powered flight and gliding to provide the pilot with an aeroplane more efficient than a light aircraft.

The retractable propeller means that the sailplane is launched via aerotow or cable ground and uses the engine to perform the 'saw-tooth' method of flight, where the engine is used to gain height and switched off to allow descent. Once in the descent, the motor is switched off and retracted. Other motor glider flight forms include those equipped with a sustainer engine, which helps the sailplane climb to its desired height. Initially, the engines were of two-stroke variety. Still, by the end of the first century of flight, electric motors started to appear in growing numbers. This allows the sailplane to be equipped with a smaller, lighter, nose-mounted engine featuring full feathering propellers. Known as a Front Electric Sustainer (FES), this development in electrically powered flight is joined by self-launching sailplanes equipped with electric motors. These engines were developed by electric engine specialists Lange Aviation.

Besides the TMG there are also self-launching sailplanes, which feature either a traditional internal combustion type or an electric motor mounted within the sailplane's fuselage and the propeller. This then drives a two- or single-blade propeller, which is raised and powered via a belt-driven mount.

A final method used to achieve self-launch is the jet engine. This method first used the two-seat Caproni Vizzola Calif A-21SJ sailplane, which offered pilots a 0.16 or 0.20 kN (36 or 45lbf) thrust engine, which helped the A-21SJ achieve a modest 10,000-metre (33,000-foot) ceiling. As unorthodox as it sounds, this idea is featured in more contemporary aircraft such as the Jonker JS-1 Revelation. This sailplane, first flown on 12 December 2006, features the latest aeronautical innovations, including a more flexible wing that aids a more effective dihedral angle stabilizing the light fuselage roll axis. The wings are then capped with vertical extension winglets. These increasingly common devices help reduce the aerodynamic drag associated with the airborne vortices that develop at the wingtips in flight, improving range and fuel efficiency in the case of powered aircraft. In addition, the JS-1C features a retractable MD-TJ42 turbojet. As a result, this diesel-powered jet can be used to gain height or return to the home airfield at a modest 124mph (200km/h).

The other gliding disciplines, hang gliding and paragliding, are represented by the FAI Hang Gliding & Paragliding Commission (Commission Internationale de Vol Libre (CIVL)). Hang gliding features the pilots flying a fabric-covered rigid frame steered by using their weight in opposition to a control frame. Hang gliders are capable of travelling great distances, as well as performing aerobatics.

Paragliding is a recent development that owes its origins to Canadian Domina Jalbert, who invented the ram-air airfoil in 1957. This was followed by a period of development which saw Jalbert refine the design to create the ram-air double-surfaced fully flexible airfoil in 1963 and patented as the Multi-cell Wing Type Aerial Device.

In 1961, the French engineer Pierre Lemoigne produced a radically redesigned parachute which featured removed panels at the rear and sides of the canopy. These cutouts enabled the modified parachute to be towed and steered into the air from the ground. The design was called the ParaCommander, and it generated considerable interest, especially from The National Aeronautics and Space Administration (NASA), who sought to use it as part of the safety package in use with the Gemini 6 programme. However, by the decade's end, the ParaCommander was in use as a recreational tool.

At the same time, David Barish was developing the sail wing, a single-surface wing parachute. The former USAF pilot had been working on his gliding parachute design since 1964. By 1966 he was testing his work with his sons and hang gliding expert Dan Poynter. The testing took place around Hunter Mountain, New York State. Soon Barish was offering summer slope-soaring experiences in the otherwise quiet ski resort.

In 1971 the first of Jalbert's new double-skinned parachutes was built by Theodore Hulsizer. There followed further design refinement by Jalbert and Hulsizer to enable the parachute airfoil to be inflated safely as well as increase overall control. Another designer keen to develop the idea of paragliding was Mark McCulloh, whose views revolutionized the sport. His list of achievements included a host of inventions such as the Winchboat, which was developed alongside several maritime architects, and handling techniques, including the first set of parasail operating guidelines. Over the next thirty years, McCulloh, who was an expert witness for parasailing accident investigations, developed the Winchboat to include tandem and gondola options. Other achievements would see him working with NASA and Disney, creating a series of safety devices. However, McCulloch's entrepreneurial spirit remained guided by ethics, and in 1998 he established the Parasail Safety Council (PSC), which represents the interests of the parasailing public. The PSC's mission is to raise the safety standards of commercial parasailing while focusing on education, certification and regulation.

In 1984, Jalbert's completed parafoil design was tested in China, confirming its potential; after that, equipment continued to improve, and the sport's popularity increased. As a result, the first (unofficial) Paragliding World Championship took place in Verbier, Switzerland, in 1987. This was followed in 1989 by the first officially sanctioned FAI World Paragliding Championship held in Kössen, Austria.

Further development of the paraglider is paramotoring or Powered Paragliding (PPG). Here a harness-worn motor can help the pilot get airborne in still air and on level ground, opening up opportunities. The engine does not provide an increase in airspeed: it is used to either climb or descend, with controls akin to that used on unpowered paragliders. The speed of the PPG is between fifteen and 50mph, with the motor capable of propelling the pilot up to 18,000 feet (5,487 metres). The PPG is seen as the ultimate expression of powered flight due to its low cost, slow speed and use of an aerofoil. However, their low weight makes PPGs susceptible to windy weather and turbulence. Both microlights and paramotoring are represented by the FAI Microlight and Paramotor Commission (Commission Internationale de Microaviation (CIMA)).

Another unpowered method of air travel is skydiving which the FAI International Skydiving Commission (ISC) represents. The first recorded skydiving incident was in 852 BCE, when Abbas ibn Firnas, a Berber polymath, attempted to fly. Wearing a cloak, Firnas leapt from a tower in Córdoba, Spain, and, instead of flying, fell. Thankfully there were folds in his cloak to soften his fall enough for him not to be too severely injured. Over the coming millennia, there followed numerous experiments. By the turn of the twentieth century parachutists were often performing daring leaps from tethered balloons to the amazement of the gathered crowds. One such performer was Dolly Shepherd, who would ascend to 2,000 feet (610 metres) before leaping from a bar and free-falling for 200 feet (61 metres) while her parachute inflated. This was not a sport for the faint-hearted, accidents often occurred, with fatal consequences. On one occasion Shepherd leapt to the rescue of one such parachutist on 9 June 1908, saving them both but paralysing herself in the process. Despite medical opinion, the inspiring Shepherd rehabilitated herself and was parachuting two months later.

The First World War saw the parachute come of age as a vital tool for aviator safety. Innovations led to the Type-A parachute, which featured a rip-cord, and a pilot chute, which drew the main chute from the individual's backpack. Development has continued, and today, there are two distinct types of parachute design: ascending and descending canopies. Ascending canopies are used in paragliders, designed to stay aloft for as long as possible, with all other types considered descending types. Numerous parachutes are now used, from those designed for aerobatic work to wingsuit flying and BASE jumping where participants jump from buildings, antennas, spans (bridges) and earth (clifftops).

Pushing the Envelope

The history of aviation has been one of constant experimentation. The 1919 Paris Convention introduced airworthiness certificates and competency to manufacturers and aviators. This was vital to ensure the safety of passengers, crew and bystanders. In a post-war world, converting military aircraft was essential to help establish a vibrant aviation scheme.

Far from stifling innovation, the convention kick-started the imagination. As the principal manufacturers began to experiment, so did the enthusiastic amateurs. For some, the idea of simply building their own aircraft was spur enough. In contrast, others were content to assemble an aircraft from a kit. The middle ground was filled by those who were experienced enough to build from plans but not confident or competent enough to build from scratch. Added to this mix were the experimental aircraft designers and builders.

Brazilian Alberto Santos-Dumont was one of the early pioneers who offered plans for his Demoiselle high-wing monoplane accessible to readers of *Popular Mechanics* in June 1910. The First World War saw many experience flight and gain a rudimentary understanding of its mechanics. While some could finance their aviation dreams in the immediate post-war years, taking advantage of the numerous and cheap surplus aircraft, others wanted to test their newly learned skills.

In the early 1920s, American barnstormer and circus performer Orland Corben, who had learned to fly at fourteen in a war-surplus Curtiss JN-4 Jenny, designed an aircraft for the Ace Aircraft Manufacturing Company. The first of three models was the single-seat Baby Ace, an aeroplane that could be built at home from either a plan or later by assembling kits. The Baby Ace featured a parasol wing and was made primarily from wood and fabric, with a small amount of welding for the mounts of the modified motorcycle engine. Corben's approach was purely egalitarian, and his desire was to make flying accessible to all.

Despite needing the time and space to build the aeroplane, which featured a fuselage length of eighteen feet (5.5 metres) and a wingspan of twenty-six feet (eight metres), Corben's design proved popular. The Baby Ace was followed by a two-seat version featuring a tandem arrangement, called the Junior Ace and a single-seater powered by a Ford Model A automotive engine. The development of further models by Corben was hampered by the Great Depression and changes in state and federal laws that banned homebuilding and flight in uncertified designs by 1938. It was not until 1948 that experimental aircraft, including the Baby Ace, were allowed to be constructed and flown once again.

Such is the popularity of this aeroplane that updated kits and plans remain available from Acro Sport for all three aircraft, nearly a century after Corben's initial design first flew.

Corben's Ace family of aircraft was followed by numerous other kits covering heavier- and lighter-than-air aircraft. Again, the imagination of thousands of hobbyists, builders and designers was caught. Unfortunately, while many of these aircraft were safe, one was so bad that it ended up being banned by two governments due to its appalling safety record.

The Flying Flea family of monoplane aircraft was designed by Frenchman Henri Mignet. Mignet had an early interest in aviation and produced several prototypes between 1920 and 1928, including the M.8 Avionnette. This featured as part of a self-produced book *Comment j'ai construit mon avionnette* (*How I built my avionnette*), which contained the plans and instructions for making the parasol-winged aeroplane. Despite encouraging others to fly, Mignet found controlling flight awkward. His lack of confidence in his coordination drove him to seek a solution to provide an easy-to-fly aircraft. In 1929, he began refining his ideas, and in 1933 developed the HM.14. The plans for the HM.14 were released for publication in 1934; hundreds of amateur builders bought copies. The HM.14 was a unique aircraft which employed a tandem-wing design, which placed the branches staggered, one behind the other, with both contributing to lift. Unfortunately, the tandem wing also does away with the elevators. As a result, the foot controls were absent, with all controls handled via the stick. The materials used in its construction were the same as many of the early self-build aircraft: wood, metal, linen and enthusiasm. While seemingly revolutionary, the HM.14's design was unstable in certain situations. By 1936, flights in the HM.14 had been grounded by four nations due to a series of fatalities. Subsequent wind-tunnel tests by French and British authorities found that control could be lost in an unrecoverable shallow nose-down attitude. This could be worsened by incorrect adjustment of the centre of gravity. Despite design changes, including a pivoting rear wing, the HM.14 remained grounded in the United Kingdom due to flight-safety concerns.

The Second World War saw private and commercial civil aviation grounded in most countries. Still, the post-war years witnessed a resurgence in home-built aircraft. The Stits SA-3A Playboy was unveiled in the United States by designer Ray Stits. Stits had served as a mechanic during the war and had a prolific design portfolio of fifteen aircraft. The single-seat, strut-braced, low-wing SA-3A was a success for Stits, who had gone from design to prototype in a mere twelve weeks. Stits also designed and built the remarkable-looking experimental SA-2A Sky Baby biplane, which was created to claim the world's smallest aircraft title. The SA-2A could only be flown by a pilot weighing 77kg (170lb) to maintain the aircraft's centre of gravity and was very much a novelty project. Despite this, the SA-2A would gain 25 hours of flight time. Stits would remain influential in the home-build and experimental aircraft circuit, becoming the Experimental Aircraft Association (EAA) founder in 1953, alongside fellow aircraft designer and veteran aviator Paul Poberezny.

In the United Kingdom, the low-wing Taylor J.T.1 and the J.T.2 Titch monoplanes designed by John Taylor were intended to be built with ease in around 2,000 hours at low cost and by someone with reasonable skills. The J.T.1 owes more in appearance to the sports tourers of the 1930s, while the J.T.2 is a sportier-looking aeroplane, with the pilot enjoying an enclosed cockpit. Both remain popular, with plans still available for the adventurous aviator. Like the US-based EAA, a similar body existed in the United Kingdom, founded in 1946 and known as the Ultra-Light Aircraft Association (ULAA). The remit of the ULAA was to issue a flying permit for aircraft not exceeding 1,000lb (453kg) in weight, carrying an engine generating no more than 40hp and having a landing speed of no more than 40mph. ULAA inspectors would ensure all stages of construction, including the final airworthiness inspection, were mechanically sound before issuing the permit to fly. In 1949 the ULAA became the Popular Flying Association (PFA), which changed its name in 2011 to the Light Aircraft Association (LAA). The purpose of the LAA remains the same as its predecessors but also encourages flying activities and the operation of classic and vintage factory-built aircraft that can no longer hold a full certificate of airworthiness.

Slowly more contemporary designs began to emerge, with materials used in their construction becoming more sophisticated. Home-built aircraft and kits began to include aluminium skins, shaped composite elements and the development of poly-fibre vinyl to replace linen. In 1973, one of the most influential home-build aircraft design and kit manufacturers was formed by designer Richard VanGrunsven, Van's Aircraft. All aircraft produced feature a monocoque aluminium skin to create a stressed skin covering over formers. This approach removes the need for stringers, saving money, aircraft weight and construction time. Van's aircraft are considered Experimental Amateur-Built (EAB) by several nations, including the United Kingdom and New Zealand. The Van's low-wing monoplane aircraft range includes the RV-three single-seat aerobatic aeroplane and the tandem-seat RV-14, which can operate as a training and aerobatic aircraft. The larger twin-seat aircraft of the Van's range are considered Experimental Light Sport Aircraft (ELSA) or particular Light-Sport Aircraft (SLSA). With over 10,000 aircraft made from kits and plans and a host of aftermarket spares and support, including construction classes, VanGrunsven, would go on to form the Aircraft Kit Industry Association (AKIA) in 2012. All of these came together to provide a one-stop shop for the home builder.

In 1971 the home-build market entered the turbojet age, with the diminutive

Bede Aviation BD-5J Micro, designed by James Bede, taking to the skies. The Microturbo TRS-18 powered BD-5J was sold as fibreglass and aluminium kits and would go on to feature in the 1983 Bond film *Octopussy*. It also holds the record for being the world's smallest jet aircraft.

Around the same time, pilot and aerospace engineer, Burt Rutan began designing and experimenting with his unique-looking aircraft and the materials to construct them. The first of his experiments was the VariViggen, which had a rear wing and forward canard with the engine in a pusher configuration. The unorthodox-looking VariViggen was far from unique as a forward canard aeroplane, especially as the 1902 Wright Flyer was constructed with a forward canard. However, it did prove that the concept was viable as a home-build option. In 1975 the prototype VariEze home-build aeroplane made its maiden flight. This was followed by more designs, including the Rutan model 68 Amsoil Racer. Rutan innovations, especially composite construction materials, would find their way into mainstream aircraft production. Rutan and brother Dick would design the Rutan Model 76 Voyager, which would become the first aircraft to fly around the world without stopping or refuelling, in 1986. Voyager would be followed in 2005 by the Virgin Atlantic GlobalFlyer, flown by Steve Fossett; he became the first pilot to complete a solo, non-stop, non-refuelled flight around the world in February 2006.

Rutan also developed sub-orbital spaceflight technology, completing a series of designs. In 2004, Rutan's sub-orbital spaceplane design SpaceShipOne became the first privately funded spacecraft to enter space, winning the Ansari X-Prize that year for achieving the feat twice within two weeks. This was followed by a series of projects with Richard Branson and Virgin Galactic, which saw the development of SpaceShipTwo and the White Knight One and Two carriers in 2008. This marked a new dawn in the evolution of civil aviation, with recent competitions and challenges such as the Ansari X-Prize pushing designers, investors and manufacturers to push the envelope of air travel once more.

Even with some reaching for the stars, the realm of home-building aircraft remains as strong as ever, with several bodies supporting those wishing to go it alone. Another such body is the FAI Amateur-Built Aircraft Commission (Commission Internationale des Amateurs Constructeurs d'Aéronefs (CIACA)). The CIACA has several functions under its remit: to promote the design, construction and operation of amateur and home-built aircraft and the restoration of vintage aircraft.

Restoration of vintage aircraft by organizations such as museums and specialist companies or individuals, either as static displays or as airworthy types, remains a vital part of civil aviation. Not only does it help preserve the physicality of these old aircraft, but it ensures the engineering skills required to build and maintain them are kept alive. For many aviators and enthusiasts, the opportunity to fly a historically significant aircraft is often a dream come true. There is also the possibility of flying full and scaled-down replicas of famous aircraft. These can either be home built or as kits, with companies such as Kip Aero and Airdrome Aeroplanes offering the experienced home builder a range of aircraft types, including the Blériot XI Monoplane. Some companies also provide aviators with the ultimate in custom-build options of old warbirds, such as the German aviation company Scalewings which offers a P-51 scaled replica, the SW-51. This 70 percent true-to-scale aircraft features contemporary design and build techniques, as well as the opportunity for the buyer to configure their purchase in various finishes. Scalewings also offers the SW+51 as a 51 percent complete aircraft ensuring registration as a home-built aircraft, but with the critical build areas professionally finished.

Human-Powered Aircraft (HPA) are also covered by the CIACA, with prizes for development in the field offered by the Royal Aeronautical Society (RAeS). Known as the Kremer Prizes, established in 1959, these are a series of challenges set and inspired by the industrialist Henry Kremer to break speed and distance records. Another prize is the Robert Graham Competition for students involved in experimental research or engineering design. In addition, the RAeS established the Icarus Cup 2012, to stimulate the development of human-powered flight into a popular sport.

Space exploration is a further development of civil aviation, with its roots in the 1920s, and large-scale rocket programmes in Germany by Fritz von Opel and Max Valier. Their work was influenced by the Russian rocket scientist Konstantin Tsiolkovsky, who in turn influenced Wernher von Braun. The space race between the Soviets and the United States accelerated in the years after the Second World War. Both sides used German wartime experimentation and scientists to help build their space programmes. The Soviet space programme was the first to achieve the critical milestones of getting the first man, Yuri Gagarin, and woman, Valentina Tereshkova, into space on 12 April 1961 and 16 June 1963 respectively. National Aeronautics and Space Administration (NASA), founded on 29 July 1958, was startled by how quickly the Soviets had achieved this and the race was on to make the first human-crewed moon landing. This was achieved on 20 July 1969 when Apollo 11 astronauts Neil Armstrong and Buzz Aldrin became the first men to set foot on the moon. The formation of NASA was followed on 13 December 1958 by the establishment of the United Nations Office for Outer Space Affairs (UNOOSA), responsible for ensuring the orderly conduct of activities in outer space, based on the Outer Space Treaty of 1967. This treaty provided the freedom of exploration and use of space for the benefit and interest of all, the prohibition of the deployment of weapons of mass destruction in outer space, and the non-appropriation of outer space, including celestial bodies. The Outer Space Treaty was followed by four further treaties which included the most important, the Rescue

Agreement of 1968, which requires member states to assist an astronaut in case of accident, distress, emergency or unintended landing.

The Apollo 11 landing achieved the ultimate goal of the space race. After that, the theme increasingly became cooperation between all parties. The 1970s and 1980s were a gradual step away from military-trained crews to specialist civil astronauts marking a change in how space was used. The Soviets launched the Salyut programme in 1971 which established four crewed scientific research space stations, known as long-duration orbital stations, abbreviated to DOS. These were joined by two-crewed Almaz military reconnaissance space stations, all launched in a period of fifteen years, from 1971 to 1986. The Salyut programme achieved numerous records, including various spacewalk records. Aside from the apparent military reconnaissance missions of the Almaz space stations, the DOS space stations focused on astronomical, biological and Earth resources experiments. Both were replaced by the modular *Mir* low-Earth orbit space station, which operated from 1986 to 2001 and was replaced by the International Space Station (ISS).

The collapse and fragmentation of the Soviet Union in the late 1980s and early 1990s saw a new agency formed in Russia on 25 February 1992, the State Space Corporation, known as Roscosmos. There was a further restructuring in 2004, with Roscosmos becoming the Russian Aviation and Space Agency and Federal Space Agency (Roscosmos). A final merger in 2015 between the Federal Space Agency (Roscosmos) and the state-owned United Rocket and Space Corporation created Roscosmos as a national body. In the post-Soviet years, Roscosmos struggled with a lack of state financing in the chaotic financial period that immediately followed the collapse of the USSR. However, 31 October 2000 witnessed a new period of cooperation with the ISS as crews began to leave the Baikonur Cosmodrome in Kazakhstan on a Soyuz launch vehicle, a practice that continues to this day. Other projects included developing and managing the Proton series of heavy booster rockets used primarily for delivering commercial loads, including satellites and developing the Angara range of launch vehicles.

Around the same time, NASA began its space station mission with Skylab. Skylab was to act as an orbital and observational platform for a range of scientific experiments. Unlike the fifteen-year lifespan of the Salyut programme, Skylab operated in orbit between May 1973 and February 1974 before lying unmanned and slowly falling into a decaying orbit which ended with re-entry and destruction on 11 July 1979. This would be the last time that NASA would use a space station as a platform for experimentation until the ISS launch in 1998. Until then, NASA moved on to use the reusable low Earth orbital Space Transportation System (STS) known as the Space Shuttle programme. Starting in 1972 and running until 2011, the Space Shuttle gave NASA an edge in space exploration. Over its lifetime, the programme launched its five shuttles on 135 missions between 21 April 1981 and 8 July 2011. Fourteen astronauts lost their lives in the *Challenger* and *Columbia* disasters which occurred on 28 January 1986 and 1 February 2003. The *Columbia* disaster led to the establishment of the Constellation programme, which sought to complete the ISS and see a crewed return to the moon by 2020 using the Ares launch vehicles. Fiscal constraints led to the programme being replaced by the Space Launch System (SLS) in 2011. The super heavy-lift expendable launch vehicle co-developed by NASA with Boeing and Northrop Grumman would not enter service until the latter part of 2022 as part of the new Artemis programme. Furthermore, it is hoped that the SLS, designed to return man to the moon by 2024, will use elements designed by external partners such as SpaceX and the European Space Agency (ESA).

The ESA, founded on 30 May 1975, covers the space exploration interests of twenty-two nations and operates the Centre Spatial Guyanais (CSG) spaceport northwest of Kourou in French Guiana. Built in 1964 as the spaceport of France, CSG has become home to the famous Ariane series of heavy-lift space launch vehicles that can be used for various tasks. These can include satellite delivery to either Geostationary Transfer Orbits (GTO) or Low Earth Orbits (LEO). The placing of the CSG is also essential as it sits approximately 310 miles (500 kilometres) north of the equator, at a latitude of five degrees, which is the best location for launching a space vehicle. Launching as closely to the equator as possible, any launch vehicle can take full advantage of the land moving at about 1,035mph (1,670km/h), adding almost 310.7mph (500km/h) to the vehicle post-launch. In 2011, CSG was the site for the first launch of Russian-built Soyuz-2 rockets from a dedicated site known as the Ensemble de Lancement Soyouz (ELS). This marked a new era of cooperation between commercial rivals for the benefit of all.

The most significant expression of cooperation in space exploration came in 1998 with the launch of the modular ISS. Developed by the United States, Canada, Russia, Japan and the ESA, this multinational space station serves as a microgravity and space environment and scientific research laboratory. The ISS evolved from the NASA Space Station Freedom project of the early 1980s and combined the knowledge and experiences of operating the early space stations to benefit all. Crewed by up to seven astronauts, this unique space station is split into two pieces. Russia's six-module Russian Orbital Segment (ROS) handles the guidance, including orbital station keeping, navigation, and propulsion control systems for the entire station. The eleven-module United States Orbital Segment (USOS) is run by the United States and other partner states. It is responsible for operational control of the gyroscopes to provide attitude control to control the station's orientation. Power from the USOS solar arrays is also transferred to the ROS to supplement their power needs. In addition, both segments are responsible for generating oxygen and scrubbing the carbon dioxide generated in the ISS interior.

This new age of cooperative space exploration also saw an increase in civil aeronautical businesses stepping up to work with space agencies and begin their own work. In the new millennium, established names like Northrop Grumman and Arianespace were joined by module maker Bigelow Aerospace and Space Exploration Technologies Corp (SpaceX), founded in 2002 by South African-born entrepreneur Elon Musk. Musk would go on to develop a range of spacecraft, including the privately financed super-heavy *Starship* reusable interplanetary spacecraft. Musk had joined an exclusive club of billionaire space explorers, including Richard Branson and Jeff Bezos (who had established Blue Origin in 2000).

With the advent of crewed space travel, the inevitable prize-giving and record-breaking came thick and fast. The FAI established the Astronautic Records Commission (International Astronautic Records Commission (ICARE)) to award astronautic medals and diplomas and keep records of human spaceflight. For ease of management, the FAI defined the limit between Earth's atmosphere and outer space at sixty-two miles or 330,000 feet (100 kilometres) above sea level. The line became known as the Kármán line, named after the aeronautics and astronautics engineer and physicist Theodore von Kármán. ICARE has also created a complete set of definitions and computations to help individuals and organizations establish whether their efforts have broken a record and define the fields in which absolute records can be awarded. There are other awards available, including the Gold Space Medal, established in 1963 and awarded to those who have contributed significantly to the development of astronautics through their activities, achievements, initiative or devotion to the cause of space exploration. The Vladimir Komarov Diploma was established in 1970 to honour the memory of Cosmonaut Vladimir Komarov who lost his life while undertaking a cosmic flight. It may be awarded annually to astronauts and members of multi-seater crews for their outstanding achievements in the field of exploration of outer space in the previous year.

Chapter 9

Icarus Has Fallen:
War, Disaster and Terror

The world of flight is not without its risks and dangers, which can come from various natural and manufactured sources. The early years of regulated flight saw the steady demise of the giant transatlantic airship crossings. These aircraft were replaced by more reliable and safer aircraft, which were allied to international conventions and governance from organizations such as the ICAO.

Despite the increasing sophistication and development of aviation, aeronautics and training, the perils of flight have remained. Flight remains subject to the forces of a great many unknowns. Regardless of the range of inventions and interventions that have developed over the years to promote and ensure safe air travel as far as possible, they occasionally fail in overcoming the human factor. Moreover, not all disasters have been human in origin; mechanical disasters have also played their part with aircraft disappearing over water and land.

Birds at War

While most civil aviation activities were grounded during times of war, there were still occasions where aviators were active in a variety of supporting roles. Even in a non-combatant role, civil aviation remains vulnerable to attack by the military forces of hostile nations, with numerous reports of civil aircraft being attacked and downed, despite clearly being in commercial service. Such attacks were, somewhat surprisingly, not subject to an explicit treaty law which covers the subject of firing on civilian aircraft. Although such acts are now considered both a crime against humanity and a war crime, especially when viewed alongside Article 3 of the Universal Declaration of Human Rights (UDHR).

The Second World War saw wide-scale reductions in civilian aviation operations, first in the European theatres of operations, followed by the Pacific theatre of operations. Initially, the German flag carrier Deutsche Luft Hansa (DLH) continued to operate flights to friendly nations, albeit under Luftwaffe's command and control, maintaining national short-haul routes. But as the war continued, the availability of aircraft and staff was reduced, especially from 1942, as the tide was slowly turning against the Axis forces in Europe, the Mediterranean and North Africa. DLH was joined by the Italian national carrier Ala Littoria S.A. in providing air services to what had become the fifth largest air route of the time in 1940. As Italy's fortune in the war changed, Ala Littoria became increasingly involved in her war effort before disbanding.

In the Far East, Japan Airways Co. Ltd had supported the Second Sino-Japanese War since 1937 as Japan Air Transport (JAT), providing charter services for the Imperial Japanese military. In December 1938 the Japanese government bought a fifty percent share in JAT, renaming it Dai Nippon Kōkū (DNK). The airline continued to serve as a charter airline to the military. It offered flights around the west and central Pacific areas. In 1942, DNK became wholly owned by the government, and its fleet and services were split between the army and navy. Despite heavy wartime losses, DNK continued to operate until its eventual disbanding in October 1945. Japanese civil aviation would remain dormant until 1951 with the founding of Japan Air Lines (JAL). Up until that point, post-war air international services had been provided by Pan Am and Northwest Orient Airlines (NWA).

In Europe, most civil aviation activities were halted with the declaration of war, and airlines such as Polskie Linie Lotnicze LOT (LOT Polish Airlines) were evacuated to safe locations. In addition, airlines were merged once the initial disturbance of war had settled, as in the case of British Overseas Airways Corporation (BOAC). BOAC would have a busy war ferrying diplomats between continents and running the gauntlet of Axis air defences in Northern Europe in de Havilland Mosquitoes carrying vital ball bearings from neutral Sweden to the United Kingdom. Other roles included maintaining the vital Poole, UK, to New York via Montréal route, and taking important military, political and diplomatic personalities to North America in its Short Empire flying boats. Maintaining this route would be the key to the success of the Atlantic Ferry Organization of the Ministry of Aircraft Production (MAP (UK)), which had begun operating the North Atlantic return ferry service in May 1941. BOAC would take over the returning of third-country aircrews from MAP later that September. This service was joined by Montréal Trans-Canada Air Lines (TCA) that operated the Canadian Government Trans-Atlantic Air Service (CGTAS) between 1943 and 1947. This service provided transatlantic military passenger and postal delivery services using the Avro Lancaster-based Model 691 Lancastrian aeroplane. The Lancastrian could carry nine passengers a maximum of 4,100 miles (6,600 kilometres) and took an average of thirteen hours to make the flight. The service also helped establish regular commercial transatlantic crossings by landplanes.

BOAC operated three former Pan Am Boeing 314 Clippers alongside their Short C-Class flying boats on the early transatlantic flights. Boeing Clippers also flew to the Middle East. BOAC airlifted 469 military personnel to the safety of Alexandria, Egypt, from the island of Crete in April and May 1941. In addition, BOAC flew civilianized Consolidated Liberators, among other aircraft, to Africa via Lisbon, Portugal.

There would be attacks against BOAC aircraft by the Luftwaffe. One notable loss on 1 June 1942 would be a former Koninklijke Luchtvaart Maatschappij (KLM) DC-3 attacked in the Bay of Biscay. The scheduled flight, operated by a Dutch crew, took off from Lisbon headed for Bristol and was attacked by eight Junkers Ju 88 C-6 maritime fighter aircraft. The attack was saddening because it occurred in an area considered safe for air passage by civilian aircraft. The attack on the DC-3 saw the loss of seventeen lives, including that of British actor Lesley Howard, who was returning from a diplomatic mission. After returning to base, the Luftwaffe pilots were notified of the flight's nature leading to understandable shock. The fact that the Luftwaffe pilots had not been informed of the aircraft's identity, despite being camouflaged, only added to the tragedy.

A Royal Air Force (RAF) Short Sunderland dispatched to the crash site was also attacked by eight Ju 88s. However, its crew managed to fight the attackers off, downing three and causing significant damage to three others. After that, flights from Lisbon were re-routed and only flown in darkness. First, Oberleutnant Herbert Hintze, the *Staffelkapitän* of Bordeaux-based 14 Staffel, V./Kampfgeschwader 40, which was responsible for the downing of Howard's flight, would later state that had his pilots known the nature of the flight then the DC-3 would have been forced to land at Bordeaux and its passengers interned. Howard's obituary would be published in *The Times*, in the same edition that reported the death of Major William Martin, whose persona was used by Allied intelligence to mislead their Axis counterparts regarding the 1943 invasion of Sicily. Between BOAC's foundation in 1940 and the war's end, the airline flew approximately 19,000,000 miles (30,577,536 kilometres) a year, performing a vital wartime role as a merchant air service.

Another critical development in British civil aviation during the war was the Air Transport Auxiliary (ATA), founded on 15 February 1940 and staffed by civilian pilots. The ATA's essential tasks would be the conveyance of new, repaired and damaged military aircraft between factories and airfields, maintenance units (MUs), and occasionally scrapyards. The ATA was also responsible for maintaining the transatlantic delivery points and ferrying service personnel, including the injured. The ATA was organized into pools; at its height, it was operating fourteen such pools across the United Kingdom, flying any one of 147 aircraft types. By war's end, the ATA, led and administered by Commodore Gerard d'Erlanger, a former director of British Airways Ltd, had delivered more than 309,000 aircraft and accumulated an impressive 415,000 flying hours.

Pilots were recruited by their ability to fly, which led to pilots who had been turned down by the authorities on the grounds of age, gender or physical fitness being recruited. The ATA developed a unique talent pool of male and female pilots, often with disabilities and occasionally of a particular vintage. The ATA also recruited pilots from outside the United Kingdom, with twenty-eight nations represented. The ATA had adopted the sobriquet of the 'Ancient and Tattered Airmen', which sat alongside the unofficial motto of 'Anything to Anywhere'. The ATA attracted much press coverage due to its use of women, of which 166 served with the service. These included Amy Johnson, who would lose her life in a flying accident and the American Jacqueline Cochran, who would return to the United States to establish the Women Airforce Service Pilots (WASP). One hundred and seventy-four ATA pilots lost their lives during the war.

Air France had a varied wartime experience, especially after the 1940 Franco-German armistice. Air France operated three services until the end of the war: Occupied Air France, Free Air France – also known as the Civil Air Liaison Services, which fell under the authority of the Free French Government – and the Lignes Aériennes Militaires, the transport branch of Free Air France, which was managed by Lionel de Marmier. This service ran alongside the Civil Air Liaison Services. In French North African territories Air France continued to run a reduced operation while Aeromaritime, which covered the West African air routes, ceased operations. The two airlines were merged in 1944 as part of the Réseau des Lignes Aériennes Françaises (Network of French Military Air Lines), which was managed by de Marmier. All services would return to the Air France brand upon nationalization on 26 June 1945.

Aeroflot had grown a sizeable fleet and airway network in the east, closing the 1930s as the world's largest airline. In January 1941, training pilot squadrons for the Red Army Air Force (Voyenno-Vozdushnyye Sily (VVS)) were established. These were run by the regional administrative boards of civil aviation. Following the Axis invasion of the USSR in June 1941, the Civil Air Fleet (CAF) was subordinated to the People's Defence Commissariat (Narodnyy komissariat oborony Sovetskovo Soyuza). This would see the establishment of six CAF special groups in the regions most threatened by invading forces: North, Baltic, Byelorussian, Kyiv, South-Western, and Moscow. Three squadrons were also established to provide air support for Soviet Naval (Voyenno-morskoy flot SSSR (VMF)) units: North, Baltic and the Black Sea. The first CAF to be given a dedicated role was the Moscow CAF special group, tasked with daily supply flight of vital war material to Leningrad.

In 1942 the CAF was developed further, with several independent regiments raised and organized, becoming frontline air troops. This was followed by establishing the Krasnoyarsk–Uelkal air route on 30 January 1943. This east-to-west air route would help to connect the USSR to continental North America via Alaska, Canada and the northwestern United States. As Soviet forces reclaimed territory lost in the Axis advances of 1941 and 1942, Aeroflot began to rebuild damaged and destroyed airfields and infrastructure in preparation for the return of peacetime civil aviation operations. This was notable as it signalled the intent to return to operating the domestic network as soon as possible, once hostilities had ceased. 9 May 1945 saw the end of the Great Patriotic War. In recognition of the CAF's significant contribution, Aeroflot's six front sub-divisions were reorganized into guard units, ten were awarded orders and twelve received honorary titles. In addition, honours, orders and medals were awarded to 12,000 CAF aviators for their efforts and bravery.

Qantas continued to cover British Empire interests in the Pacific. Alongside BOAC, it established what would become known as the Horseshoe Route between Sydney and Durban via India, Singapore and Egypt. This route, which ran between June 1940 and March 1942, enabled the passage of mail carried by ship from the United Kingdom to South Africa. After the fall of Singapore, the route was shortened from Durban to Calcutta. On 3 March 1942, twenty-two aircraft, including fifteen flying boats that had serviced the Horseshoe Route, were among a mixed group of military and civilian aircraft attacked at the Australian port of Broome. The attack by aircraft of the Imperial Japanese Navy Air Service killed some eighty-eight civilians, including four Dutch refugees who had arrived on a KLM DC-3.

Imperial Japan showed little regard for respecting national neutrality and civilian lives. This was proven by the Kweilin incident, which took place on 28 August 1938, in which a KLM DC-2 jointly operated by China National Aviation Corporation (CNAC) and Pan American World Airways was attacked by Imperial Japanese aircraft. The attack forced down the aircraft, with the loss of eighteen lives, and hinted at how ruthless Japanese military pilots could be. As the war progressed, the Imperial Japan air forces attacked and destroyed the DC-2 that had survived the Kweilin incident, on 29 October 1940, with the loss of nine lives. Finally, on 30 January 1942, a Qantas Short S.30 Empire flying boat was attacked off the coast of West Timor, Dutch East Indies and shot down by the Imperial Japanese Naval Air Service. Thirteen lives were lost. However, three survivors managed to swim to shore.

Despite these attacks, Qantas returned to flying on 10 July 1943, using Consolidated PBY Catalina flying boats between Swan River, Perth, and Koggala Lake in southern Ceylon (now Sri Lanka). These flights linked up with BOAC from London and were an extraordinary feat of airmanship, with the crew and three passengers flying an astonishing 4,027 miles (6,482 kilometres). This made the flight the longest in commercial service, taking an average of twenty-eight hours. This marathon flight was completed 271 times between June 1943 and July 1945. Passengers completing this journey received a certificate of The Rare and Secret Order of the Double

Sunrise because they witnessed two sunrises due to the duration of the flight. In 1944 the PBYs were joined by Consolidated B-24 Liberators, which also flew an extended route to Karachi, then in India via Ratmalana, Ceylon. This new route brought its own award for passengers: The Elevated Order of the Longest Hop. In addition, the Liberators were the first Qantas aircraft to carry the well-known flying kangaroo logo.

New Zealand airline Tasman Empire Airways Limited (TEAL) did not start operations with its two Short S.30 Empire Flying Boats until April 1940. The weekly 1,340-mile (2,157-kilometre) route between Auckland and Sydney was part of the Empire Air Mail Scheme (EAMS) which was cut off by Italy's entry into the war against the British Empire. Slowly the service was built up, linking up with Pan Am's transpacific flights until Imperial Japan entered the war. By 1944 TEAL was providing a thrice-weekly service between Auckland and Sydney.

The continent of North America, except for the Japanese Fu-Go balloon bomb campaign of 1945 and the Aleutian Islands campaign, escaped serious harm during the Second World War. It was also the only region that saw a growth of civil aviation air routes, with the United States taking full advantage of growing its civil air network to support the nation's war effort. For the major airlines, the pre-war purchases of DC-3s were well timed, as these would be used to provide services to the government. At the same time, the engineering growth of the aeronautical industry would benefit aviation beyond its wildest dreams in the post-war world. However, regardless of the air route growth, the opportunities to develop new civil airlines were non-existent as all eyes turned to win the war. After all, the Allies would be fighting a war on numerous fronts from December 1941.

There was the inevitable loss of staff, especially aircrew, to various domestic training programmes. These included continued support of the Civilian Pilot Training Program (CPTP) and the British Flying Training School Programme, established under the 1941 Lend-Lease Bill. The CPTP would be split into the War Training Service (WTS), or Civil Aeronautics Authority (CAA) War Training Service, acting as a screening service for potential pilots between 1942 and 1944. By the time of the programme's closures, a remarkable 435,165 pilots had been trained. Former barnstormers and air racers such as Opal Kunz taught many of these pilots. In addition, Kunz established the women-only Betsy Ross Air Corps in 1931, intended to support the US Army Air Corps in times of emergency. The Betsy Ross Air Corps also trained women to build a reserve that could be called upon at a moment's notice.

Another service that emerged during the United States wartime experience was the Women's Flying Training Detachment (WFTD), founded by record-breaking Jacqueline Cochran and approved by General Henry Arnold on 15 September 1942. The purpose of the WFTD was to take over male ferry pilots' tasks, freeing them up for flying offensive operations. Entirely civilian in nature, the women of the WFTD still had to pass an exacting twenty-three-week course which helped students build a portfolio of 115 hours of flight time. However, the course was soon extended to thirty weeks, and flying time increased to 230 hours. To apply, the student needed to be between eighteen and thirty-five years of age and have a minimum of 200 hours of flying time, although this later dropped to thirty-five as the need for pilots grew. Many of the pilots trained by the WFTD would serve in the Women's Auxiliary Ferrying Squadron (WAFS), founded by Nancy Harkness Love. Like Cochran, Love was an experienced pilot and had been flying since she was sixteen, working as a test pilot and air racer. As part of her duties, Love ensured she could fly any aircraft her subordinates were expected to fly.

On 5 August 1943, the WFTD and the WAFS were combined and renamed the Women Airforce Service Pilots (WASP). At this point Love was in command of four squadrons and remained in charge of all WASP ferrying operations. Meanwhile, Cochran remained in control of selection and training and was made overall organization director. This put immense socio-political pressure on Cochran, and the politics of segregation meant that African American candidates were unsuccessful in their application to join the WASP. One potential candidate was Janet Bragg, who in 1942 became the first African American woman to hold a commercial pilot licence. Bragg would go on to qualify through the CPTP. Regardless of the racial barriers, Ola Mildred Rexroat was successful in being accepted by the WASP and, in doing so, became the only known Native American woman to complete the training. Rexroat, an Oglala Sioux, would tow targets for air gunnery students. After the disbanding of the WASP, Rexroat would transfer to the air force where she worked as an air traffic controller, a trade she maintained after service.

During its short life, the WASP profoundly affected the war effort, freeing up 900 male pilots. However, of the 25,000 applicants to the programme, only 1,074 graduated to earn their wings. Between September 1942 and December 1944, all three organizations delivered 12,652 aircraft of seventy-eight different types, including the B-29. Since the WASPs were employed as federal civil service employees, they did not qualify for military benefits. However, they would finally be awarded recognition as a military unit and all the benefits that it entailed in 1979.

Soon the national airlines were providing services overseas. Pan Am crews drew on their pre-war experiences of providing long-haul flights in unique ways. The Japanese attack on Pearl Harbor also bombed the Pan Am shoreline base, leaving a single Boeing 314 the Herculean task of flying from Wake Island to New York to escape the attack. Soon aircrews joined the military, and aircraft lost their shimmering silver finishes to matt camouflage schemes. It was not just the aircrews and aircraft the military wanted from Pan Am;

it was also their experience in transoceanic flight and access to the systems that supported it.

Pan Am soon became heavily involved in the war, helping evacuate Hong Kong and providing an air bridge alongside the China National Aviation Corporation (CNAC) over the The Hump between India and southern China. This service provided crucial logistical support to Chiang Kai-Shek's Chinese forces fighting Japanese troops in the region. In addition, Pan Am helped prevent the Japanese Imperial General Headquarters from moving the region's forces into the eastern and South Pacific areas by providing this service. Pan Am was also responsible for the safe carriage of President Roosevelt, taking him to the important Casablanca Conference of January 1943 via Brazil and Gambia. Another critical service provided by Pan Am during the war was the transportation of around seventy-five percent of the uranium ore from the Belgian Congo (Congo-Kinshasa) needed for use in the Manhattan Project.

Other airlines soon undertook long-haul transoceanic flights. Trans World Airlines (TWA) contracted its five Boeing 307 Stratoliners to the US Army Air Forces' Air Transport Command. These aircraft made over 3,000 flights between the continental United States, Europe and Africa. They also ushered the Lockheed C-69 Constellation into service in 1944, which joined its four-engine Douglas C-54 Skymasters on long-haul flights.

Another American transoceanic airline was American Export Airlines (AEA), which operated between New York and Foynes, Ireland, crossing with the Sikorsky VS-44 flying boat. The AEA service ran from 26 May 1942. By the war's end, AEA offered connections to Northern European destinations.

Neutral nations continued to fly during the war, though this was not without its perils, as KLM had learned in the Far East. Before the Japanese invasion of the Dutch East Indies, KLM had its operations curtailed by the onset of war in Europe on 3 September 1939, with flights to Germany and France forbidden. KLM painted most of its fleet orange to prevent loss of life and continued to operate routes to the United Kingdom, Scandinavia and Belgium. After the German invasion of the Netherlands on 10 May 1940, European-based KLM DC-2s and DC-3s evacuated to the United Kingdom, or in the case of the Dutch East Indies, aircraft remained operating. The European-based aircraft maintained the UK–Portugal route, while the KLM operations in the Pacific continued as long as possible, culminating in airlifting Dutch evacuees to Australia's Northern Territory. KLM continued to fly in America's expanding routes to the Caribbean during the war.

The Swiss wartime experience was perhaps the most complex, given the nation's geographical location and political situation. Initially, the flag carrier Swissair had to suspend its operations due to France and Germany closing their airspace on 27 August 1939. This led to the airline initially ceasing operations before slowly starting to re-run flights to southern Germany, northern Italy and Spain. The destruction of a Swissair DC-2 during an Allied air raid on Stuttgart saw the airline cease all external flight operations in August 1944. As a result of the slow contraction of services, groundcrew and aircrew found themselves in military service, which occasionally saw them fly interned aircraft such as Boeing B-17s to dedicated storage sites. Swissair resumed normal air operations on 30 July 1945.

The neutral countries of Scandinavia had a mixed wartime experience. After the German invasion of Denmark and Norway on 9 April 1940, Det Danske Luftfartselskab (DDL) and Det Norske Luftfartselskap (DNL) flight operations were paused. DDL was granted permission to operate domestic services and services to Berlin in June. Other air routes were permitted, which included flights to Munich, Vienna and Malmo; Sweden followed soon after. However, DNL operations were restricted to covering routes in the north of the country, which in turn helped the German occupying forces. On 21 March 1941, after most DNL pilots had fled to the United Kingdom to join the resistance, the occupying Germans cancelled all DNL routes and flights, signalling the end of domestic flights until liberation.

AB Aerotransport (ABA), the national airline of Sweden, faced several challenges throughout the war; despite being the flag carrier of a neutral country, it soon became isolated by regional events. This isolation was not helped when a Finnish civilian Junkers Ju 52-3/mge was shot down by two Soviet Ilyushin DB-3 bombers killing all nine occupants, on 14 June 1940. Eventually the RAF opened a route between Leuchars, Scotland, and Stockholm, which helped pass on airmail and diplomatic staff between the Allies and Sweden. In addition, in 1942 Sweden began operating its own service between Sweden and Scotland, further developing into an intercontinental service.

On 22 February 1943 Svensk Intercontinental Lufttrafik (SILA) was established to run the service. However, despite sending delegates to the United Kingdom and the United States, the purchase of aircraft, while agreed in principle, would not occur until the war's end. This decision was probably influenced by the loss of a DC-3 on 22 October 1943, on an Aberdeen-to-Stockholm flight which was attacked by a Ju 88 over Hållö Island. The aeroplane crashed and thirteen of the fifteen occupants were killed.

Sweden had several interned B-17s in its possession. As part of the Swords to Ploughshares agreement, the B-17s were kept by the Swedish government in return for the crews' release from internment. SAAB was then tasked to convert seven of the bombers, known as Felix – a nod to the help given to the project by the American defence attaché Lieutenant Colonel Felix M. Hardison – into fourteen-seat passenger aircraft. Five went to ABA and SILA on 6 May 1944, with the remaining two passed on to DDL. The final B-17 Felix retired from service in 1948.

In December 1944, the Chicago Convention on International Civil Aviation was signed and brought into the International Civil Aviation Organization (ICAO), a specialized agency of the United Nations linked to the Economic and Social Council (ECOSOC). With the ICAO came a defined organizational rationale:

Whereas the future development of international civil aviation can greatly help to create and preserve friendship and understanding among the nations and peoples of the world, yet its abuse can become a threat to the general security; and
Whereas it is desirable to avoid friction and to promote that co-operation between nations and peoples upon which the peace of the world depends;
Therefore, the undersigned governments having agreed on certain principles and arrangements in order that international civil aviation may be developed in a safe and orderly manner and that international air transport services may be established on the basis of equality of opportunity and operated soundly and economically;
Have accordingly concluded this Convention to that end.

Such a statement was groundbreaking. With it came the hope that in the post-war world civil aircraft would not be lost to attacks by the military. This was not the case. Despite the best intentions and numerous technical advances in flight safety, including the widespread fitting of Identification Friend or Foe (IFF) systems, aircraft continued to be lost.

The 1950s saw two incidents of aircraft lost to military interception, in China and Bulgaria. This was followed by the loss of a United Nations Douglas DC-6 carrying UN Secretary-General Dag Hammarskjöld and fifteen others on 18 September 1961. Initially, the crash near Ndola, Northern Rhodesia, now present-day Zambia, was thought to have occurred due to pilot error. Nonetheless, rumours have persisted of an aerial interception taking place by a third party. However, despite several witness accounts testifying, there have been no further investigations.

Despite the increasing use of radar and IFF, the losses of passengers, crews and civil aircraft continued to mount throughout the 1970s and 1980s. Losses often occurred around political or regional tensions. One of the most notable losses of life occurred on 1 September 1983, when a Boeing 747-230B of Korean Air Lines (KAL) Flight 007 was shot down by a Soviet Sukhoi Su-15 after it had made a navigational error. All 269 occupants, including Congressman Larry McDonald who was part of a party returning from celebrating the thirtieth anniversary of the United States–South Korea Mutual Defense Treaty, died.

The losses continued throughout the 1980s in regions such as South Africa and South West Asia, where the Cold War had become 'hot' and tensions were running high. The 1990s initially saw a global calming of uncertainties as the Cold War ended and Soviet influence waned. Then, on 6 April 1994, the unthinkable happened in an otherwise calm region of Africa. A Dassault Falcon 50 long-range business jet carrying Rwandan president Juvénal Habyarimana and Burundian president Cyprien Ntaryamira was shot down by a surface-to-air missile as it prepared to land in Kigali, Rwanda. Both presidents were killed in the missile strike, the origin of which has never been verified. Within twenty-four hours of the attack an explosive wave of violence, lasting four months, swept across Rwanda with the deaths of nearly a million Tutsis and moderate Hutus.

The twenty-first century saw military attacks on civil aviation continue to decline. However, occasional incidents occurred, including the accidental downing of Siberian Airlines (SBI) Flight 1812 on 4 October 2001. The aircraft involved, a medium-range Tupolev Tu-154 tri-engine jet aeroplane, was en route from Israel to Siberia when it was downed by an S-200 surface-to-air missile. The missile, operated by the Ukrainians, had been fired during a joint Ukrainian-Russian military exercise, accidentally engaging the Tu-154 as a target and killing all on board.

This region would gradually become increasingly unsettled due to Russian-backed military actions along its border with Ukraine. For example, on 17 July 2014, a Boeing 777-200ER of Malaysia Airlines (MAS) flying from Amsterdam Airport Schiphol, Netherlands, to Kuala Lumpur International Airport, Malaysia, was downed by a Buk surface-to-air missile. The missile, operated by pro-Russian Donetsk separatists, completely destroyed the aircraft and killed 283 passengers and fifteen crew. Subsequent investigations into this act led to the Netherlands and Australia seeking a prosecution of the perpetrators under Article 84 of the Convention on International Civil Aviation. On 17 November 2022 the District Court of The Hague sentenced the accused, Kharchenko, Dubinskiy and Girkin, in absentia, to life imprisonment for causing Flight MH17 to crash and for the murder of the 298 passengers on board. One defendant, Pulatov, was acquitted.

On 8 January 2020, Flight 752, a Ukraine International Airlines (AUI) Boeing 737-800 was on a flight from Tehran, Iran, to Kyiv, Ukraine, when it was shot down, killing all 176 people on board shortly after take-off, by the Iranian Islamic Revolutionary Guard Corps (IRGC) who were guarding the airport. The AUI Boeing had been unintentionally targeted when it was mistaken for a cruise missile launched by the United States.

Such losses of life are thankfully infrequent in an age where travel information is rapidly updated, and safety systems are in place to prevent all but the most unfortunate accidents. However, these incidents also show how devastating military equipment is when used against civil aircraft.

Humble beginnings. A view of Bulltofta Airport, Malmö, Sweden in the 1930s. (Scandinavian Airlines)

An Aircraft Transport and Travel de Havilland DH.16, with attending dispatch rider. (Unknown)

Well-wrapped passengers in front of a civil Handley Page O/400, of Handley Page Transport, prior to departure on the London–Paris route. (Swissair)

Handley Page W.8b of the Belgian airline Sabena on the Amsterdam–Rotterdam–Basel route. (Unknown)

Left: Airmail carriage was not restricted to the northern hemisphere. On 31 January 1921, Captain Euan Dickson (centre) flew the first regular airmail service in New Zealand. (Archives New Zealand)

Below: A BOAC Lockheed Lodestar flying over Cairo before landing at Heliopolis, Egypt, during the Second World War. (IWM)

An Armstrong Whitworth AW.15 Atalanta preparing for take-off at Talangbetoetoe Airport near Palembang, Indonesia. (Unknown)

An early Qantas booking office. (State Library of Queensland, Australia)

Left side view of Zeppelin LZ 129 Hindenburg burning at Lakehurst, New Jersey, 6 May 1937. (Arthur Cofod Jr)

Above: All hands on deck as QEA staff push an aircraft into a hangar for safe keeping. (State Library of Queensland, Australia)

Right: Croydon Airport, London, would see many passengers arrive at its terminal, including German Foreign Affairs Minister Joachim von Ribbentrop on 18 March 1936. (ETH Library)

A four-seat Blériot-SPAD S.56 at rest. This aircraft formed the backbone of more than one early French airline. (Unknown)

A Junkers F 13 floatplane acting in the medevac role has its patient deplaned by medics and a suitably attired pilot at Reykjavík, Iceland. (Reykjavík Museum of Photography)

A Sabena flight after arriving in Léopoldville, Belgian Congo, 1935. (Unknown)

A KLM Douglas DC-2 was the first plane to land at Oslo Airport, Norway, during its official opening on 1 June 1939. (Anders Beer Wilse)

Lockheed 3 Air Express in Western Air Express service. (Unknown)

Student of the Civilian Pilot Training Program under instruction at Fort Worth, Texas. (Library of Congress)

Juan Trippe, seen here with a Boeing Stratocruiser, would create an American icon in Pan Am. (Unknown)

Opposite page, 2nd from bottom: An Aer Lingus ATL-98 Carvair at Southend, United Kingdom. (Richard Goring)

Opposite page, bottom: India's Prime Minister, Pandit Nehru, arrives in Zurich-Kloten, Switzerland, aboard an Air India (AIC) Lockheed L-749A Constellation. (Swissair)

A Scandinavian Airlines, or SAS, Fokker F.XXII in flight. (Scandinavian Airlines)

A TWA Boeing Model 307 Stratoliner readying itself for take-off. The influence on wartime heavy bomber design is apparent in this excellent study photo. (San Diego Air and Space Museum Archive)

An Aeroflot ANT-20 bus at an unknown location in the Soviet Union. (Unknown)

A Sociedad Colombo Alemana de Transportes Aéreos, or SCADTA, Junkers W 34 pulled up to a riverbank attracts the attention of the local population. (Unknown)

War-surplus B-17s, B-25s and B-26s at Ontario, California in May 22, 1946. Many of these would end up in the hands of private owners with big ideas. (Bill Larkins)

An Avro York of Dan-Air at Manchester (Ringway) Airport in 1960. (Ruth AS)

A Douglas DC-6 operated by BCPA, at Brisbane Airport, Australia. (Unknown)

The prototype Bristol Type 167 Brabazon, which made its first flight on 4 September 1949. (HMSO)

The Società de Agostini e Caproni Stipa-Caproni designed by engineer Luigi Stipa. (Unknown)

Above: An Ansett-ANA Vickers 747 Viscount. Note the rounded doors fore and aft. (Chris Finney)

Left: An Aeroflot (AFL) Tupolve TU114 at Budapest Airport, Hungary. (Fortepan)

A Frontier Airlines (FFT) Convair CV-340 at Phoenix Sky Harbor International Airport, Arizona. (Jon Proctor)

An Aeroflot (AFL) Ilyushin Il-12 at Budapest Airport, Hungary. (Fortepan)

Right: A Sikorsky VS-44 flying boat. (Harold Martin)

Below: In the 1930s there was a requirement for flight attendants to be qualified nurses, should the worst happen, hence the clinical garb. (San Diego Air and Space Museum Archive)

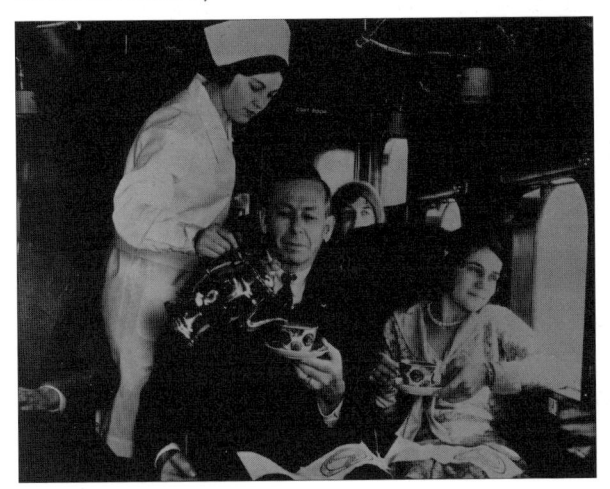

A Pan Am Boeing 314 Clipper attracts a crowd. The flying boats were glamour personified. (Tudor Washington Collins)

Ever since the Wrights, aircraft design has been a collaborative process. Finnish designer Martin Vainio discusses a design point with test pilot Esko Halme. (Journalistinen kuva-arkisto JOKA)

Above: Louise Thaden wearing the essentials for all aviators of the pioneering age: riding boots and jodhpurs. (San Diego Air and Space Museum)

Left: The interior of a Dornier Do X flying boat. These beautifully appointed aircraft led the way in opulence and grace. (German Federal Archive)

Big is beautiful. The Boeing 747 is displayed to the public for the first time. (Scandinavian Airlines)

Élise Raymonde de Laroche (1882–1919) was the first woman to be awarded a pilot's licence, in 1910. (George Grantham Bain Collection, Library of Congress)

Lady Mary Bailey, pictured in her de Havilland DH.60 Cirrus II Moth biplane in which she set the world record for altitude of 5,268 metres (17,283 feet). (Unknown)

Above left: Amy Johnson pictured with Jason, a de Havilland DH.60 Gipsy Moth, after their arrival in Australia on 24 May 1930. (National Library of Australia)

Above right: Mary Anita 'Neta' Snook (84) emerges from the flight simulator for advanced aircraft at Ames Research Centre, California. Snook was one of the first women pilots and, in 1920, was Amelia Earhart's flight instructor. (NASA)

Left: Amelia Earhart posing in a US Department of Commerce's Bureau of Air Commerce Stearman Hammond Y-1 in 1936. Earhart would inspire a generation with her exploits. (Picryl)

Bessie Coleman with her aeroplane, a Curtiss Jenny JN4, in 1922 at Curtiss Field, New York. Coleman would become the first female African-American pilot and earned the nickname 'Queen Bess'. (Cradle of Aviation Museum)

Louise Thaden prepares for an altitude flight, date unknown. (San Diego Air and Space Museum)

Looking prepared, Bobbi Trout and Gladys O'Donnell ready themselves before the 1929 Women's Air Derby, Santa Monica, California. (*Los Angeles Times*)

Louise Thaden with Frances Marsalis in their Curtiss Thrush named I.J. Fox after their sponsor, in which they made a record endurance flight of 196 hours 5 minutes in 1932. (San Diego Air and Space Museum)

Jackie Cochran, head of the Women Airforce Service Pilots (WASP) in the cockpit of a Curtiss P-40 Warhawk. (USAF)

Ninety-Nines members Bobbi Trout and Ruth Elder with bouquets of flowers christening a Pickwick Airways Bach 3-CT-6 Air Yacht at Grand Central Air Terminal dedication on 23 February 1929. (*Los Angeles Times*)

Pancho Barnes with her Mystery Circus of the Air female aviators. (Unknown)

Women of the Air Transport Auxiliary (ATA). Commandant of the Women's Section of the ATA Pauline Gower (far left) chats with fellow pilots (from left) Mrs Winifred Crossley, Miss Margaret Cunnison, The Hon Mrs Margaret Fairweather, Mona Friedlander, Miss Joan Hughes, Mrs G. Paterson, Miss Rosemary Rees and Mrs Marion Wilberforce. (IWM)

WASP Elizabeth L. Remba Gardner of Rockford, Illinois, at the controls of a Martin B-26 Marauder at Harlingen Army Airfield, Texas. (US National Archives and Records Administration)

A promotional photograph of Lee Ya-Ching on her 1937 goodwill tour to raise money for refugee care. (Unknown)

Jean Ross Howard received her helicopter rating in 1954, the thirteenth woman in the world to do so. (Texas Women's University)

Polina Osipenko, Valentina Grizodubova, and Marina Raskova in front of their Tupolev ANT-37 bis, Rodina. (Boris Yeltsin Presidential Library)

Betty Miller steps out of her Piper Apache in Brisbane, Australia, on 13 May 1963. (Barry Pascoe/Courier-Mail Photo Archives)

Captain Lynn Rippelmeyer became the first international B-747 female captain. (Lynn Rippelmeyer)

Barbara Cartland with her glider. Cartland's work with long-distance glider towing (to their designated release points) would be used during the Second World War, notably on D-Day. (Unknown)

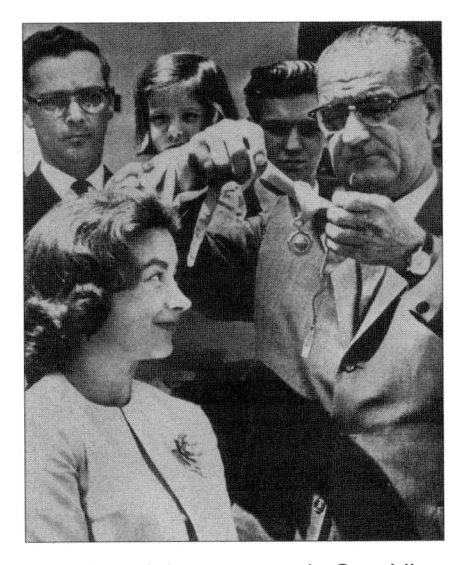

President Johnson awards Geraldine Mock the Federal Aviation Agency Gold Medal for Distinguished Service in 1964. (Associated Press)

Early airfield infrastructure was rudimentary to say the least, as witnessed in this wonderful photograph of the first great air meet in the United States in 1910 at Dominguez Field near Los Angeles. (Charles Pierce)

Preparing a civilian Consolidated Catalina at Rose Bay, Australia, 1939, a far cry from contemporary health and safety precautions. (*Sydney Morning Herald*)

Caterers prepare for Scandinavian Airlines' inaugural Stockholm–New York service from Bromma International Airport, Sweden on 17 September 1946. (Scandinavian Airlines)

The Apollo 11 spacecraft command module (CM) is loaded aboard an Aero Spacelines Super Guppy Aircraft at Ellington Air Force Base for shipment to the North American Rockwell Corporation at Downey, California. (NASA)

Juan de la Cierva with his autogyro at Lasarte Airfield, Spain, 1930. (Pascual Marín)

A retrieved Fu Go balloon showing various parts of the balloon and its mechanics. (NARA)

French aeromedical pioneer Marie Marvingt would champion the humanitarian use of aircraft throughout her life. (Library of Congress)

A suitably camouflaged BOAC Boeing Model 314A Clipper, lands on Lagos Lagoon in Nigeria for moorings at the West African flying boat terminal at Iquoi. (IWM)

A RAF trainee takes a flying lesson in a Vultee BT-13 of the US Air Force Basic Pilot Training School based at Gunter Field, Montgomery, Alabama. (IWM)

Like its contemporaries, Pan Am was quick to answer the call to arms. Here President Franklin Roosevelt shares a lighter moment celebrating his 61st birthday aboard the Dixie Clipper, 1943. (Pan Am Historical Foundation)

Nancy Harkness Love preparing for take off. (NARA)

A DC-4 of Near East Air Transport airlifting Habbani Jews from the South Arabian Peninsula during Operation Magic Carpet, August 1950. (Fritz Cohen)

United Nations Secretary General Dag Hammarskjöld arrives at Lydda Airport in Israel in a UN DC-3 in 1956 on his way from Beirut to Cairo. Hammarskjöld was en route to negotiate a ceasefire in the Congo when his Douglas DC-6 airliner SE-BDY was shot down near Ndola, Northern Rhodesia (now Zambia). (Israel Government Press Office)

Svensk Intercontinental Lufttrafik (SILA) pilots gather in front of a B-17 Felix. (Scandinavian Airlines)

Inspecting a BEA Vickers Vanguard nose after a bird strike on approach to Schiphol Airport, Netherlands. (Joop van Bilsen/ Anefo)

NASA were keen to start the ball rolling wit its space station concepts very early on in the spce program. 60 years on from these ideas a new commercial space race is gearing up, and ready to push the boundaries of lfight once more. (DVIDS).

Four members of the United States Women's Airforce Service Pilots (WASPs), who belong to the first class of WASPs to complete the nine-week flight training course on the B-26 Marauder receive final instructions as they chart a cross-country course. (National Archives at College Park)

Disaster and Calamity

Airlifts
The airlift is a vital element of humanitarian aid, enabling the delivery of medical evacuation services carried out by civilian and military aircraft. There have, however, been examples of civil aviation taking on the task of airlifts, aiding authorities in times of need.

One of the first instances of a civil airline arranging airlift activities occurred with Israeli flag carrier El Al Israel Airlines Ltd (ELY), which airlifted over 160,000 Jewish immigrants. The first airlift occurred between June 1949 and September 1950 as part of Operation Magic Carpet, which saw some 49,000 mainly Yemenite Jews brought to Israel via Saudi Arabia and involved 380 flights. This was followed in March 1951 by Operations Ezra and Nehemiah, which saw a further 130,000 Iraqi Jews initially airlifted from Cyprus, then Baghdad, ending in early 1952.

Fifteen years later the Biafran Airlift, the largest civilian airlift of its kind, took place between 1967 and 1970. Around thirty non-governmental organizations (NGOs) and governments provided non-military direct and indirect aid in support of the operation, including the International Committee of the Red Cross (ICRC), a host of church organizations known as Joint Church Aid (JCA) and Oxfam. Many aircraft were used in the airlift, from Lockheed C-130s to Boeing C-97 Stratofreighters, which took part in ferrying in foodstuffs in a desperate attempt to stave off starvation. Despite early issues with the Nigerian government, the NGOs and the JCA, in particular, were able to slowly build an operating base on the island of São Tomé in early 1968. This Portuguese overseas department lay 646 miles (1,040 kilometres) south of Nigeria in the Gulf of Guinea and had several ports which could unload supplies from donors. Moreover, the island's governor, Silva Sebastio, fully supported the airlift. By September 1968 a system was in place which saw ten aid flights take off every night for Biafra. The same flight would return with those, mainly children, who needed medical assistance.

The island airstrip was joined by a second, established at Umuchima village, Uli, which lay on the western border of the breakaway Biafra. This strip would soon be code-named Annabel Airport, becoming the second busiest airport in Africa after Johannesburg. Operating from the Annabel Airport required professionalism and bravery as Nigerian forces often attacked the site. Aircraft were also targeted when they were most vulnerable – during landings. These flights were soon joined by French aircraft operating from Libreville in Gabon, south of Nigeria. Like the flights from São Tomé, the flights from Annabel Airport and Libreville took place at night to avoid Nigerian military aircraft, which maintained air superiority.

Come 1969 and the airlift delivered an average of 250 metric tons of food each night to Annabel Airport. The food deliveries were joined by cargos of fuel, basic medical supplies and supplements. By the time the missions ended, the JCA had overseen over 5,300 flights by ten carriers, lifting 60,000 tons into Annabel Airport. Twenty-nine lives were lost during the operation, with the ICRC losing three staff after Nigerian forces shot down a DC-7.

Civil aviation's single largest airlift occurred between 13 August and 20 October 1990. Air India (AIC) and Indian Airlines (IAC) evacuated 170,000 Indian nationals and their families who had been stranded in Kuwait due to the Iraqi invasion. The first leg of the repatriation was a bus journey via Basra and Baghdad to Amman, Jordan. From there, the evacuees were flown to Bombay, 2,558 miles (4,117 kilometres) away by one of the 488 flights.

These acts have proved that civil aviation is more than capable of stepping up to the mark. It is more than equal to military aviation. It is capable of remarkable feats of preparation and logistics; it pulls together diverse teams from across the globe to help make the difference between life and death.

Disasters

Despite year-on-year improvements in aircraft design, as proved by the Boeing 707 barrel roll on 7 August 1955 made by test pilot Tex Johnston, disaster can strike any time. The act of flight, as demonstrated by the tales of Icarus and Eilmer of Malmesbury, is strewn with hazards.

There is a sad inevitability that the development of powered flight would also see a corresponding increase in fatalities, especially as the numbers of aircraft built and passengers carried increased. Many early accidents occurred due to aircraft being unable to stand sudden stresses, mostly from deflecting forces generated by engine cranks creating instability, known as the gyroscopic effect. The experiences of the First World War eliminated such incidents as the need to build aircraft capable of withstanding hard and heavy handling was paramount. However, as well as the gyroscopic effect, the issue of eradicating stalling remained, leading to an increase in aviator skills and engineering developments. These incidents gave aviators, designers and engineers vital early lessons upon which they built their knowledge; a little over a decade later, the figures showed a vast improvement.

Between 1919 and 1924, the number of aircraft miles flown per accident resulting in death or injury to occupants was 357,000 miles (574,535 kilometres) to one. As the art of aeronautics continued to develop, the number of aircraft miles flown per accident resulting in death or injury to occupants had risen to an unbelievable 2,336,000 miles (3,759,400 kilometres) to one between 1930 and 1933. It was not until the release of the 1944 Chicago Convention's Annex 13 Aircraft Accident and Incident Investigation on 11 April 1951 that a legal definition of an accident was described. The wording is understandably lengthy but essential in identifying what would become a vital piece of aeronautical equipment: the Flight Data Recorder (FDR) system. This is defined as 'Any type of recorder installed in the aircraft for the purpose of complementing accident/incident investigation' with the specifications laid out, including what data should be recorded. Also included were the physical specifications and the standards any design must reach. These can vary nationally, with FDR systems used on British-registered aircraft meeting the obligatory British Civil Airworthiness Requirements (BCAR), for example. However, regardless of national specifications, all FDR systems must meet the standards made by the ICAO.

The first designs of what would become an essential piece of equipment first appeared in 1939. Designed by engineers François Hussenot and Paul Beaudouin in 1939 and used at the Marignane flight test centre, the early recording systems used photographic film to record a latent image. The image was made by a beam of light which in turn was controlled by a tilting mirror to record the physical aspects of an aircraft flight. This FDR was followed by a British design made during the Second World War, engineered to withstand the rigours of a crash. Developed at Farnborough for the Ministry of Aircraft Production by Len Harrison and Vic Husband, this new FDR used copper foil indented by various-sized styluses to periodically record the aircraft's instrument readings and control settings. At the same time, Finnish aviation engineer Veijo Hietala developed the first modern FDR, known as the Mata Hari in 1942. The name was chosen because FDR recorded an aircraft's essential dynamic details during flight, a nod to the First World War Dutch dancer and alleged spy. A year later the United States Army Air Forces (USAAF) was developing another vital piece of equipment that would join the FDR – the Cockpit Voice Recorder (CVR).

Post-war developments of both systems continued unabated, with companies like Royston Instruments launching their Midas 270 channel FDR. Australian research scientist David Warren developed the Aeronautical Research Laboratories (ARL) flight memory unit. The ARL recorded both instrument readings and cockpit conversations as a combined unit. After refinement of the design, supported by industry and Sir Robert Hardingham, the secretary of the British Air Registration Board, the first ARL was used in 1965. The familiar red colour of the recorder bodies was introduced at this time. By 1966 all British aircraft were required to carry an FDR. The following year a Royston Midas was instrumental in investigating the 1967 Stockport air disaster in which seventy-two people lost their lives.

In the United States development had also continued post-war and mechanical engineer Professor James Ryan patented his Ryan flight recorder in 1953. Ryan's FDR was approved for use in 1960 and was followed by a second design, known simply as a Coding Apparatus For Flight Recorders. Like other FDRs, Ryan's plan measured as much telemetry and dynamic movement of the host aircraft as possible, including exhaust temperature, fuel flow and control surfaces position. In 1961 the Cockpit Sound Recorder (CSR) was designed and patented by aeronautical engineer Edmund Boniface. Boniface redesigned his CSR to include a delete switch that would erase flight recording on completion of a safe flight after fears about cockpit privacy. In addition, the CSR was fitted with a thirty-minute continuous erasing/recording loop capable of picking up any sound that could be heard in the cockpit.

A further device that has proved invaluable to in-flight safety and overall operational efficiency has been the Quick Access Recorder (QAR). This unit can be accessed at any time to provide raw flight data fed to it by the Flight Data Acquisition Unit (FDAU) which receives its information from several sensors and onboard avionic systems. The FDAU also provides information to the FDR. Their data feeds to the QAR can be removed from the host aircraft to help engineers and crews build a picture of fight performance and possible technical issues.

FDRs and CSRs/CVRs remained analogue until the early 1990s when solid-state recorders became commercially viable. This led to the use of the combined digital Cockpit Voice and Data Recorder (CVDR), which is virtually maintenance-free and offers extended voice recording. With access to these devices, crash investigators can understand what has happened and piece together the often-catastrophic last moments of flight. Despite training and flight safety devices, losses still occur.

Terror
One of the most significant security fears for crews remains the threat of hijacking, also known as air piracy, terrorism and sabotage, which can be carried out for political gain. It remains the final challenge to the safety of the aircraft and its crew and passengers in flight.

The first recorded hijacking at gunpoint took place in spring 1919, the exact date has not been recorded. The perpetrator was an Austro-Hungarian aristocrat and pioneering palaeontologist Baron Franz Nopcsa von Felső-Szilvás who had fled the short-lived Socialist Federative Republic of Councils in Hungary, founded on 21 March 1919, to Vienna, Austria. This hijacking was followed by several minor incidents, although it was not until 21 February 1931 that a major incident occurred. A group of armed insurgents approached Ford Trimotor pilot Byron Richards at the airport in Arequipa, Peru, demanding the use of his aircraft and services to help them leave the county. Richards stood his ground and the revolutionaries refused to let him go. However, ten days later, news reached the group that the revolution had succeeded and Richards was allowed to leave after taking one group member to Lima.

A surge of hijackings occurred with the conclusion of the Second World War, all politically motivated except one, a mid-air robbery on 16 July 1948 when a Cathay Pacific Catalina crashed into the Pearl River Delta near the South China Sea after the pilot was shot dead resisting the hijackers. Of the twenty-six passengers, there was only one survivor, Huáng Yù, one of the hijackers.

On 24 Match 1950 the first mass hijacking occurred as three DC-3s were simultaneously hijacked by former Czech Royal Air Force pilots, some of whom were also flying the DC-3s. The pilots flew the aircraft to a United States Air Force base near Erding, West Germany, where they sought political asylum from the Communist regime in Czechoslovakia. The theme of escaping political oppression behind the Iron Curtain continued when, on 13 July 1956, seven Hungarian students hijacked a small aircraft leaving Budapest, forcing a landing, again in West Germany.

The 1960s and 1970s saw hijacking increasingly used as a political tool with many hijackers demanding flights to the newly established Republic of Cuba. The 1960s also saw the start of the first government-backed hijacking intervention agency when the United States Federal Aviation Administration (FAA) started its peace officer programme in 1962.

The role of the armed peace officer was to travel incognito on those flights deemed to be in peril of hijacking, especially from politically motivated individuals keen to reach Cuba. By the decade's end, the role had expanded; the peace officer programme became the sky marshal programme in 1968, which changed again in 1982 to the Federal Air Marshal Program. The programme changed its name to the Federal Air Marshal Service in January 2002 and management was transferred to the Transportation Security Administration (TAS). This project was followed by several other countries, including Pakistan, which introduced its Airports Security Force (ASF) in 1976, Austria, which introduced its service in 1981 and India in 1999. After the terrorist attacks of 9/11, more nations would establish similar programmes, including the UK's Aircraft Protection Operations (APO) programme, run by the Metropolitan Police and the Canadian Air Carrier Protection/Protective Program (CAPP). The CACPP is a multi-agency programme run by the Royal Canadian Mounted Police (RCMP) with Transport Canada, which oversees Canadian aviation security and the Canadian Air Transport Security Authority (CATSA).

Not all hijackings are necessarily political. On 31 October 1969, 19-year-old US Marine, Raffaele Minichiello, hijacked a TWA Boeing 707 on a scheduled flight from Los Angeles to San Francisco. After stewardess Tracey Coleman negotiated the release of all the passengers and most of her fellow flight attendants in Denver, Minichiello, three pilots and a stewardess made the longest hijacking in history. Crossing the continental United States, the Atlantic and Northern Europe, Minichiello would finally finish his hijacking in Rome, Italy. There he would endeavour to visit his dying father but was caught by Italian authorities hiding in a church before he could make contact.

After flying 6,900 miles (11,104 kilometres), Minichiello was tried by an Italian court and sentenced to seven years for the hijacking. He was released after serving eighteen months and remained in Italy.

The 1970s would see a rise in terrorist hijackings. The high-profile hijacking of four aircraft started the decade by the Popular Front for the Liberation of Palestine (PFLP). On 6 September, what was to be the second mass hijacking in aviation history took place with the aircraft involved assembled at Dawson's Field, a desert airstrip near Zarqa in Jordan. It was a former Royal Air Force station, which had become PFLP's 'revolutionary airport'. The first aircraft that arrived was TWA Boeing 707 and Swissair (SWR) Douglas DC-8. On 9 September, a third aircraft, a BOAC Vickers VC10, joined the 707 and DC-8. A PFLP sympathizer had hijacked the VC10 to put pressure on the British authorities to free the hijacker, Leila Khaled. She had been detained after a failed hijacking in which her partner, Patrick Argüello, had been killed by sky marshals in an El Al (ELY) Boeing 707 on the 6 September. Khaled was a member of the PFLP while Argüello was a member of the Sandinista National Liberation Front, which was operating in support of the hijacking in exchange for guerrilla warfare training. A fourth, a Pan Am Boeing 747, was also hijacked by the PFLP after security prevented two of the group from boarding the El Al flight. However, due to the 747's size, the hijackers landed the 747 in Cairo as it was too large to land at Dawson's Field.

The PFLP segregated the flight crews and fifty-six Jewish hostages before releasing the remaining hostages. These were transferred to Amman, Jordan's capital, and subsequently freed on 11 September. The following day the PFLP used explosives to destroy the empty planes. A photograph of the exploding VC10 would become a world-famous image, synonymous with the event.

The use of Dawson's Field was also a significant move by the PFLP in terms of Arab-Palestinian activity within Jordan, which was increasingly threatening the country's stability. King Hussein had grown weary of the threat posed by organizations like the PFLP and the Palestine Liberation Organization (PLO), which had come together as the Fedayeen, bent on establishing a state within a state. On 16 September Jordan declared martial law which triggered a civil war, as Jordanian forces sought to secure areas under Arab-Palestinian control. This short-lived war would become known as Black September. The Jordanians secured a rapid victory, which enabled a 30 September deal that saw the remaining PFLP hostages released in exchange for Khaled along with three PFLP members imprisoned in Switzerland.

The following year one of the strangest hijackings took place, on 24 November 1971, when a cross-country flight from Portland, Oregon to Seattle, Washington, flown by a Northwest Orient Airlines (NWA) Boeing 727, was hijacked. Shortly after take-off, a man identified as Dan Cooper passed Florence Schaffner, a flight attendant, a note announcing that he had a bomb. Showing Schaffner the bomb, Cooper demanded a ransom of $200,000 (worth $1.3 million in 2020). He then began to issue detailed instructions, first to Schaffner and then to another fight attendant Tina Mucklow. Mucklow, who would remain next to Cooper throughout the flight, passed on Cooper's demands to the captain William Scott, which included four parachutes. Scott then relayed Cooper's demands to the authorities, including the Federal Bureau of Investigation (FBI), as he continued the flight. To ensure there was no panic among the thirty-five other passengers, Donald Nyrop, NWA's chief executive officer, arranged to meet Cooper's demands. Scott was instructed to circle Puget Sound, a Pacific inlet in Washington State. Scott did this under the pretence of a 'minor mechanical difficulty' at Seattle's airport, preventing landing.

When the aircraft finally landed in Seattle, Cooper let the passengers alight, sending Mucklow to collect the ransom money and four parachutes, while the crew remained behind. Cooper then shared a technical series of instructions with the crew on how he would like the aircraft to fly to his desired location, Mexico City, via a fuel stop at Reno, Nevada. Cooper, Mucklow and Scott, along with First Officer William Rataczak and Flight Engineer Harold Anderson, took off at 19:40 local time with the 727 flying at the minimum airspeed possible, as per Cooper's instructions. The flight was shadowed by two USAF Convair F-106 fighters and an Air National Guard (ANG) Lockheed T-33.

Five minutes after take-off Mucklow saw Cooper for the last time. He had instructed her to help lower the 727's rear ventral stairs, but she told Cooper that she was scared of being pulled from the aircraft once the stairs were lowered. So Cooper decided to confine her to the cockpit with the rest of the crew. Mucklow would be the last person to see Cooper. Around 20:00, the aft ventral stairs indicator alarm activated as Cooper opened it. Using the intercom Scott asked if Cooper needed assistance, to which he replied a simple 'no'. At 23:02, the 727 landed at Reno. After a thirty-five-minute search by Scott, it was confirmed that Cooper had gone, parachuting into the pitch black of night over the Washington wilderness.

Later forensic analysis of Cooper's discarded cigarette butts would show traces of chemicals used in the aviation industry. Combined with his detailed flight plan and demands on how the aircraft was to be configured for its flight from Seattle to Reno, investigators considered Cooper was possibly a Boeing employee. This, given Cooper's knowledge of the area and the 727, was not beyond the realms of possibility, though no employee matching Cooper's description was ever found. Nevertheless, the FBI pursued the case, buoyed in 1980 by a camper's discovery of three bundles of banknotes with serial numbers given to Cooper; this discovery would only fuel the mystery. Regardless of the numerous leads, suspects and time invested into the crime, the FBI closed the unsolved case in 2016.

Cooper's hijacking spurred several copycat acts in 1972, leading to major aviation security (AVSEC) changes. These included the installation of metal detectors, mandatory baggage inspection and extra scrutiny of passengers who paid for their tickets on the day of departure with cash. As a result of these improvements in AVSEC, aircraft hijacking for financial gain decreased markedly by 1973.

Aircraft such as the Boeing 727, equipped with ventral stairs, were fitted with Cooper vanes due to the incident. The Cooper vane is a simple mechanical device consisting of a spring-loaded paddle fitted to the fuselage side of an aeroplane equipped with ventral stairs. While on the ground, a spring keeps the paddle perpendicular to the fuselage, and the attached plate does not block the stairway. However, once airborne, the airflow pushes the paddle parallel to the fuselage and the corresponding plate is moved underneath the stairway, preventing its use.

While hijackings for ransom decreased due to the AVSEC measures, the 1970s saw a sharp rise in increasingly violent acts on aircraft. Most of these acts had their roots in regional political and religious upheavals, culminating in the events of 11 September 2001.

The rise of Middle Eastern-based terrorism and growing resentment against Israel and her supporters would soon begin to spill over into commercial aviation. One of the most notable was the Rome–Fiumicino Airport attacks and hijacking on 17 December 1973, shortly after the conclusion of the Yom Kippur War. Five members of the PLO had smuggled firearms and grenades into the terminal building. Once inside all weapons were uncovered and the terrorists began firing, shattering glass and killing two people. The group then made their way toward Pan Am 707-321B Flight 110, bound for Tehran, Iran. The group attacked the aircraft, which was still being prepared for flight, using grenades, including a phosphorus type. Despite the crew's efforts to evacuate the aeroplane, the attack killed thirty, including the purser, Diana Perez.

The five terrorists then took several hostages, which included Lufthansa (DLH) ground crew and boarded a DLH Boeing 737, but not before killing Antonio Zara, an Italian border police officer. After several stops the Boeing landed in Kuwait where the terrorists negotiated 'free passage' to an unknown destination in exchange for releasing their hostages.

The hijackings continued, but one sinister thread that began to appear among threats used by hijackers was that of turning an aircraft into a missile and crashing it into a critical landmark. However, by the late 1970s, civil anti-terrorism units and special operations forces (SOF) had built up sufficient skills to meet the new threats posed by terrorism in aviation.

The Israeli Sayeret Matkal (General Staff Reconnaissance Unit) launched an ambitious rescue mission known as Operation Thunderbolt on 4 July 1976.

The task was to free 248 passengers and crew from an Air France (AFR) Airbus A300, which two members of the Popular Front had hijacked for the Liberation of Palestine – External Operations (PFLP–EO). The A300, which had been hijacked on 27 June, had been flown to Entebbe Airport, as Uganda's Idi Amin was sympathetic to the PLO cause. The hijackers then moved all the occupants to a disused building. They separated the Israeli passengers from the others, who were subsequently released. Ninety-four passengers, along with the twelve Air France crew, were under constant threat of death. Working with Mossad, Sayeret Matkal instituted Operation Thunderbolt, flying 2,500 miles (4,000 kilometres) from Israel to Uganda, arriving in four C-130 Hercules aircraft at 23:00 on 3 July. The ensuing battle saw the Sayeret Matkal return all but three of the hostages to Ben Gurion Airport and kill all the hijackers.

The following year the PFLP hijacked a Lufthansa Boeing 737-230C on 13 October 1977 to secure the release of imprisoned Red Army Faction leaders. The aircraft journey terminated at Aden Adde airport, Mogadishu, Somalia. On 18 October, the West German Federal Police Grenzschutzgruppe 9 (GSG9) tactical unit, which had flown in to secure the passengers, stormed the aircraft a little after 02:00. The operation saw the team kill three of the four hijackers, rescuing the eighty-six passengers and four crew with no further loss of life.

Hijackings continued throughout the remainder of the twentieth century, with political and religious ideologies remaining the reason behind the acts. On 14 June 1985, TWA Boeing 727-231 Flight 847 from Cairo to San Diego was hijacked shortly after take-off from Athens, one of its first stop-offs. A horrific experience for passengers, including Greek singer Demis Roussos followed. The hijackers' identities remain unknown, but due to their demands for the release of Lebanese Shia prisoners held by Israel, many analysts have considered the group members of Hezbollah. Over the next seventy-two hours the hijackers would be joined by a dozen accomplices, as the hijackers began to single out military personnel. A United States Navy (USN) diver, twenty-three-year-old Robert Stethem was murdered and his body thrown out of the aircraft onto the apron at Beirut Airport. Shortly after seven Americans were forced off the aeroplane and taken as hostages to the Shia Prison in Beirut. Throughout the hijacking, hostages were released in dribs and drabs. The final hostages were released after a joint effort by American and Lebanese officials on 30 June, seventeen days after their ordeal began. One of the critical ramifications of the hijacking was the extension of the Federal Air Marshal Program and its use on international flights. Another role was the establishment of regular security assessments of third-country airports. Other nations' security agencies soon picked up on this important task.

The hijackings would continue and the death toll of innocents caught up in events would continue to rise. On the morning of 9 September 2001, nineteen men carried out four coordinated hijackings with the intent of carrying out a series of suicide terrorist attacks. Two of the aircraft, an American Airlines (AAL) Boeing 767-223ER and a United Airlines (UAL) Boeing 767-200, were flown into the Twin Towers of the World Trade Center (WTC) in New York City. Not long after, an AAL Boeing 757-223 ploughed into the United States military's headquarters at the Pentagon in Arlington County, Virginia. A fourth aircraft, a Boeing 757-222 belonging to UAL, was forced down as passengers and crew attempted to retake it from the hijackers after learning of the attacks on the WTC earlier that morning. The Boeing crashed sixty-five miles (105 kilometres) southeast of Pittsburgh, Pennsylvania.

The actions of the hijackers, later identified as members of the militant Islamic extremist network Qaeda al-Jihad, otherwise known as al-Qaeda, led to the deaths of 2,996 people. The FAA responded quickly with FAA National Operations Manager Ben Sliney, on his first day in the role, closing down American airspace to all international flights. At the same time, the Department of Defense (DoD), FAA and the Federal Communications Commission (FCC) initiated their joint plan for the Security Control of Air Traffic and Air Navigation Aids (SCATANA) for the first and only time. SCATANA had been designed to control air traffic and air navigation aids effectively under emergency conditions, such as an aerial attack by a foreign power. In the instance of these attacks, which would become known as 9/11, the emergency plan was only partially implemented as the DoD left command and control of the air traffic system with the FAA. This allowed all American-operated radio navigational aids to remain operational so that the process of controlling and landing aircraft still flying in domestic airspace could be facilitated.

The twenty-first century continued to see the use of hijacking as both an act of terror and a political tool. As a result, however, changes to the aircraft's internal security arrangements, including the universal adoption of security doors, are controlled via the cockpit. A safety override code is available when dealing with an emergency in the cockpit, such as a fire.

Bombings

The most problematic and indiscriminate terror act remains the use of a bomb on board an aircraft. Despite often striking without warning, bombings remain isolated incidents. Still, their effects can result in a horrific loss of life in the air and on the ground.

The first recorded use of a bomb as a sabotage device on an airliner occurred on 10 October 1933, on a UAL Boeing 247 carrying seven on a cross-country route between Newark, New Jersey, and Oakland, California. The wreckage exhibited some strange damage to the area around the toilet at the rear of the aircraft. Subsequent chemical analysis by the Crime Detection Laboratory at

Northwestern University, Illinois, concluded that Boeing had been blown up.

The next bombing would not occur until 9 September 1949, when a Canadian Pacific Air Lines (CPC) Douglas DC-3 was blown up on a cross-country flight from Montréal to Baie-Comeau. The bombing was planned by Joseph-Albert Guay, Généreux Ruest and Marguerite Pitre using a timed dynamite bomb. The intention had been to kill Guay's estranged wife, who was on the flight and collect the life insurance he had taken out on her. This would also leave him free to marry a 17-year-old waitress with whom he was having an affair. Aided by Ruest, a clockmaker, and Pitre, who purchased the dynamite from a hardware store, Pitre then delivered the bomb as an item of airmail for carriage on the DC-3.

The bomb, timed by Guay to go off as the DC-3 crossed the Saint Lawrence River, would make any forensic investigation almost impossible. However, the take-off was delayed by five minutes. When the bomb exploded the DC-3 was over the Cap Tourmente National Wildlife Area and twenty-three lives were lost. Thankfully a full forensic investigation took place into North America's first bomb attack against a civil aircraft. The ensuing investigation would see Guay, Ruest and Pitre charged with murder and sentenced to death. Marguerite Pitre would be the last woman to hang in Canada.

Thankfully the occurrences of bombing an aircraft remained relatively low throughout the 1950s despite the growth in commercial aviation. However, 1955 would see two attacks, one on 11 April on an Air India (AIC) Lockheed L-749A Constellation, which was later declared an assassination attempt on Chinese Premier Zhou Enlai by the Chinese Nationalist Party or Kuomintang (KMT). Zhou had missed the flight due to a medical emergency; only three of the nineteen on board survived. The second occurred on 1 November when Jack Graham packed a timed bomb in his mother Daisy's suitcase to claim against life insurance policies he'd purchased at the airport before she started her journey. Ms Graham was travelling to see her daughter in Alaska. Part of the journey was flying on a UAL Douglas DC-6B cross-country flight between Denver, Colorado, and Seattle, Washington. Shortly after take-off the DC-6B exploded with the loss of all forty-four on board. Graham, who had a prior conviction for embezzlement, was soon apprehended. The case against him was strengthened by forensic evidence later found at the crash site and at his home. He was executed on 11 January 1957.

While most bombings were deadly, there were instances of aircraft surviving explosions, a testament to improvements in design and materials. In such cases, the bombs were often detonated in the lavatory, the physical location of which often helped the aircraft survive. The 1960s saw an increase in politically motivated bombings with the loss of an Air France (AFR) Lockheed L-1649 Starliner over Algeria, killing all seventy-eight on board on 10 May 1961. The following year, on 22 May 1962, the first jet aircraft, a Boeing 707-

124, was blown up near Centerville, Iowa, in what would later be described as a suicide bombing committed as insurance fraud.

The 1970s saw the rise of the bomb as a terror weapon, with the Popular Front for the Liberation of Palestine – General Command (PFLP–GC) claiming responsibility for the destruction of a Swissair (SWR) Convair CV-990-30A-6 on 21 February 1970. What made this attack unique was the use of a barometric-triggered bomb in the cargo bay and the fact that the bomb itself did not destroy the CV-990. The aircraft was not brought down by the initial explosion, and the crew tried to land at Zürich. However, smoke filled the cockpit and prevented the instruments from being read. This led to the aircraft deviating from its flight path, crashing a short time later in a wooded area at Würenlingen, near Zürich, killing all forty-seven on board. On the same day, a bomb exploded on an Austrian Airlines (AUA) Sud Aviation SE 210 Caravelle after it took off on a Frankfurt–Vienna flight. On this occasion, the aircraft was able to land safely.

The 1970s saw a steady increase in bombings from political groups, and the trend continued into the 1980s. The decade would witness two of the deadliest bomb attacks against civil aviation, involving Boeing 747s. The first took place on 23 June 1985 when Air India (AIC) Flight 182, flying from Montréal to London; the aircraft suffered structural decompression leading to mid-air break-up due to a bomb exploding in the forward cargo hold. The 747-327B disintegrated over the Atlantic Ocean close to the coastline of southwestern Ireland. The bomb, which had been planted by Sikh militants, claimed the lives of 329. However, it was not until 2003 that Inderjit Singh Reyat, who had assembled the bomb that downed Flight 182, and another that had detonated at Tokyo airport killing two baggage handlers, was imprisoned for fifteen years. This closed what was to become the most expensive Canadian criminal case in history, and which cost nearly Can$130 million ($99 million).

Three years later, on 21 December 1988, Pan Am Flight 103, which had left London Heathrow for John F. Kennedy International Airport, New York, would explode over the United Kingdom. What became known as the Lockerbie Bombing would see the 747-121 initially break up into three distinct parts: nose, mid-section, and wing and tail. Large pieces of the disintegrating 747-121 would fall on the small town of Lockerbie, Scotland, killing eleven as well as the loss of all 259 on board. The mid-section and parts of the wings and engines fell almost vertically, hitting a petrol station and engulfing several homes in aviation fuel. The nose would be found near Tundergarth, along with a trail of debris constituting what was to be called the Southern Wreckage Trail, some two miles away from the site of the mid-section crash. The rest of the aircraft, including the tail, would constitute the Northern Wreckage Trail.

Numerous organizations claimed responsibility for the bomb, although it would not be until 1991 that a joint Dumfries and Galloway Constabulary and FBI investigation identified the key suspects. On 13 November 1991, indictments for murder were issued against Abdelbaset al-Megrahi, a Libyan intelligence officer, and Lamin Khalifah Fhimah, the Libyan Arab Airlines (LAA) manager at Luqa Airport, Malta. Protracted negotiations finally saw both men handed over by Libyan authorities to face justice at the neutral venue of Camp Zeist in the Netherlands on 5 April 1999. In 2001 al-Megrahi was convicted of murder by a panel of three Scottish judges and sentenced to life imprisonment; Fhimah was acquitted. In 2003, Libyan President Muammar Gaddafi took responsibility for the act and paid compensation to the victims' families. Megrahi served just over ten years of his sentence, which was considered to start on 5 April 1999, before being released on compassionate grounds, on 20 August 2009. Six years later, Scottish prosecutors identified two Libyan nationals as persons of interest in their investigation.

As a result of improvements in AVSEC technology and overall procedures, the number of bombings fell significantly after 9/11. However, the effects of these attacks had barely been processed when on 22 December 2001, Richard Reid, a self-proclaimed al-Qaeda operative, attempted to detonate explosives hidden in his shoes on an AAL Boeing 767-300ER which was on a transatlantic flight from Paris to Miami. Thankfully the device was not viable but showed the lengths to which a new generation of terrorists was willing to go in achieving their aims. The world reacted accordingly to these latest threats, including a plot to use liquid explosives on a transatlantic aircraft was discovered after a series of raids by the British Metropolitan Police on 9 August 2006.

On 25 December 2009 Umar Farouk Abdulmutallab, a member of al-Qaeda in the Arabian Peninsula (AQAP), attempted to detonate chemical explosives he had sewn into his underwear. He boarded an NWA Airbus A330-323E flight from Amsterdam to Detroit Metropolitan Airport to detonate his bomb. But, like Reid's shoe bomb, Abdulmutallab's underwear bomb was not viable. He was quickly subdued by fellow passenger Jasper Schuringa; the aircraft made an emergency landing at Detroit Metropolitan Wayne County Airport.

The bombings lessened as post-9/11 international intelligence became more sophisticated, and global politics changed due to the War on Terror. By the mid-2010s, a new threat had emerged from post-Hussein Iraq, the Islamic State of Iraq and the Levant (ISIL). On 31 October 2015 ISIL planted a bomb on board an Airbus A321-231 operated by the Russian airline Kogalymavia Meroject (KGL) at Sharm El Sheikh International Airport, Egypt, returning holidaymakers to Pulkovo Airport, Saint Petersburg. At 06:13 local time, the device detonated while the aircraft flew over North Sinai, costing 214 lives. The bombing became the worst disaster in Russian civil aviation history.

The last recorded bombing incident occurred on 2 February 2016 when an Harakat al-Shabaab al-Mujahideen, (al-Shabaab) member, Abdullahi Abdisalam Borleh, detonated a device on an Airbus A321-11. Operated by

Somalia Daallo Airlines (DAO), the Airbus was on a short-haul flight between Mogadishu and Djibouti City, carrying eighty-one occupants. Twenty minutes after take-off the bomb, which had been placed in a laptop, blew a hole in the side of the fuselage close to the fuel tanks. Due to a delay in taking off, the bomb detonated before the cabin was fully pressurized enabling the aircraft to return to Aden Adde International Airport, where it made an emergency landing. Borleh's body had fallen from the plane and was later found near a village close to the airport.

In the case of in-flight bomb alerts, which are often hoaxes used to manipulate a crew into following a hijacker's demands, special procedures have been developed to prevent the loss of an aircraft. All of these acts would play a role in further refining the AVSEC, which would offer increased protection for all involved in flight.

Security
By the end of the first century of flight, the security management of flying, especially in the financially lucrative commercial sphere, had developed into a sophisticated process. Each nation established its own rules, protocols and systems, all guided by ICAO, which eventually evolved into the Global Aviation Security Plan (GASeP). GASeP also brought member states together to achieve priorities designed to meet the needs of the UN Counter-Terrorism Committee Executive Directorate (CTED).

The physical security aspects have seen airfields and airports going from open areas offering little security and shelter to town-sized enclosed spaces awash with architectural wonders explicitly designed for use in the aviation industry. But, as with all things, it is the details that matter. Over a century AVSEC has developed various technical solutions for potential issues. Often borrowing from other sectors such as medicine and the military, today's security processes are a world away from those passengers first encountered in the 1950s as the network of commercial routes expanded.

The simple visual bag checks and uniform questions were eventually backed up with machinery with its own rules, regulations and procedures. These items soon became accepted pieces and processes of the journey to be endured to enable the pleasure of the outward journey. These items have become gatekeepers to the start of the flying experience, literally and figuratively.

The first security gateway to flight after passing through ticket and passport control is the Walk-Through Metal Detector (WTMD). This requires the passenger to remove all removable metal objects such as belts and coins. The WTMD is now supplemented by the next generation of scanners known as Advanced Imaging Technology (AIT), which uses electromagnetic radiation to make a full body scan of the individual inside. These have proved invaluable in detecting suspicious non-metallic items concealed on the person. These simple archways indicate to the supervising security officer any presence of metallic or otherwise undeclared objects. The nature of these objects can be confirmed either by using an auxiliary Hand-Held Metal Detector (HHMD) or visually. At the same time, hand, baggage, footwear and any object worthy of closer inspection, such as a walking stick, is passed through an X-ray machine. This scans items in real-time, allowing the trained operator to identify any objects which may be suspicious and warrant closer inspection in the presence of the owner.

There are also two systems available for detecting explosive vapours and residues. The first is the Explosives Trace Detector (ETD) which uses swabs to detect explosive residues. The second is the Explosives Detections Systems (EDS), specifically designed to scan hold luggage as it passes along a conveyor. The system uses a combination of X-ray and image analyses to create a Computerized Tomography (CT) scan of each item. This system will automatically alert the operator should it detect any object that displays the characteristics of explosives. A bigger EDS machine is also available for scanning unaccompanied cargo packed in a Unit Load Device (ULD). As a result of the 2010 al-Qaeda campaign, which saw explosive devices concealed in printer cartridges, the United States initiated the Air Cargo Advance Screening (ACAS) programme. The ACAS programme, led by the United States Customs and Border Protection (CBP) and the Department of Homeland Security (DHS), went into effect on 12 June 2018 and is a simple administrative intervention that operates much like a customs declaration. This process gives all agencies involved in overseeing the shipment of products by air into United States territories a clear audit route back to the exporter. This system allows the CBP, and more importantly, the DHS, to judge the sender's authenticity.

There remains a final piece of highly sensitive equipment that can only be used by specialist personnel: the detection dog. The use of dogs for AVSEC purposes was pioneered by the Royal Air Force Police during the Second World War. Often chosen from hunting breeds, the dogs are used to locate illicit narcotics, explosives and firearms, as well as being a visible deterrent in public areas. Dogs may also be used in a security role, with walking patrols taking place in sensitive areas such as bonded warehouses. Another vital security device is the Closed-Circuit Television (CCTV) system which monitors all non-sensitive areas in an airport, including the secure departure area. Airport operators also enforce the need for security awareness among staff, from bird control to boarding gate greeters. This gives passengers, operators and external law enforcement agencies peace of mind.

Unavoidable Disasters

By its very nature, the act of flight invites a host of possible calamities, from pilot error to structural failure, there have been multiple losses since the 1919 Paris Convention. Over the next century these losses would be categorized into types which would help designers, pilots and investigators, among others, find solutions to prevent situations from reoccurring. Perhaps the most common damage to an aircraft is FOD. Potential hazards include a range of objects that can affect the passage of an aeroplane. These include bird strikes and artificial or natural debris such as volcanic ash or hail. FOD can also occur in the interior of an aircraft, with objects as significant as loose fastenings fowling controls, moving parts or shorting out electrical connections. As aircraft designs became more complex, the risks posed by FOD increased, especially with the arrival of the jet engine.

The results could be catastrophic and dramatic in the case of external FOD, especially that ingested by jet engines. Designers have sought to mitigate these effects, especially for aircraft, such as helicopters and transport types that would be expected to operate on rough surfaces or for aircraft vulnerable to bird strikes. Tests of both situations are easy to simulate and include the infamous chicken test, where a dead chicken to fired into a running engine from the aptly named chicken gun. The chicken test is also used to test cockpit windows, nose panels and body sections where a bird strike is likely to occur. While rare, bird strikes can occur and are capable of bringing down an aircraft. In such incidents, the professionalism and training of the crew can make the difference between life and death, as shown by the actions of Chesley 'Sully' Sullenberger and Jeffrey Skiles. The pair glided their powerless US Airways (AWE) Airbus A320-214 onto the Hudson River, without loss of life, after experiencing a bird strike not long after taking off from New York City's LaGuardia Airport on 15 January 2009. The ditching saw all crew members awarded the Master's Medal of the Guild of Air Pilots and Air Navigators in recognition of their 'heroic and unique aviation achievement'.

Whether caused by a bird strike, mechanical failure, fuel, oil and coolant exhaustion or human error, the outcomes of engine failure can be catastrophic. The vital preventative interventions remain up-to-date training on emergency procedures, following the correct operating and engineering practices, and overall quality control around engine management. This last factor has been built on experience. It covers various subjects, including fuel quality checks, improved material quality and overall design progress.

Pilots flying piston-engine or propeller-driven aircraft can either feather the failed engine or allow it to free spin. This allows the pilot to glide while seeking a suitable landing site without obstructions that is easy to access should a rescue be required. Gliding is a well-used technique in finding a safe landing area and can be maintained over a considerable distance. For example, on 21 August 2001, an Air Transat (TSC) Airbus A330-243 on a transatlantic flight between Canada and Portugal ran out of fuel due to a leak as the result of improper maintenance. Luckily the aeroplane's captain, Robert Piché, an experienced glider pilot, and First Officer Dirk de Jager glided the A330 to the safety of the Azores, a distance of a staggering seventy-five miles (121 kilometres). Piché was awarded the Superior Airmanship Award by the Air Line Pilots Association, International (ALPA) for his airmanship.

For rotorcraft, engine failure is not necessarily disastrous; the pilot will enter autorotation, where the main rotor system is turned by the action of air moving up through the rotor. This process is a standard emergency procedure taught to pilots as part of their training. It uses the unpowered rotor to slow the aircraft's descent and gives the pilot a measure of control, allowing for a safe emergency landing. Thankfully, engine failures are rare and aircrew will follow the procedures that best cover the situation they find themselves in, including an emergency landing, as demonstrated by Sullenberger and Piché.

While many military aircraft and helicopters have some form of FOD deflectors or design features, such as vents or gauze over intakes, such design features are simply not practicable for civil aircraft. This has led to the development of the gravel kit, which is placed on the bottom of the leading edge of a low-hanging engine nacelle. Resembling a pitot tube, it uses compressor bleed air to act as vortex dissipaters. It pushes the debris away from the engine intake. A common addition is that of shields, which resemble giant skis, fitted over wheels; such devices can be seen in use by airlines such as Air Inuit (AIE) on their Boeing 737-200s.

FOD can also be transferred to aircraft manoeuvring between areas by vehicles which have passed through several areas around an airfield, including tanker vehicles. As a result various methods have been developed to prevent and clean up FOD. From the simple expedient of drivers checking their tyre treads before moving across aircraft aprons and runways, to using runway sweepers which clean up a wide range of FOD and are fitted with magnetic bars. In the twenty-first century, the addition of specially textured FOD prevention mats known as a Trackout Control System (TCS) fitted at access points has helped prevent the carriage of FOD.

Despite the best efforts of airport workers and authorities, FOD-related disasters still happen. The most notable FOD incident occurred on 25 July 2000. An Air France (AFR) Concorde, Flight 4590, hit a piece of the engine cowl that had fallen from a Continental Airlines (COA) McDonnell Douglas DC-10-30, which had taken off five minutes previously. The titanium alloy strip shredded an undercarriage tyre as it passed over it, leading to its disintegration. This began a chain of catastrophic events as fragments of the tyre struck the underside of the wing and damaged the undercarriage preventing its retraction. The debris also ruptured the wing fuel cells causing the integral fuel tank to crack, leading to large amounts of fuel igniting as it spilt out of the wing. The rapid loss of fuel resulted in engines one and two losing power. The loss of engine power, high drag and fire damage to flying controls combined to make the Concorde uncontrollable. With flames bellowing from its port wing, Concorde crashed into a nearby hotel, killing all 109 on board and six hotel guests. Despite a much-publicized fleet-wide safety retrofit, the AFR and British Airways (BAW) Concorde fleets were retired on 10 April 2003, with the last flight, made by a BA Concorde, taking place on 26 November 2003.

The environment of the airport has grown to be increasingly busy. For example, in 2019, Hartsfield–Jackson Atlanta International Airport, Georgia, handled a staggering 110,531,300 passengers from 895,000 aircraft movements. With such a busy environment, runway incursions inevitably occur despite developing numerous safeguards and devices. The deadliest of these incursions occurred when two Boeing 747s collided on the runway at Los Rodeos Airport, Canary Islands on 27 March 1977. At the time Los Rodeos was acting as a relief airport for aircraft, including the two 747s that diverted from Gran Canaria Airport after a terrorist attack by the Canary Islands Independence Movement. Due to the increase in traffic Los Rodeos resorted to using the taxiway as a temporary apron to park the diverted aircraft. This had a knock-on effect of having to make departing aircraft taxi on the runway. To add to an already chaotic situation, the weather had closed in and fog banks were drifting across the runway, hampering visibility.

A KLM Boeing 747-206B captained by Jacob Veldhuyzen van Zanten used the runway to taxi before taking off while a Pan Am Boeing 747-121 captained by Victor Grubbs was making its way to the taxiway/apron. What followed was a collapse in communication. The KLM 747-206B began its take-off as Van Zanten believed he had been given take-off clearance from air traffic control. Instead, he was headed directly for Grubbs and his 747 turning across the runway length to access the temporary apron. Grubbs, realizing what was about to happen, tried to steer his aircraft off the runway to avoid colliding with Van Zanten's 747. The KLM 747 managed to get airborne, but its left-side engines, lower fuselage and main landing gear struck the upper right side of the Pan Am's fuselage as it took off. The KLM 747 flew for approximately 150 metres before hitting the ground and sliding 300 metres, igniting its full fuel load. The resulting inferno was so hot it melted parts of the stricken KLM 747. The collision killed 583 people, making it the deadliest accident in aviation history.

The disaster highlighted the importance of standardized phraseology in radio communication. Alongside these changes cockpit procedures were also reviewed, contributing to establishing Crew Resource Management (CRM) as a fundamental part of airline pilots' flight safety training.

Another flight hazard remains the Mid-Air Collision (MAC), occurring for various reasons, including navigational error, poor flight planning and a lack of situational awareness. The first recorded collision of the post-Paris Convention era occurred in poor visibility on 7 April 1922 at Picardy, France, when a Daimler Hire Ltd de Havilland DH.18A Mailplane collided with a Compagnie des Grands Express Aériens (CGEA) Farman F.60 passenger plane. Seven people were killed in the collision, which prompted a series of precautions. The first was aircraft designers ensuring pilots had a clear view ahead, the second was the carriage of radios aboard airliners, and the third was the adoption of the 'keep to the right' becoming the universal rule of the air. The need for disciplined air planning was also discussed, including establishing defined air routes between Northern Europe and the British Isles.

Collisions continued despite improved procedures. On 4 July 1948, a mid-air collision over Northwood, London, between an RAF Avro York C.1 and a Scandinavian Airlines System (SAS) Douglas DC-6 saw another fundamental change. Stacking aircraft 500 feet (152 metres) apart in holding patterns was nothing new, especially when dealing with aircraft operating at a slow flying speed. It is also used when managing aircraft working in poor weather, as was the case on this occasion. The SAS DC-6 was maintaining its stack position below the RAF York when the captain decided to leave the stack and divert to Amsterdam. The DC-6 asked permission to leave the stack and, minutes later, the York was ordered to descend and take the DC-6's place. Neither aircraft acknowledged their instructions from ATC, which were given at approximately 15:00 local time. Three minutes later, the two aircraft collided in Britain's worst mid-air collision, killing thirty-nine passengers and crew. As a result of the ensuing investigation, one of the recommendations was to increase the vertical separation distance between stacked aircraft to 1,000 feet (305 metres).

As the skies became busier, near misses and collisions continued. On 30 June 1956 a UAL Douglas DC-7 collided with a TWA Lockheed L-1049 Super Constellation over Grand Canyon National Park, Arizona; 128 lives were lost. The Grand Canyon collision proved to spur the modernization of civil ATC safety and management systems, including the introduction of radar. Another fundamental change was the passing of the Federal Aviation Act of 1958. This act dissolved the Civil Aeronautics Administration (CAA) and created the Federal Aviation Agency (FAA), renamed the Federal Aviation Administration in 1966. The FAA was also given complete authority over United States airspace, including all military activity. It also oversaw the upgrading of ATC safety and management systems and procedures.

One of the critical technical developments was the on-board Traffic Collision Avoidance System (TCAS) which entered service in 1987. TCAS monitors the airspace around an aircraft for other aircraft equipped with a corresponding active transponder. The TCAS acts independently of air traffic control and warns aircrew of the presence of aircraft which may present a threat of collision. A similar system was developed for ground use by ATC, named the Air Traffic Control Radar Beacon System (ATCRBS), which is used to enhance surveillance radar monitoring and help identify aircraft by their individual aircraft-mounted transponders, which contain the necessary information to allow for unique identification and control.

Over the years, the TCAS was developed further still, belonging to a range of Airborne Collision Avoidance Systems (ACAS), including the Portable Collision Avoidance System (PCAS). The PCAS transponder/receiver system was developed in the late twentieth century by pilot Zane Hovey and is a simple plug-in device which has seen success in the general and light aviation industries. This was joined by FLARM, developed by Urs Rothacher and Andrea Schlapbach in 2003, and has been in commercial service since 2004. FLARM uses GPS, a barometer, and software to help identify and predict potential obstructions.

All of these changes have seen a steady drop in the frequency of mid-air collisions and the accompanying loss of life towards the end of the first century of regulated civil flight. The last significant loss of life occurred over the dense jungle of western Brazil on 29 September 2006 when a Boeing 737-800, on a domestic flight flown by Gol Linhas Aéreas Inteligentes (GLO), collided with an ExcelAire (XLS) Embraer Legacy 600 on a delivery flight. This incident, which saw the loss of 154 lives, was caused by a combination of pilot and ATC error. This led to considerable changes in Brazil's civil aviation systems, including ATC, which had been in technological and managerial decline since the 1980s.

The final challenge for any aircraft operator is structural fatigue. Although rare, it remains a concern. The losses of the early de Havilland Comets showed that even the most cutting-edge designs could not beat the laws of physics. Most pre-war disasters were often a result of natural causes such as ice build-up, pilot error in poor visibility or reckless airmanship, and sabotage, such as bombs. However, there are also unexplained occurrences, such as the case of the German passenger airship LZ 129 *Hindenburg* on 6 May 1937, in which best guesses have included a bomb and the ignition of leaking hydrogen.

Cases of fatigue have increased as aircraft have become bigger, and the technological advances steadied. The early days of civil aviation, especially in the airline industry, saw change happening so rapidly that aircraft were rarely in use beyond fifteen years before they were replaced. By the end of the first century of flight, the average age of the retiring aircraft was twenty-five years. This is a testament to acting on those early lessons, and engineers sharing their information, and the quality of the maintenance teams worldwide. Incidents still occur for various technical and mechanical reasons, including metal fatigue and human error. In a modern aircraft such as an airliner with a pressurized fuselage, the effects of any integrity to its structure can lead to three types of decompression: gradual, rapid and explosive.

Gradual decompression can be as deadly as explosive decompression, especially in an airliner climbing to its operating altitude, as occupants will eventually lose consciousness from hypoxia or oxygen deficiency. Gradual decompression can include an engine flameout as experienced by a BWA Boeing 747-200 on 24 June 1984, or pilot error such as that on a Pinnacle Airlines (FLG) Bombardier CRJ-200 on 14 October 2004. Rapid decompression is experienced within and typically takes more than half a second. This can occur for various reasons, such as the cargo-door malfunction experienced by an AAL DC-10-10 on 12 June 1972.

Explosive decompression is by far the most terrifying and can have a catastrophic effect on the integrity of the aircraft, as in the case of the de Havilland Comets lost in 1954. Explosive decompression occurs at a rate swifter than when air can escape from the lungs, which is around 0.1 to 0.5 seconds. There is also a risk of physical injury to the aircraft's occupants, as any unsecured object can be turned into a projectile due to the explosive force of decompression. While many aircraft have been lost due to explosive decompression, there are instances of aircraft able to land after experiencing such a catastrophic event.

On 24 February 1989 a UAL Boeing 747-122 on a Los Angeles–Sydney flight experienced a forward cargo-door failure after departing from a Honolulu stop-off. The resultant explosive decompression tore off part of the fuselage's starboard side skin to the rear and below the cockpit, blowing out several rows of seats and leading to the deaths of nine passengers. Miraculously, the 747 was able to return to Honolulu without further incident. The United States National Transportation Safety Board (NTSB), responsible for civil transportation accident investigation, recommended all 747-100s in service replace their cargo-door latching mechanisms with new, redesigned locks. This was followed by a sub-recommendation which suggested replacing all outward-opening cargo doors, such as the one which failed, with inward-opening doors. This change would equip the aircraft with a door that could not open in flight due to pressure differential.

On 8 March 2014 a Malaysia Airlines Boeing 777-200ER Flight 370 (MH370) took off for a routine international flight from Kuala Lumpur International Airport in Malaysia to Beijing Capital International Airport, China. The MH370, which carried a total of 239 passengers, disappeared from ATC radar contact round forty minutes after taking off at 01:22 Malaysian Time (MYT). Military radar subsequently tracked MH370 for a further hour as the aircraft deviated westwards from its planned flight path. At 02:22 MYT the military lost contact with it as it flew over the Andaman Sea. A subsequent report in October 2017 by Malay authorities stated that the aircraft distress systems were not activated. Furthermore, there was no indication of foul play, extreme weather, or technical factors that led to the loss of MH370.

What followed was a vast and much publicized multinational search for the lost aircraft and passengers, which would become the most expensive in aviation history. Military and civilian agencies worked together during political friction between the Chinese and Malay governments. Chinese concerns were mainly focused on the Malaysians slow sharing of information during the early stages of the search, which they felt hindered results. In China the government faced similar criticisms at home from the families of those on the lost aircraft.

From October 2014 to January 2017 a 46,000mi^2 (120,000km^2) seabed search, led by the Australians, proved fruitless despite items of debris belonging to MH370 washing up on East African shorelines. Between January and May 2018, led by tidal drift modelling, the US marine exploration company Ocean Infinity began to search a new area for debris. Despite mapping 43,000 mi^2 (112,000 km^2), Ocean Infinity found no further indication or evidence of the lost Boeing. Despite this, the Malay government remains resolute in locating and retrieving the lost aircraft and passengers.

There have been many theories on what occurred to MH370 including the possibility of uncontrolled decompression. However, until key data is retrieved from flight recording systems, these remain theories. Quick to respond to the disaster, the ICAO and IATA both began to look into ways to locate a downed aircraft in deep water. By November 2018 the ICAO had introduced a new standard for all aircraft flying over large bodies of water to report their positions every fifteen minutes. This was followed by an amendment to the Chicago Convention requiring aircraft manufactured after 1 January 2021 to be fitted with autonomous tracking devices. These would send location information at least once per minute in distress circumstances. Another key safety feature was provided from May 2014 by Inmarsat, who offered their tracking service for free to all aircraft equipped with Inmarsat satellite connection equipment. This covered nearly all commercial aircraft in service, and to further enhance flight safety the 'handshake' or connection between the Inmarsat satellite would change from hourly to fifteen minutes.

Another act that can lead to the loss of life and aircraft is sabotage. The first aircraft lost to an act of sabotage is believed to be that of an Imperial Airways Armstrong Whitworth Argosy II on 28 March 1933. The Argosy was on Imperial's regular London–Brussels–Cologne route when it crashed in flames near the city of Diksmuide, Belgium. Witnesses reported someone jumping without a parachute shortly before the Argosy crashed, and the body of German dentist Albert Voss was subsequently recovered. The investigation into the crash concluded a fire had been started at the aircraft's rear. At the same time, an inquest into Voss's death concluded that an investigation by the Metropolitan Police into Voss and drug smuggling had led him to light the fire with the intent of downing the aircraft. However, his estranged brother claimed that Voss planned to use the immediate confusion to slip away from the crash site and hope his death would be recorded, thus letting him escape justice.

Sabotage is thankfully rare, and the many safety protocols evolved over the years – including the physical and mechanical pre-flight checks developed by Boeing during their development of the B-17 –can pick up on any oddities. This, combined with the employment of a professional workforce, and stringent personal security checks, means that most acts of sabotage remain firmly in the realm of terrorism. That's not to say the occasional act does not occur. Thankfully these are quickly picked up on and dealt with by authorities. As recently as 2019 an AAL mechanic, Abdul-Majeed Marouf Ahmed Alani, glued a piece of foam into the inlet of a Boeing 737's air data module. This device is an essential piece of flight equipment which primarily measures the plane's pitch and speed. Thankfully the flight crew noticed an issue with their instrument readings when they increased engine power. This shows how well-trained flight crew are and how well they know their aircraft. This event occurred on 17 July, a date which has gained a reputation in aviation circles for being the worst day in aviation, with 902 fatalities from sixteen airliner crashes on this one day alone. For his sabotage, Alani, was sentenced to thirty-seven months.

Thankfully, for all the thousands of flights that take place every day incidents of calamity and terrorism are extremely few and far between. Indeed, over the past century, aircrews have remained calm, often under intense pressure, whether artificial or natural. This character trait has seen many potentially deadly situations resolved without loss of life. Coupled with a highly trained cabin crew, refined emergency procedures and equipment on the ground and in the air, numerous lives have been saved over the past century. The steady development of the multiple safety features, a history of continuous type development and a keenness to act on lessons learned has meant that air travel has delivered the safest transportation available today. By the end of the first century of regulated flight, the fatality figure sits at a mere 0.07 deaths per billion passenger miles travelled, which contrasts with all other forms of travel.

Conclusion

Into the Second Century: The Future is Now

On 13 October 2019, a century after the signing of the Paris Convention, Air Malta (AMC), one of over 5,000 airlines operating today, retired its only A319-100. The A319 was a family of aircraft whose design ethos would have been familiar to the early pioneers such as Junkers and de Havilland a century earlier. The reliable short-haul, narrow-bodied aeroplane was the backbone of many smaller operators, such as AMC, which had provided sterling service with a safety record par excellence. However, in contrast to Airbus's success, Boeing found its fourth-generation 737 MAX grounded over safety concerns regarding its new Manoeuvring Characteristics Augmentation System (MCAS), which had led to the loss of two aircraft. Nevertheless, this showed that although civil aviation remained potentially dangerous, it had also built robust safety systems.

The second century of civil aviation is already witnessing massive changes as it comes out of the Coronavirus pandemic of 2020–2022. The most notable change was the retirement of the Boeing 747-400 passenger aircraft. These giants of the skies are now being replaced by the next generation of equally capable two-engine, wide-bodied aircraft like the Boeing 777 and Airbus 350. However, regardless of the passing of such a recognizable icon as the 747, the danger that aviation would follow other transport industries in producing products which look 'same old, same old' remains very slim.

Aviation and aeronautics are very much at the cutting edge of technical innovation. As we welcome the second century of regulated flight, we do so with a generation of new ideas and ideals. Engine and airframe material technologies are constantly developing, driven by a desire to lessen the environmental impact of air travel while improving efficiency and lowering costs. Meanwhile, the physical and software design processes are striving to work together to improve overall efficiency.

The second century of regulated flight will bring the same passion and invigoration that accompanied the first, delivering new and exciting designs that match our desire for a cleaner, more efficient world. The global financial downturn that affected the investment in large-scale aeronautical development at the start of the twenty-first century is now easing. The return on this investment has seen a range of new and exciting innovations within civil aviation. Supersonic Transport (SST) is back on the agenda and developing at a pace, with Boom Technology and Spike Aerospace developing SSTs. Smaller than the awesome Concorde and Tupolev Tu-144s, this next generation SST will offer airlines and private owners the opportunity to cruise at speeds over 1,000mph (1,600km/h).

There is also a renaissance for the airship as creating more environmentally sound travel systems remains at the forefront of innovation. In the United Kingdom, Hybrid Air Vehicles (HAVs) have been developing the Airliner series of helium-filled airships to fulfil various needs, from commercial passenger operations to humanitarian relief cargo. Airliner will be powered by electrical engines often developed by partnerships between industry and academia. These developments are also encouraged with competitions for innovation from organizations such as the British Royal Aeronautical Society (RAeS), and the rebirth of the 1920s air races with the four-day Pulitzer Electric Aircraft Race planned for the mid-2020s. Other power sources are being developed by American Honeywell Aerospace using their expertise gained from building aviation products, including Auxiliary Power Units (APUs), to create hybrid and hydrogen engines. There is also a drive to increase the use of sustainable aviation fuel (SAF) sourced from a range of biomass, including waste cooking oil.

Many of these innovations will see aircraft designs tailored to get the most from their power sources, with engineers, designers and end users working to shape the next generation of aircraft and aviation solutions. Such innovations will include the zero-emission Ducted Electric Vectored Thrust (DEVT) single-stage, electrically powered jet engines used on the German Lilium Jet aeroplane. The mounting of DEVT engines on the elevators, ailerons and flaps also gives the Lilium capacity to operate as a VTOL aircraft. In addition, the development of hydrogen-powered aircraft continues, albeit at a slower pace, though there remains interest in its use to power rotorcraft and personal light aircraft.

The future of Unmanned Aerial Vehicles (UAVs) is also assured as the technology has constantly improved since the turn of the twenty-first century. The UAV, particularly electrically powered types, has already replaced rotorcraft and light aircraft in several roles, such as image gathering. Furthermore, the increased use of artificial intelligence will see UAVs develop into autonomous craft capable of operating Beyond Visual Line of Sight (BVLoS). Another development could well see e-Urban Air Mobility (UAM) providing a safe and efficient system, transporting passengers or cargo at low altitudes within urban and suburban areas. Already companies such as American mobility service provider Joby Aviation, alongside Japanese automotive company Toyota, are designing a multi-engine Electronic Unmanned Air Vehicle (EUAV) that will fulfil a UAM role in the continental United States.

Space travel will also become a defining feature of civil aviation's second century as more private investors strive to make space travel accessible with the continued use of private sector investment. The planned return of human-crewed moon missions by NASA will be achieved by collaborative working with companies such as SpaceX's Starship Human Landing System (HLS). These missions will provide vital data and understanding for continued exploration beyond the moon, and it is hoped that a human-crewed mission to Mars will take place by the 2040s. Space tourism will also grow, with companies such as Space Adventures Ltd continuing to develop its range of leisure activities, including zero-gravity flight experiences, and stays on the International Space Station (ISS). With so many new challenges ahead, it is fair to say that regulated civil aviation will make its second century as inspiring and dramatic as its first.

Bibliography

Print Media

Davies, K., *Aviation's People Movers: Ninety Years of Passenger Flight*, Freshwater, Regional Publications, 2001

Dobson, A., *A History of International Civil Aviation: From Its Origins Through Transformative Evolution*, Abingdon, Routledge, 2019

Donald, D. (General Editor), *The Encyclopaedia of World Aircraft*, Enderby, Blitz Editions, 1997

Falcus, M., *50 Airliners That Changed Flying*, Stroud, The History Press, 2018

Gayden, G., *Commercial Aviation 101*, Greg Gayden, 2019

Gero, D., *Aviataion Disasters; The World's Major Civil Airliner Crashes Since 1950*, Stroud, The History Press, 2017

Grant, R. G., *Flight: The Complete History of Aviation*, London, Dorling Kindersley Ltd, 2017

Greenwood, J, (Editor), *Milestones of Aviation: Smithsonian Institution Nation Air and Space Museum*, Fairfield, Hugh Lauter Levin Associates, Inc. 1995

Gunston, B., (Editor-in-chief), *The Illustrated Encyclopaedia of Propeller Airliners,* Leicester, Winwood Imprint, 1980

Gwynn-Jones, T., *The Air Racers: Aviation's Golden Era 1909–1936*, London, Pelham Books Ltd, 1984

Hale, J., *Women in Aviation*, Oxford, Bloomsbury Publishing Plc, 2019

HMSO, *Airlines at War: British Civil Aviation 1939–1944. An Official History*, (London 1946), Barnsley, Air World Books, 2018

Inglis, T., *Mid-century Modern Graphic Design*, London, Batsford, 2019

Johnson B., *Classic Aircraft*, Basingstoke, Macmillan/Channel 4 Books, 2000

Konemann L. (Editor) & Fecker, A., *The Complete Book of Flight: Facts Figures and the Story of Airports, Airlines and Aircraft*, Bath, Parragon, 2010

Marck, B., *Women Aviators: From Amelia Earhart to Sally Ride, Making History in Air and Space*, Paris, Flammarion, 2013

Marriot, L., *80 Years of Civil Aviation*, Devizes, Select Editions, 1997

Mondey, D., (General Editor), *Aviation: The Complete Book of Aircraft and Flight, London*, Book Club Associates, 1980

Olins, W., *Corporate Identity: Making Business Strategy Visible Through Design*, London, Thames & Hudson, 1989

Rendall, I., *Reaching for the Skies: The Adventure of Flight*, London, BBC Books, 1988

Rendgen, S. & Weidemann, J. (Editor), *Information Graphics*, Cologne, Taschen, 2009

Stewart, S., *Flying the Big Jets*, Shrewsbury, Airlife Publishing Ltd, 1986

Thetford, O., *Aircraft of the Royal Air Force 1918–57,* London, Putnam, 1958

Woodley, C., *Flying Boats; Air Travel in the Golden Age*, Stroud, The History Press, 2018

Woolford, S. & Warner, C., *Flight: The Evolution of Aviation,* London, André Deutsch, 2019

Web Resources

https://airlinehistory.co.uk/
https://airracee.com/
https://applications.icao.int/postalhistory/index.html
https://aviationbenefits.org/social-development/rapid-disaster-response/
https://boomsupersonic.com/
https://centreforaviation.com/
https://data.worldbank.org/indicator/IS.AIR.GOOD.MT.K1
https://data.worldbank.org/indicator/IS.AIR.PSGR
https://lilium.com/
https://panchobarnes.com/about-the-pancho-barnes-archive/
www.aerodacious.com/
www.aerosociety.com/
www.aerosociety.com/media/11409/2-wendy-pritchard.pdf
www.airbus.com/en
www.airplaneboneyards.com/post-wwii-military-airplane-boneyards.htm
www.airracinghistory.freeola.com/the%20great%20races.htm
www.aviationfanatic.com/ent_list.php?ent=21&CL_CollectionID=36
www.avjobs.com/history/index.asp
www.britishairracing.org/
www.boeing.com/
www.britishairways.com/travel/home/public/en_gb/
www.caa.co.uk/
www.claimcompass.eu/blog/biggest-busiest-airports-in-the-world/
www.deltamuseum.org/home
www.embraer.com/
www.esa.int/
www.faa.gov/air_traffic/flight_info/hurricane_season
www.fai.org/
www.flyingdoctor.org.au/
www.hugojunkers.bplaced.net/junkers-luftverkehrs-ag.html
www.hybridairvehicles.com/
www.icao.int/
www.iata.org/
www.jobyaviation.com
www.nasa.gov/
www.ninety-nines.org/

www.qantas.com/us/en.html

www.statista.com/chart/20788/aircraft-models-with-the-highest-estimated-production-figures/#:~:text=Since%20it%20first%20flew%20in,rolling%20off%20the%20production%20line.

www.si.edu/

www.thefamouspeople.com/aviation.php

www.un.org/Depts/ptd/search/node/aviation

www.wai.org/100-most-influential-women-in-the-aviation-and-aerospace-industry

www.weforum.org/agenda/2022/10/supersonic-flight-sustainable-aviation/

www.worldrecordacademy.org/transport/aviation-world-records

INDEX

9/11, 112, 123, 213, 218, 221

AB Aerotransport (ABA), 206
Abbott, Robert S., 132
Abbott-Baynes Scud 3 sailplane, 187
Abdulmutallab, Umar Farouk (al-Qaeda bomber), 221
Abeles, Peter, 90
Ace Aircraft Manufacturing Company, 190–91
  Baby Ace, 190
  Junior Ace, 190
Acosta, Bert, 44
Acro Sport, 191
Ader, Clémant, 11
  and *Ader Éole*, 11
adventure and sports flying, 15, 31–32, 34, 36, 38, 44, 146, 148, 166, 174, 178, 184, 186, 192
Aer Lingus, 94, 97, 117
Aerial Experiment Association (AEA), 67
Aerial Medical Service (AMS) (Australia), 170
Aerial Navigation Act 1911 (UK), 21
Aerlínte Éireann *see* Aer Lingus
Aero Club of America, 158–59
Aero Spacelines, 110, 168
  Conroy Virtus, 168
  Guppy, 168–69
  Pregnant Guppy, 168
  Super Guppy, 110, 168–69
aerobatics, stunt flying, 44, 46, 133, 135, 137, 140–41, 146–47, 179–82, 186, 188, 190
Aéro-Club de France, 158
Aeroflot (Public Joint Stock Company – Russian Airlines), 62–64, 92–93, 201–2
aeromodelling, 178, 181–82
Aeronautica Macchi, 33–34, 36
  M.16 sports, 34
  M.39, 34
  M.52, 34
  M.52R, 34
  M.7bis biplane seaplane, 34

INDEX

M.C.72, 34
M.C.73, 34
Aeronautical Commission of the Peace Conference (ACPC), 21
Aeronautical Research Laboratories (ARL), 211
Aeronca Aircraft, 142
  Super Chief 65 LB monoplane, 142
Aérospatiale *see also* Airbus, 109–11, 172
Aerovías Nacionales de Colombia (Avianca), 76
African Americans, 101, 103, 132–33, 147, 204
  and racism in aviation, 132–33, 204
Agnello, Francesco, 34
Agusta, 36
AIC Flight 182 bombing, by Sikh militants, 220
Air Canada (ACA), 86, 122
Air Commerce Act 1926 (US), 66
Air France (AFR), 30, 60, 83, 85–86, 96–99, 112, 123, 155, 201, 217, 219, 225
Air India (AIC), 210, 219–20
Air Inter, 99
Air Line Pilots Association, International (ALPA), 225
Air Mail Act 1930 (US), 69–70
Air Mail Scandal (1934), 69–70, 74
Air Mail Service Beacon System (US), 64–65, 68
Air Marking Group (in Bureau of Air Commerce), 134
Air Ministry (UK), 23, 35, 55, 140–41
Air Navigation (Restriction in Time of War) Order 1939, 55
Air New Zealand, 118
Air Orient *see also* Air France, 60
Air Race World Championship, 46
Air Safety Board (US), 74
Air Tahiti Nui (THT), 123–24
air traffic control (ATC), 74, 121, 152, 156–58, 178, 204, 226–29
Air Transat (TSC), 224
Air Transport Auxiliary (ATA) (UK), 95, 129, 138–40, 200
Air Transport Command (USAAF), 205
Air Transport Licensing Board (ATLB), 96–97
Air Union *see also* Air France, 57, 60
Airbus Commercial Aircraft, 103, 109–10, 112, 120–21, 123, 149, 161, 165, 169, 217, 221, 224, 232
  A300, 103, 109–10, 217
  A300-600ST Beluga, 110, 169
  A320, 109, 121, 224
  A321-231, 221–22

A330, 110, 123, 221, 224–25
A330-743L Beluga XL, 110
A340, 110, 123
A350, 123
A380, 109–10, 149, 161, 165
Aircraft Industries Association (AIA), 142
Aircraft Kit Industry Association (AKIA), 192
Aircraft Manufacturing Company Limited (Airco), 51–52
Aircraft Protection Operations (APO) (UK), 213
Aircraft Transport and Travel (AT&T), 51, 61
Airdrome Aeroplanes, 194
Airline Deregulation Act 1978 (US), 75, 120
airmail, 11, 26, 54, 56–57, 62, 64–67, 70–71, 73–75, 77, 105, 148, 156, 162, 206, 219
airships, 18–19, 21, 23, 43, 49, 58, 149, 171, 198, 228, 233
   LZ 127 *Graf Zeppelin*, 58
   LZ 129 *Hindenburg*, 228
   R33, 171
   R34 and first return crossing of the Atlantic, 18–19
Airspeed, 96, 129
   Oxford, 129
   Ambassador, 96
Akaflieg Darmstadt D-36 Circe sailplane, 186
Akaflieg Stuttgart fs24 'Phoenix' sailplane, 186
Aktiengesellschaft für Luftverkehrsbedarf (Luftag), 90–91
Ala Littoria S.A., 199
Alcock, John, 17–18, 43
Aldrin, Buzz, 194
All Women's Transcontinental Air Race, 142–43
All-American Flying Derby, 28, 40
Allen, Cecil, 39
   and R-1/2 *Spirit of Right*, 39
Almaz space stations, 195
al-Megrahi, Abdelbaset *see also* Lockerbie Bombing, 221
al-Qaeda, 218, 221, 223
al-Shabaab bombing, by Borleh, Abdullahi Abdisalam, 221–22
American Airlines (AAL), 28, 71, 102–4, 151, 218, 221, 229, 231
and Mercury skysleeper service, 28, 71
American Airways (AA), 69
American Export Airlines (AEA), 100, 205
American Helicopter Society, 143
American International Airways (AIA), 72

American Overseas Airlines (AOA), 100
American Red Cross, 142
Amsterdam Airport Schiphol, 61–62, 208
Annabel Airport, 209–10
Ansari X-Prize, 193
Ansett Airways (AAW), 89–90
Ansett, Sir Reg, 89–90
Antonov, 92, 167, 169
   An-2, 92, 169
   An-24, 93
   An-225 Mriya, 167–68
Antonov Airlines (ADB), 167
Argüello, Patrick (Sandinista National Liberation Front), 214
Armistice (11 November 1918), 16, 23
Armstrong Flight Research Center (AFRC) (NASA), 137
Armstrong Whitworth, 24–25
   Argosy, 97
   Argosy II, 230
   AW.15 Atalanta, 25, 54
   AW.27 Ensign, 25
Armstrong, Neil, 194
Arnold, Gen Henry 'Hap', 139, 203
Arnold, Henry, 72
Associated Motion Picture Pilots' (AMPP) Union, 137
Atlantic Ferry Organization (of MAP), 199
Auriol, Jacqueline, 140
Australian National Airways (ANA), 89–9
Automobile Cup, 42
Avenger Field, Texas, 142
Aviation Corporation of the Americas (ACA), 72
Aviation Traders, 83–84
   ATL-98 Carvair, 83–84
Aviatrissa (women's aero club), 147
Avro, 85, 115, 172
   504K, 50
   Lincoln,
   Model 671 Rota, 172
   Model 691 Lancastrian, 199
   Tudor, 83
   York, 85, 227
Ayulo, Ernesto, 76

Babb, Charles, 39
Bacon, Roger, 10
  and *History of Art and Nature*, 10
Baikonur Cosmodrome in Kazakhstan, 195
Bailey, Lady Mary, 126–27
Bailey, Sir Abraham, 126
Ballin, Albert, 58
ballooning, 42, 125, 179, 183–84
Barish, David, 188
Barlow, Howard, 41–42
Barnes, Florence 'Pancho', 44, 133, 135–38, 179
barnstorming, 40, 129, 135, 137, 162, 179, 190, 203
BASE jumping, 190
Baumeister, Al, 145
Bayles, Lowell, 38–39
BEA Airtours, 98
Beaverbrook, Lord, 95
Bede Aviation BD-5J Micro, 193–94
Beech, Walter and Olive, 134, 174–75
Beechcraft/Beech Aircraft Corporation, 27, 99, 134–35, 174–76
  Beech Bonanza, 176
  Beechcraft C17R Staggerwing, 142, 175
  C17R biplane, 134
  Model 18, 27
  Model 80 Queen Air, 175
Beijing Capital International Airport, 229
Beijing Daxing Airport, 155
Beirut Airport, 217
Bell Aircraft, 172
  UH-1A Iroquois, 172
Bell Helicopter School, Fort Worth, 143
Bell, Dr Alexander Graham, 67
Bell, Larry, 143
Bendix Trophy, 39, 45, 134–35, 138
Bendix, Vincent, 45
Benítez, Merino, 77
Bennett, James Gordon Jnr, 42
Bennett, Jim and Shiers, Wally, 19
Berlin-Brandenburgisches Luftfahrtunternehmen GmbH (Berline), 91
Bermuda Agreement (1946), 80
Berry, Albert, 14
Berta, Ruben Martin, 76

Besançon, Georges, 156
Bessie Coleman Aero Club, 133
Betsy Ross Air Corps (BRAC), 137, 203
Bezos, Jeff, 197
Biafran Airlift, 209
Big Four (American, Eastern, TWA and United), 69, 71–72, 100, 103
Bigelow Aerospace, 197
bird strikes and FOD, 35, 149, 157, 224–25
Birmingham Small Arms Company (BSA), 52
Black September war, 214
Blanchard, François, 11
Bleecker, Maitland, 33
Blériot, 29
  SPAD S.27, 29
  SPAD S.56, 57
  XI monoplane, 13, 106, 194
  XI-2 monoplane, 15
Blériot, Louis, 13, 43, 57, 113, 158, 181
Bloch (Société des Avions Marcel Bloch), 30
  MB.120, 30
  MB.161 Languedoc, 30, 98
  MB.220, 30
  MB.60, 30
Blue Origin, 197
Boardman, Russell, 39
Boeing, 26–27, 30, 37. 59, 66–67, 70–72, 84, 92, 94, 97, 99, 102, 107–10,
  115, 119, 121–24, 152, 161, 163, 165, 168, 196, 215, 231–32
  200/221 Monomail, 26
  247, 26–27, 30, 61, 70, 218
  307 Stratoliner, 27, 66, 72
  314 Clipper, 74, 107, 115, 199, 204
  377 Stratocruiser, 83–85, 94, 168
  40, 26
  40A, 67
  707, 88, 91–92, 94, 99–101, 104, 210, 213–14, 216, 219–20
  720, 92, 102, 104
  727-100, 90, 214, 216
  727-231, 217
  737-100, 167, 216
  737-200, 97, 225
  737-230C, 217
  737-800, 208, 228

737 MAX, 232
747-100 (Jumbo Jet), 66, 95, 99, 101–4, 109–12, 119–20, 123, 147, 149, 157, 161, 166–67, 214, 226, 229, 232
747-121, 220, 226
747-122, 229
747-200, 229
747-206B, 226
747-230B, 207
747-327B, 220
747-400, 169, 232
747-400 Large Cargo Freighter (LCF), 166, 169
757-222, 218
757-223, 218
767, 123
767-200, 218
767-223ER. 218
767-300ER, 221
777, 123, 208, 229–30, 232
787 Dreamliner, 123, 168–69
80 Trimotor, 26
B-17 Flying Fortress, 82, 206
B-47 Stratojet, 103
B-52 Stratofortress, 103, 168
C-97 Stratofreighter, 209
Shuttle Carrier Aircraft (SCA), 167–68
VC-25 (*Air Force One*), 109
VC-137 (*Air Force One*), 112
Boeing 737 MAX groundings, 232
Boeing Air Transport (BAT), 67, 161
Boeing School of Aviation, Oakland, 141
Boeing 'Sky Girls', 161
Boeing Vertol, 172
  CH-47 Chinook, 172
Bombardier Aerospace/Aviation, 33, 177
  CRJ-200, 229
  Global 7500, 177
bombings, 218–222
Bond, William Langhorne, 73
Boniface, Edmund and CSR, 212
Bowlus, William, 65
Brabazon Committee, 86, 96
Brackley, Herbert, 54

branding and PR, 32, 58, 59, 61, 91, 113, 115–17, 121–23, 130–31, 134, 142, 201
Braniff International Airways (BIA), 72, 83, 118
Braniff, Paul, 71
Braniff, Thomas, 71, 74
Branson, Richard, 120, 193, 197
Brazil mid-air collision, 228
Breguet, Louis Charles, 57, 171
Bréguet Deux Ponts, 98
Bréguet-Richet Gyroplane, 171
Breitling Orbiter 3 circumnavigation, 183–84
  and Brian Jones and Bertrand Piccard, 183
Bristol Aeroplane Company, 84, 97, 111
  Type 170 Freighter, 84, 166
  Brabazon, 86
  Type 175 Britannia, 94
Bristow Helicopters Ltd, 173
Bristow, Alan, 173
Britannia Trophy, 127
British Aerobatic Association (BAMA), 181
British Air Racing Championship, 43
British Air Registration Board, 211
British Aircraft Corporation (BAC) *see also* Concorde, 36, 97, 103, 109–11
  One-Eleven, 97, 103
British Airways (BA), 98, 109, 112, 118, 120, 226
British Airways Board, 98
British Airways Limited (BAL), 55–56
British Commonwealth Pacific Airlines (BCPA), 88, 90
British Continental Airways (BCA) *see also* British Airways Limited (BAL), 56
British Eagle, 97
British European Airways (BEA), 94–90
British Marine Air Navigation Company Ltd, 52
British Overseas Airways Corporation (BOAC), 55–56, 62, 83–85, 88, 93–98, 112, 199–200, 202, 214
British South American Airlines (BSAA), 94
British United Airways (BUA), 97
British Women's Pilots' Association, 140–41
Brittin, Louis, 71
Brown, Arthur Whitten, 17–18, 43
Brown, Jill E., 147
Brussels Haren Airport, 61

Bureau of Air Commerce, 74, 134
Bureau of Safety Regulation, 74
Burgess, Carter, 101
Burton, Roland, 48
business jets, 175–77, 208
Butler, Alan, 36
BWIA West Indies Airways Limited, 7

Cain, Yola, 147
Canadair Sabre 3, 139
Canadian Air Carrier Protection/Protective Program (CAPP), 213
Canadian Air Transport Security Authority (CATSA), 213
Canadian Alpine Helicopters, 173
Canadian Government Trans-Atlantic Air Service (CGTAS), 199
Canadian Pacific Air Lines (CPC) Douglas DC-3 bombing, 219
Cape Canaveral, Florida, 168
Caproni Vizzola Calif A-21SJ sailplane, 187
Cardiff Municipal Airport, 44
Cartland, Dame Barbara, 148
Casablanca Conference (1943), 205
Castoldi, Mario, 34
Cathay Pacific (CPA), 122
Cathay Pacific Catalina hijacking, 213
Caudron G.3, 126
Caudron, René, 57
Cavendish, Henry, 10
Cayley, Sir George, 185
Central Committee of the Communist Party of the Soviet Union, 63
Centre Spatial Guyanais (CSG) spaceport, 196
  Ensemble de Lancement Soyouz (ELS), 196
Cessna, 99–100, 174–76
  120, 100, 175
  140, 100, 175
  150, 175
  152, 175
  172 Skyhawk, 100, 175
  172P, 175
  180, 144
  Airmaster, 174–75
  AT-17 Bobcat/T-50, 174
  Citation, 175
  SR-3 racer 174

INDEX

Cessna, Clyde, 174
Chadwick, Roy, 85
Challenge International de Tourisme, 127
Chambers, Reed, 68
Changi Airport, Singapore, 155
Channel Islands Airways Ltd *see also* British Airways Limited (BAL), 56
Charles de Gaulle Airport, Paris, 112
Chermayeff, Ivan, 117
Chiang Kai-Shek, 141, 205
Chicago Convention on International Civil Aviation (1944), 79–81, 153, 155, 206, 210, 230
Chicago Defender, 132
Chicago International Air Race, 41
Chicago-O'Hare Airport, 104
Chief Directorate of the Civil Air Fleet see *also* Civil Air Fleet *see also* Aeroflot, 63
China Airways, 73
Chinese National Aviation Corporation (CNAC), 73, 202, 205
Church, Ellen *see also* Boeing 'Sky Girls', 161
Cia de Aviacion Peruanas SA, 76–77
Cia Mexicana de Aviación (CMA), 78
CIAM Flyer, 182
Cirrus Aircraft, 178
  Vision Jet, 178
Cirrus Airframe Parachute System (CAPS), 178
Cirrus Hi-Drive aero engine, 40
Civil Aeronautics Act 1938 (US), 74
Civil Aeronautics Administration (CAA), 75, 134, 143, 227
Civil Aeronautics Authority (CAA), 74–75, 136, 203, 227
Civil Aeronautics Authority War Training Service, 203
Civil Aeronautics Board (CAB), 75, 100, 120
Civil Air Fleet (CAF) (USSR), 63, 146, 201
Civil Air Guard (UK), 140
Civil Air Liaison Services (Free Air France), 201
Civil Air Patrol (CAP), 135, 138, 142
Civil Aviation (Licensing) Act 1960 (UK), 96
Civil Aviation Authority (CAA) (UK), 97, 179
Civilian Pilot Training Program (CPTP), 75, 142, 159, 203
Cleveland Air Races, 38
Cleveland National Air Show, 44–45
Cobham, Sir Alan, 53
Cochran, Jacqueline, 39, 138–42, 201, 203–4

Coleman, Alfred, 58
Coleman, Bessie *aka* 'Queen Bess', 132–33
Coleman, Tracey, 213
Collins, Eileen Marie, 147
Colonial Air Transport (CAT), 66, 69
Colonial Airlines, 102
Comité d'Étude pour la Navigation Aérienne au Congo (CENAC), 60
Commonwealth Aircraft CA-28 Ceres, 174
Compagnie des Grands Express Aériens (CGEA), 57, 227
Compagnie des Messageries Aériennes (CMA), 66–67
Compagnie Franco-Roumaine de Navigation Aérienne (CFRNA) *see also*
Compagnie Internationale de Navigation Aérienne (CIDNA), 57
Compagnie Générale Aéropostale *see also* Air France, 60
Compagnie Internationale de Navigation Aérienne (CIDNA) *see also* Air France, 57, 60
Compañía de Aviación Faucett, 76
Compania Mexicana de Transportación Aérea (CMTA), 77–78
Concorde (Aérospatiale/BAC), 17, 87, 95, 104, 106, 109–12, 123, 225–26, 233
Concorde Flight 4590 (AFR) disaster, 225–26
Conroy, John, 168–69
Consolidated
　B-24 Liberator, 200, 203
　Convair 880, 101
　Convair CV-340, 91
　Convair CV-990-30A-6, 220
　Convair F-106, 215
　LB-30, 93
　PBY Catalina flying boat, 89, 202
Continental Airlines (COA), 66, 225
Contract Air Mail Act 1925 (US) *aka* Kelly Act, 65, 68, 70
Contran École d'Aviation, 141
Cooper, Dan and NWA Boeing 727 hijacking, 214–16
Corben, Orland, 190–91
Cornu, Paul, 171–72
Cosyns, Max, 183
Coupe d'Aviation Maritime Jacques Schneider *see* Schneider Trophy
Coupe Femina, 126
　and Barrault, M., 126
COVID-19 pandemic, effects of, 46, 124, 167
Crawford, Frederick, 44
Crimson Fleet Air Circus, 140

crop spraying/dusting, 92, 173
Croydon Airport, 53, 61, 157
Curtis, Eleanor, 140
Curtiss Aeroplane and Motor Company *see* Curtiss Aeroplane Company
Curtiss, Glenn, 11–14, 33, 158
Curtiss-Wright Corporation *also* Curtiss Aeroplane Company, 14, 29, 33, 66–67, 134, 175
   AT-32C Condor II, 29, 31, 71, 161
   Autoplane, 33
   Carrier Pigeon biplane, 67
   CR-1, 44
   Curtiss-Bleecker SX-5-1 helicopter, 33
   JN-4 'Jenny', 129, 133, 135, 190
   June Bug, 11
   NC flying boat, 16–17
   R3C, 33
   Reims Racer, 106
   Thrush J monoplane, 134–35
Curtiss-Wright Aviation School, 129
Curtiss-Wright Technical Institute, Glendale, 163
Czechoslovak Airlines, 92

Da Vinci, Leonardo, 10, 171
Daedalus, 9
Dahl, Alexander, 183
Dai Nippon Kōkū (DNK), 199
Daily Mail, 13, 17–18, 43, 46, 128, 162, 185
Daimler Airway, 52
Damon, Ralph, 101
Darcy, Jimmy, 170
Dassault,
   Mirage IIIR, 140
   Falcon 50, 208
Dassault Falcon 50 SAM shoot-down (Rwanda/Burundi), 208
Dawson's Field and PFLP hijackings, 214
De Aresti, José Luis, 181
De Assis Brasil, Érico, 76
De Bernardi, Mario, 34
de Havilland Aeroplane Hire Service, 53
de Havilland Aircraft Company Ltd, 23–25, 29, 36–38, 51–53, 66, 84, 86, 88–89, 92, 96, 232
   DH.4, 44, 64

DH.4A, 23, 52, 106
DH.14, 51
DH.16, 52, 61
DH.18, 23, 26
DH.18A Mailplane, 227
DH.29, 23
DH.34, 23, 52
DH.37, 36–37
DH.50,
DH.50 floatplane, 53, 170
DH.53 Humming Bird, 37
DH.60, 30
DH.60 Cirrus II Moth, 127
DH.60 Gipsy Moth, 127, 140
DH.66, 24, 53
DH.80 Puss Moth, 37, 56, 128
DH.82 Tiger Moth, 37
DH.83 Fox Moth, 56, 117
DH.84 Dragon, 24, 128
DH.86, 24, 114
DH.87 Hornet Moth, 37
DH.88 Comet, 37, 47, 140
DH.89 Dragon Rapide, 24–25, 56, 76
DH.91 Albatross, 25
DH.95 Flamingo, 25–26
DH.104 Dove, 86
DH.106 Comet, 37, 86, 88–92, 94, 128, 228–29
DH.106 Comet 2, 88, 90
DH.106 Comet 4, 88, 92
DH.106 Comet 4B, 88, 96
DH.114 Heron, 86
DH.121 Trident, 96
Nimrod (military Comet), 88
de Havilland Canada
DHC-2 Beaver, 174
de Havilland, Geoffrey, 37, 51–52
De Jager Dirk, 225
De la Cierva, Juan and his Cerva C.30 rotorcraft, 172
De la Meurthe, Henri Deutsch, 12
De la Roche, Baroness, 14
De Marmier, Lionel, 201
decompression, 220, 229–30

Defense Advisory Committee of Women in the Service (in USAF), 135
Delta Airlines (DAL), 103, 108, 116–17, 122, 151
Department of Commerce (US), 65, 75
d'Erlanger, Gerard, 95, 200
Deroche, Élise *aka* the Flying Baroness, 125–26
Deruluft-Deutsch Russische Luftverkehrs A.G. (DDRL), 62
Det Danske Luftfartselskab A/S (DDL) *see also* Scandinavian Airlines (SAS), 62, 84, 206
Det Norske Luftfartselskap AS (DNL) *see also* Scandinavian Airlines (SAS), 84, 206
Deutsche Aero Lloyd AG (DAL) *see also* Deutsche Luft-Reederei (DLR), 59
Deutsche Luft Hansa A.G. (DLH) *see also* Lufthansa, 59–60, 90
Deutsche Lufthansa *see also* Lufthansa, 59–60, 90–91, 122, 198, 216
Deutsche Luft-Reederei (DLR), 58–59, 62
Deutsche Luftschifffahrts-Aktiengesellschaft (DELAG), 57–58
Deutsche Zeppelin-Reederei (DZR), 58
DH.88 Comet *Grosvenor House see also* Scott, Charles & Black, Tom, 37, 47
Diener, Nelly Hedwig, 161
   Tuttlingen accident, 161
Dillon, Reed & Co., 102
Dobrolyot (Russian Society of Voluntary Air Fleet), 63
Dolgorunaya, Princess Sophie Alexandrovna, 159
Doolittle, Jimmy, 39, 44, 137
Dornier, 107, 162
   Do J Wal flying boat, 76
   Do X, 107, 115
   Komet, 62
DOS space stations, 185
Double Eagle II and balloonists Ben Abruzzo, Maxie Anderson and Larry Newman, 184
Douglas Aircraft Company, 27–29, 37, 70–72, 83–85, 100, 103–4, 108, 152, 163, 168
   C-54 Skymaster, 100, 205
   Cloudster, 65
   Commercial series (DC), 28
   DC-1, 28, 70–71
   DC-2, 28, 61–62, 71, 107, 163, 172, 202, 205–6
   DC-3 (C-47), 26, 28, 62, 64, 71–72, 77, 82–84, 86, 89, 95, 98, 103, 105, 107, 117, 146, 152, 166, 200, 202–3, 205–6, 213, 219
   DC-4, 28, 83–84, 89, 98, 100, 108
   DC-4E, 72
   DC-6, 48, 84, 89–90, 174, 207

DC-6B, 219, 227
DC-7, 84, 103, 210, 227
DC-7C, 100
DC-8, 100, 102, 104, 214
DC-9-40, 97
Douglas Sleeper Transport (DST), 28 71
M-2 biplane, 62–67
Douglas, Donald, 70
drones (UAVs), 173, 182, 233

Earhart, Amelia, 27, 128–33, 135, 137, 144
Eastern Air Lines (EAL), 102, 118, 179
  Shuttle, 102
Eastern Air Transport (EAT), 69–70
easyJet, 84, 165
Eckener, Dr Hugo, 58
École de pilotage Caudron du Crotoy, 132
Economic and Social Council (ECOSOC) (UN), 207
Edsel Ford Air Tour, 13
Edwards Air Force Base, 137–38, 167
Edwards, Sir Ronald, 97–98
Egyptair, 141
Egyptian Aviation Club, 141
Egyptian Organization of Aerospace Education, 141
El Al Israel Airlines Ltd (ELY), 209, 214
Elmer of Malmesbury, 9–10
Elnadi, Lotfia, 141
Embraer, 178
  Legacy 600, 228
Emirates (UAE), 110
Empire Air Mail Scheme (EAMS), 203
Empresa de Transportes Aéreos Aerovias Brasil S/A, 7
English Electric Aviation Ltd, 97
  Canberra, 48
Entebbe Airport, 217
  Sayeret Matkal (Israeli General Staff Reconnaissance Unit), 216–17
Entebbe Airport, Israeli raid on, 217
Enterprise for Friends of the Air Fleet (ODVF), 63
EoN Olympia sailplane, 186
European Aviation Safety Agency (EASA), 110
European Space Agency (ESA), 196
Eustace, Alan, 183

INDEX

Everett Plant (Boeing), 108
ExcelAire (XLS), 228
Experimental Aircraft Association (EAA) (US), 192

Fahie, Bill, 141
FAI Amateur-Built Aircraft Commission (Commission Internationale des Amateurs Constructeurs d'Aéronefs (CIACA)), 193–94
FAI Astronautic Records Commission (International Astronautic Records Commission (ICARE)), 197
FAI Ballooning Commission (Commission Internationale de l'Aérostation (CIA)), 185
FAI CIVA (Commission Internationale de Voltige Aérienne) aerobatics commission), 179–80
FAI CIVA World Aerobatic Championships (WAC), (Lockheed Trophy),180
FAI Hang Gliding & Paragliding Commission (Commission Internationale de Vol Libre (CIVL)), 188
FAI Aeromodelling Commission (Comité International d'Aéromodelism, (CIAM)), 181–82
FAI International Gliding Commission (IGC), 187
FAI Microlight and Paramotor Commission (Commission Internationale de Microaviation (CIMA)), 189
FAI Rotorcraft Commission (Commission Internationale de gravitation (CIG)), 173
FAI World Paragliding Championship, 189
Fairgrave's Flying Circus, 135
Farley, James, 70–71
Farman (brothers), 60, 126, 158, 181
   F.120, 62
   F.60 Goliath, 29, 56–57, 227
   Voisin-Farman I, 12
Farman, Henri, 12–13, 29
Farnborough air base, 211
Faucett, Elmer, 76–77
Fauchille, Paul *see also* Oxford, Hague Peace *and* International Air Navigation conferences, 20–21
Federal Air Marshal Service/Program, 213, 217
Federal Aviation Act 1958, 227
Federal Aviation Administration *also* Agency (FAA), 75, 108–10, 120–21, 143, 157, 213, 218, 227
Federal Bureau of Investigation (FBI), 215, 221
Fédération Aéronautique Internationale (FAI) (World Air Sports Federation), 127, 147, 173, 178–182, 185, 187–89, 193, 197

Fhimah, Lamin Khalifah *see also* Lockerbie Bombing, 221
Fiat AS.6 V24 aero engine, 34
Fileux, Alfred, 14
Filton (Bristol) *see also* Concorde, 111
financing, 49, 52, 67, 72, 105, 149, 161–62, 164–65, 190, 195, 197
fire-fighting, 154, 161, 172–74
Firle, Otto, 117
first aeroplane fatality (1908) *see also* Selfridge, Thomas, 12
first air hijacking (by Franz Nopcsa von Felső-Szilvás, 1919), 212
first aircraft bombing (UAL Boeing 247) 1933, 218–19
first balloon cross-channel flight (1785) *see* Blanchard, François and Jeffries, Dr John, 11
first carriage of official air mail (1911) *see* Pequet, Henri, 14
first female flyer to cross the Channel (1912) *see* Quimby, Harriet, 14
first flight UK to Australia (1919) *see* Macpherson-Smith, Ross and Keith and Bennett, Jim and Shiers, Wally, 19
first flyer to cross the Channel (1909) *see* Blériot, Louis, 13
first powered flight, at Kitty Hawk, NC (1903), 11
first recorded mid-air collision, Picardy, France, 227
first transatlantic flight (1919) *see* Alcock and Brown, 17–18
first use of a parachute from an areoplane (1912) *see* Berry, Albert and Jannus, Tony, 14
first woman to qualify as a pilot *see* De la Roche, Baroness, 14
First World War, 9, 16–17, 19–20, 23, 32, 34–35, 40, 42, 47, 50–51, 53–54, 57–59, 66, 68, 76, 105–6, 112–13, 126–27, 129–30, 137, 149, 156, 159, 162–63, 172, 179, 182, 189–90, 210–11
Fitzmaurice, James, 14-
Five Freedoms agreement *see* International Air Transport Agreement (IATA)
FLARM, developed by Urs Rothacher and Andrea Schlapbach, 228
Flight 110 Pan Am 707-321B hijacking, 216
Flight 1812 (SBI) shoot-down, 208
Flight 752 (AUI) shoot-down, 208
Flight 847 TWA Boeing 727-231 hijacking, 217
Flight MH17 (MAS) shoot-down, 208
Flight MH370 (MAS) disappearance, 229–30
Flightmaster (ARS), 96
Florida Airways (FAW), 68
Flug- und Fahrzeugwerke Altenrhein FFA P-16 (Swiss) *see* Learjet
Flugzeugbau Friedrichshafen 49c floatplane, 62
Flybe (BEE), 122
flying clubs, 42, 125, 128, 146–47, 151, 162, 178, 197
Flying Doctor Service (Royal) (RFDS), 170

Flying Training School Programme (UK), 203
Flynn, Rev John *see also* Flying Doctor Service (Royal) (RFDS), 170
Focke-Wulf, 31
   A.16, 31
   A.32 Bussard, 31
   Fw 47 Höhengeier, 31
   Fw 200 Condor, 31, 62
Fokker, 26, 29, 56, 61, 63, 122
   F.II, 29, 61
   F.III, 61
   F.VII Trimotor, 61–62
   F.VIIa-3m, 66,
   F.XII, 62
   F.XVIII Trimotor, 56, 61–62
   F.XX, 29
   F.XXXVI, 29
   Fokker-Grulich F.III, 62
   F-VII/3m Trimotor, 72, 130
Fokker, Anthony, 29
Folger-Brown, Walter, 69–70
Ford Air Transport Service (FATS), 66
Ford Company, 47
   4-AT Trimotor, 29, 66–69, 78, 114, 144, 212
   Stout 2-AT, 66
   Stout 3-AT, 29
Ford National Reliability Air Tour, 47
Formula One Air Racing, 47
Fossett, Steve *see also* GlobalFlyer, 193
François Hussenot and Paul Beaudouin, 211
Frankfurt Airport, 91
Freddie Lund Trophy, 180
Freedoms of the Air, 80
Friedrichshafen G.III, 59
Frye, Jack, 101
Fysh, Hudson, 170
Fysh, Wilmot *see also* Qantas, 54

Gaddafi, Muammar *see also* Lockerbie Bombing, 221
Gagarin, Yuri, 194
Gannon, Don, 48
Garnerin, Elisa, 125
Garnerin, Jeanne-Geneviève (first female parachute jump), 125

Garros, Roland, 15
Gates Aviation Corporation/Gates Learjet Corporation, 177
Gatwick Airport (W. Sussex), 56. 153
Gee Bee (Granville Brothers Aircraft), 38
  Model A, 38
  Model D, 38
  Model E, 38
  Model Y Senior Sportster, 39
  Model Z Super Sportster, 38–39
  R-1, 39
  R-6 International Super Sportster,
  R-6H, 138
  Sportster, 38
General Aircraft Monospar ST-4, 77
General Electric aero engines, 88, 108
General Tire and Rubber Trophy, 39
Gengoult Smith, Sir Harold, 47
George V, King, 18, 43, 128
Giffard, Henri, 49
Gipsy Major aero engines, 24
  Twelve, 25
Girard, Alexander 'Sandro', 118
gliding, 33, 41, 43, 72, 145, 148, 159, 179–80, 182, 185–88, 225
Global Positioning System (GPS), 83, 228
GlobalFlyer (Virgin Atlantic), 193
Gloster, 114
Goddard, Peter, 43
Gol Linhas Aéreas Inteligentes (GLO), 228
Goodyear, 115
Goodyear Europa Blimp, 169
Goodyear Trophy, 39
Gordon Bennett Aviation Trophy, 42
Gordon Bennett, James, 42
Gorky Agit-Squadron, 145
Gorst, Vern, 68
Gower, Pauline, 140–41
Grace Shipping Company (Panagra), 76
Graham, Daisy, victim of bombing of a UAL Douglas DC-6B, 219
Graham, Jack, bombing of a UAL Douglas DC-6B, 219
Grand Canyon collision, 227
Grand Prix d'Aviation, 12
Grands Express Aériens, 19, 51, 57, 227

Granville Brothers *see also* Gee Bee, 38–40
Granville, Zantford, 38
Great Air Race (1919), 47
Great Depression, 74, 134, 137, 162, 174, 176, 190
Great Lakes 2T-1A Sport Trainer, 180
Greve, Louis, 44
Grizodubova, Valentina, 145–46
Grubbs, Victor, 226
Grumman
 G-44 Widgeon, 140
 Gulfstream II, 177
Guay, Joseph-Albert and CPC DC-3 bombing, 219
Guest, Amy, 130
 and Fokker *Friendship*, 13
Guillaume, Maxime, 86
Guthrie, Sir Giles, 48, 94–95

Habyarimana, Juvénal, 208
Hadid, Safdie and Zaha (airport) architects, 155
Hagg, Arthur, 37
Hague Peace conferences (1899 & 1907), 20
Hall, Bob, 38
 and Williams Trophy, 38
Hall, Donald, 66
Hall, Floyd, 102–3
Haller, Augustine, 185
Hamburg-Amerikanische Packetfahrt-Action-Gesellschaft (HAPAG), 58
Hamilton Cove Seaplane Base (Catalina Island), 67
Hammarskjöld, Dag, 207
Handley Page, 23, 35, 52
 H.P.42, 24, 54
 HP.115, 111
 HP.81 Hermes, 85
 W8, 23, 52
 W8f, 23, 53, 60
Handley Page Transport (HPT) *also* Handley Page Indo-Burmese Transport, 51
hang gliding, 185, 188
Hanshue, Harris, 67
Harewood Gold Cup, 49
Hargreaves, William, 35
Harmon Trophy, 85, 127, 138, 140–41, 141–45

Harmsworth, Alfred, 43
Harris, David E., 211
Harrison, Len and Husband, Vic, 211
Hartsfield–Jackson Atlanta International Airport, 151, 226
Hawker Beechcraft Corporation (HBC),
Hawker Siddeley Group, 38, 109, 111
    Harrier, 43
    HS 748, 167
Hawks, Frank, 130, 185
   and flight of 'Texaco Eaglet', 185
Hearst, Randolph, 14
Hell's Angels (movie), 41, 137, 179
Helms, Lynn, 121
Henry, Ernestine, 125
Herring, Augustus, 33
Herring-Curtiss Company *see* Curtiss Aeroplane Company
Heston Aerodrome, 56
Hezbollah, 217
Hietala, Veijo and 'Mata Hari' FDR, 211
Highland Airways *see also* British Airways Limited (BAL), 55–56
Highlands and Islands Medical Service in Scotland, 170
Hillmans Airways *see also* British Airways Limited (BAL), 56
Hinkler, Bert, 128
Hintze, Herbert, 200
Hirth, Wolfram, 185
Holt-Thomas, George, 51
Holzindustrie Meckenbeuren GmbH, 58
home-build aircraft kits, 179, 190–94
HondaJet *see* Piper Aircraft Corporation
Honeywell Aerospace, 233
Honeywell-Bendix Trophy for Aviation Safety, 45
Honourable Company of Air Pilots (UK), 160
Hounslow Heath Aerodrome, 47, 51
Howard, Lesley, 200
Hudson River emergency landing, by AWE Airbus A320-214, 224
Hughes Aircraft Company, 41
    H-1 Racer, 41
Hughes, Billy, 46
Hughes, Howard, 41, 85, 101–2, 137, 143, 179
human-powered flight, 10, 194
Humber-Sommer Biplane, 14
Hunting Aircraft, 97

Hussein, King, 214

Iberia Líneas Aéreas de España, S.A. (LAPE), 61
Ibero-American Convention on Air Navigation (1926), 22
Icarus, 9, 210
Icarus Cup 2012, 194
Ilyushin,
  DB-3 bomber, 206
  Il-12, 92
  Il-14, 93
  Il-18, 91, 93
  Il-62, 93
  Il-76, 169
Imperial Airways Ltd (IAL), 19, 24, 52, 107, 114–15, 230
Imperial German Army Air Service (Luftstreitkräfte), 29
Imperial Japanese Airways, 72
Imperial Japanese Navy Air Service, 202
incidence of clear air turbulence (CAT) UAL Boeing 747-100 (1997), 157
Indian Airlines (IAC), 210
Instone Air Line (IAL), 52
Instone, Sir Samuel, 52
Inter-Allied Control Commission, 107
Interflug Gesellschaft für Internationalen Flugverkehr m.b.H., 91
Interim Council (in ICAO), 81
International Aerobatic Club (IAC), 181
International Aeronautical Body (in ICAO), 81
International Air Freight, 29
International Air Navigation Conference (Paris 1910), 21
International Air Traffic Association (IATA), 22, 80–81, 152, 155, 167, 230
International Civil Aviation Organization (ICAO), 49, 80–81, 150, 155, 170, 198, 207, 211, 222, 230
International Commission for Air Navigation (ICAN), 9, 22, 80
International Committee of the Red Cross (ICRC), 209–10
International Space Station (ISS), 195–97, 234
ISIL (Islamic State of Iraq and the Levant) bombing, 221

Jalbert, Domina, 188–89
Jannus, Tony, 14
Japan Air Lines (JAL), 199
Japan Air Transport (JAT), 199
Japan Airways Co. Ltd, 199
Jean Luc Lagardère Airbus A380 final assembly plant, 110

Jeffries, Dr John, 11
John F. Kennedy Airport, New York, 112, 220
Johnson, Amy, 43, 127–29, 135, 140, 201
   and DH Comet *Black Magic*, 128
   and DH.84 Dragon, 128
   and Gipsy Moth *Jason*, 127
   and Puss Moth, 128
Johnston, Tex, 210
Joint Church Aid (JCA), 209–10
Jonker JS-1 Revelation, 187
Jouett, John, 72
Jubilee Cup *see also* British Air Racing Championship, 43
Junkers (Junkers Flugzeug- und Motorenwerke AG), 30, 56, 59, 63, 162, 232
   F 13, 30, 59, 75–76, 106
   G.23/G.24, 30
   G.38, 30
   J 10, 59
   Ju 13, 174
   Ju 52, 30, 61, 95
   Ju 52/3m, 31, 61, 76, 206
   Ju 86B, 77
   Ju 88, 200, 206
Junkers Flugzeugwerk AG (JFA), 61
Junkers, Hugo, 30, 61
Junkers-Luftverkehr AG (JLA), 59

KAL Flight 007, 207
Kamov (helicopters), 173
Kármán line, 197
Kasmirski, Ed, 182
Keys, Clement, 67–69, 73
Khaled, Leila, 214
Kharkiv Flight School, 145
King Fahd International Airport, Dammam, 150
King's Cup Race, 43
Kingsford-Smith, Charles, 72
Kinner Airster biplane, 130
Kinner Field, 129
Kinner, Winfield, 129
Kip Aero, 194
Kirovograd Flight School, 146
Klingensmith, Florence, 38, 45

KLM (Koninklijke Luchtvaart Maatschappij N.V.), 48. 51, 61–62, 67, 83–85, 116, 118, 200, 202, 205, 226
Komarov, Vladimir, 197
Konrad, Max, 143–44
Kooper, Captain, 48
Korchuganova, Galina, 3, 146
Korean Air Lines (KAL), 207
Kremer Prizes, 194
Kremer, Henry, 194
Kubis, Heinrich, 161
Küchemann, Dietrich, 111
Kunz, Opal, 137, 203
Kweilin incident, 202
Kyllmann, Guillermo, 76

Lady Hay Drummond Award, 143
'Lady Lindy' *see* Earhart, Amelia
LaGuardia Airport, 224
Laker, Freddie, 83–84
Latécoère (Groupe Latécoère), 57
Latécoère, Pierre-George, 57
Lathière, Bernard, 109
Le Bourget Airport, 99–100, 109, 111
League of Nations, 22, 79–80
Lear, William, 176–77
Learjet Corporation, 177
  Learjet 23, 176
  Learjet 28, 177
  Learfet 40, 177
  Learjet 45, 177
  Learjet 60, 177
  Learjet 70/75, 177
  Learjet 85, 177
Lecky-Thompson, Tom, 43
Lee Ya-Ching, 141
Lemoigne, Pierre, 188
Lend-Lease Bill (1941), 203
Lévy-Le Pen amphibious biplane, 60
Light Aircraft Association (LAA), 192
Ligne Aérienne du Roi Albert (LARA), 60
Lignes Aériennes Militaires, 201
Lindbergh, Charles, 65–66, 68, 72, 78, 129, 131, 136

Spirit of St. Louis see also Ryan, Tubal Claude, 65, 129, 131
Línea Aérea Nacional de Chile (LAN), 77
Línea Aeropostal Santiago-Arica (LASA), 77
Lincoln Standard L.S.5 biplane, 77–78
Lioré et Olivier LéO 21, 30
Lisunov Li-2 (PS-84), 64, 92, 146
Lloyd Aéreo Boliviano S.A.M (LAB), 76–77
Lockheed, 27–28, 71, 77, 85, 101, 110, 152, 176
    10 Electra, 27–28, 56, 77, 107
    10E Electra monoplane, 131
    12 Electra Junior, 28
    14 Super Electra, 28, 56, 77
    9 Orion, 27
    C-130 Hercules, 209, 217
    C-69 Constellation, 83, 85, 92, 205
    F104G Starfighter, 140
    Hudson V, 139
    L-1011 TriStar, 109
    L-1049 Super Constellation, 91, 94, 98, 100–2, 196, 227
    L-1649 Starliner, 219
    L-188 Electra, 90, 103, 114, 117
    L-300/C-130, 169
    L-329 JetStar, 177
    L-749A Constellation, 219
    Lodestar, 77, 174
    T-33A Shooting Star, 135
    Vega, 27, 71, 131
Lockerbie Bombing, Pan Am Flight 103, 220–21
London Heathrow Airport, 48, 51, 95, 97–98, 150, 220
Los Angeles International Airport, 150
Los Angeles Speedway, 129
Los Rodeos Airport, Canary Islands disaster, 226
Love, Nancy Harkness, 139, 204
Lowe, Thaddeus ('father of US military aviation'), 136
Ludington Airline (LAL), 69
Lufthansa *also* Deutsche Lufthansa Aktiengesellschaft (DLH), 59, 91–92, 122, 216–17
Luftschiffbau Zeppelin GmbH, 58
Lycoming turbine engines, 172
Lyon, Harry, 72
LVG C.VI biplane, 59
LZ 129 *Hindenburg* disaster, 228

INDEX

Macchi *see* Aeronautica Macchi
MacCracken, William Jr, 70
MacIntyre, Malcolm, 102
Macpherson-Smith, Ross and Keith, 19, 46–47
MacRobertson Trophy Air Race, 28, 37, 39, 47, 140
Maddux Air Lines (MAL), 69
Maddux, Jack, 69
Mahoney, Ben, 65
Maitland, Edward, 19
  and Naval Airship Station, 19
Malaysia Airlines (MAS), 6, 229
Mallett-Bennett, James, 47
Mallory, William, 78
Manchester (Ringway) Airport, 62
Manhattan Project, 205
Manning, Henry *see also* Earhart, Amelia, 131
Mantz, Paul *see also* Earhart, Amelia, 131
Marchetti, Alessandro, 36, 61
Marignane flight test centre, 211
Marrow, Anne, 72
Martin Company, 73, 85
  2-02-2, 101
  4-0-4, 85, 101
  M-130 *China Clipper*, 73
  M-130 flying boat, 73
Martin, William, 200
Marvingt, Marie, 170
Masefield, Peter Gordon, 95–96
Maurer, Richard, 117
Maxim, Sir Hiram Stevens, 11
McCulloh, Mark, 188
  and Winchboat, 188
McDonald, Larry, 207
McDonnell Douglas, 110, 122
  DC-10, 103, 109, 229
  DC-10-30, 225
McGinness, Paul *see also* Qantas, 54
McKay, Hugh, 170
McMaster, Fergus *see also* Qantas, 54
Menasco B6 Buccaneer air-cooled aero engine, 40
Mercedes-Benz, 115
Mercury 13 women astronaut programme, 139

263

Mermoz, Jean, 57
Messerschmitt-Bölkow-Blohm (MBB) *see also* Airbus, 109
Meyer-Labastille, Otto Ernst, 76
Miami Air Races, 40
Michelin, 149
microlights, 150
Microturbo TRS-18 aero engine, 193
Mignet, Henri, 191
  HM.14 kit, 191
  M.8 Avionnette kit, 191
Mil (helicopters), 173
Mi-10, 173
Mi-26, 173
Miller, Betty, 143–44
Miller, Chuck, 143
Minichiello, Raffaele and TWA Boeing 707 hijacking, 213–14
Minimoa glider, 185–86
Ministry of Aircraft Production (MAP (UK)), 199
Ministry of Supply (UK), 111
Mir low-Earth orbit space station, 19
Mitchell, Reginald, 35–36
Mock, Geraldine 'Jerrie', 144–45
  and Cessna 180 *Spirit of Columbus*, 144–45
Mock, Russell, 144–45
Mogadishu (Aden Abbe Airport), West German Federal Police
  Grenzschutzgruppe 9 (GSG9) raid on, 217
Moisant, John, 14
Mollison, Jim, 128
Mono Aircraft Company, 139
  Monocoupe 90, 136
Montgolfier, Joseph and Étienne, 10–11, 50, 183
Montijo, John, 130
Montréal Trans-Canada Air Lines (TCA), 199
moon landing (Apollo 11), 194–95
Moore-Brabazon, Lord John, 86
Morane-Saulnier G sports, 15
Morgan, Morien, 111
Morgan, Thomas, 73
Moscow Aviation Institute, 146–47
Mossad, 217
Mucklow, Tina *see also* Cooper, Dan, 215
Multilateral Aviation Convention (in ICAO), 81

Murdoch, Rupert, 90
Musk, Elon, 197

NASA, 111, 137, 139, 144, 168–69, 188, 194–96, 234
   Apollo programme, 168, 194–95
   Gemini programme, 168, 188
   Pioneer 10 space probe, 169
National Advisory Committee for Aeronautics, 136
National Aeronautic Association (US), 181
National Aeronautics and Space Administration *see* NASA
National Aeronautics Association of the United States, 135
National Air Communications (in UK Air Ministry), 55
National Air Races, 40, 43–45, 133
National Air Transport (NAT), 67, 69
National Capital Wing (in CAP), 132
navigation systems, 18, 21, 64, 66, 68, 78, 80–83, 97, 115, 150, 159, 196, 218
Negro Airmen International (NAI), 103
Nélis, Georges, 59
New Deal Program, 74–75
New Hampshire Air National Guard, 135
Nichols, Ruth, 133, 142
Ninety-Nines, 38, 44, 134–35, 138, 143–44, 160
Noonan, Fred *see also* Earhart, Amelia, 131–32
Norman Thompson Flight Company, 35
North American Aviation, 115
Northolt Aerodrome, 95
Northrop Grumman, 196–97
Northrop T-38A-30-NO, 139
Northwest Airlines (NWA), 71–72, 102
Northwest Orient Airlines (NWA), 199, 214–15, 221
Northwood, London mid-air collision, 227
Nott Julian, 183–84
Noyes, Blanche, 133–35
Noyes, Dewey, 133
Ntaryamira, Cyprien, 208
Nycum, J. F., 108

O'Donnell, Gladys, 133
Odekirk, Glenn, 41
Œhmichen, Étienne Edmond, 172
   Oehmichen No. 2 rotorcraft, 172

oil crisis (1973) *see also* Organization of Arab Petroleum Exporting Countries (OAPEC), 109, 112, 119, 122, 164
Oishi, Wasaburo, 157
Omlie, Phoebe, 135–36
Omlie, Vernon, 136
Oneworld Alliance, 123
Operation Ezra, 209
Operation Magic Carpet, 209
Operation Nehemiah, 209
Operation Thunderbolt, 216–17
Orbis Flying Eye Hospital (DC-10), 171
Organization of Arab Petroleum Exporting Countries (OAPEC) *see also* oil crisis (1973), 119
Organization of Black Airline Pilots (OBAP), 103
Osipenko, Polina, 146
OSOAVIAKhIM Volunteer Society for Cooperation with the Army, Aviation, and Navy, 145–47
Osoaviakhim-1 balloon disaster, 183
  deaths of Pavel Fedoseenko, Andrey Vasenko and Ilya Usyskin, 183
Outer Space Treaty (1967), 194
Oxfam, 209
Oxford conference (1880), 20

Pacific Air Transport (PAT), 68–69
Palestine Liberation Organization (PLO), 214, 216–17
Palmer, Richard, 41
Pan American (Pan Am), 72–74, 76, 78, 83–85, 88, 100, 102–3, 107–8, 115–17, 122, 152, 155, 199, 202–5, 214, 216, 220, 226
  Clipper Pacific, 115
  Honolulu Clipper, 115
Pandit, Aarohi, 148
PAO S. P. Korolev Rocket and Space Corporation Energia, 147
parachuting, 125, 135, 146–47, 189, 215
paragliding, 185, 188–89
Paramount Business Jets, 173
Parasail Safety Council (PSC), 188–89
parasailing, 188–89
Paris Convention (1919) *also* Convention Relating to the Regulation of Aerial Navigation, 20–22, 49–53, 64, 79, 149, 159, 166, 190, 224, 227, 232
Paris-Orly Airport, 98–99
Paton, Dr David (founder Orbis), 171
peace officers/air marshals, 213

Pearl Harbor, Japanese attack on, 76, 204
Peel, Clifford, 170
Pemberton-Billing Ltd *see* Supermarine Aviation Ltd
Pemberton-Billing, Noel, 34–35
Penn School of Aeronautics, 134
Penza Flying Club, 145
People's Defence Commissariat (Narodnyy komissariat oborony Sovetskovo Soyuza), 201
Pequet, Henri, 14
Percival Vega Gull, 47, 128
Perez, Diana, 216
Philadelphia Rapid Transit Air Service (PRTAS), 66–67
Piccard, Auguste, 183
Piché, Robert, 225
Pinnacle Airlines (FLG), 229
Piper Aircraft Corporation, 142–43, 176
   Cub series, 176
   HA-420 HondaJet, 176
   PA-18 Super Cub, 176
   PA-23 Apache, 143, 145
   PA-28 Cherokee, 176
   PA-47 PiperJet, 176
Piper, William, 143, 176
Pitcairn Aircraft Company (PAC), 68
   PCA-2 autogyro, 47, 131
Pitcairn, Harold Frederick, 68
Pitre, Marguerite and CPC DC-3 bombing, 219
Pitts Special S1, 179–80
Pitts, Curtis, 170
Pittsburgh Aviation Industries, 134
Poitevin, Louise, 125
Polskie Linie Lotnicze (LOT), 199
Popular Flying Association, 192
Popular Front for the Liberation of Palestine (PFLP), 214, 217, 220
Popular Mechanics, 190
Potez 560, 77
Powder Puff Derby *see also* Women's Aerial Derby, 44, 113, 130, 133, 137
Powell, William, 133
Power Jet WU (Whittle Unit),8 6
Poynter, Dan, 188
Pratt & Whitney aero engines, 108
   Hornet, 26

   R-1535 twin-row, 41
   R-1690-S1C3G, 40
   S1D1 Wasp, 26
   turboprop engine, 168
Prier, Pierre, 14
Pritchard, John, 19
Provisional Air Routes (in ICAO), 81
Pucci, Emilio, 18
Pulitzer Electric Aircraft Race, 233
Pulitzer Speed Trophy, 43
Pulitzer Trophy, 22, 43–40
Pulitzer, Ralph, 43
Putnam, George, 130–31
PZL-106 Kruk, 174

Qantas (Queensland and Northern Territory Aerial Services), 54–55, 88, 90, 118, 162, 170, 202–3
Qantas Empire Airways Limited (QEA), 55
Quimby, Harriet, 14

radio-controlled (RC) aircraft, 181–82
Railey, Hilton, 130
Railway Air Services (RAS), 55, 95
Ramensk Avionics Construction Bureau, 146
Rancho Oro Verde Fly-Inn Dude Ranch, 137–38
Rand Airport (Johannesburg), 48
Raskova, Marina, 146
Rathenau, Walther, 58–59
Read, Albert Cushing, 16–17
Reconstruction Finance Corporation (RFC), 82
Red Army Air Force (Voyenno-Vozdushnyye Sily (VVS)), 201
Red Bull Air Race, 46, 123
Red Bull Air Racing, 46
Reid, Richard (al-Qaeda bomber), 221
Relief Wings, 142
Renault, Louis, 57
Reno National Championship Air Races, 44, 46
Réseau des Lignes Aériennes Françaises (Network of French Military Air Lines), 201
Reuter, Otto, 106
Rexroat, Ola Mildred, 204
Richards, Byron and Peru hijacking, 212

INDEX

Richet, Charles, 171
Rickenbacker, Eddie, 68, 102
Ride, Sally, 139
Rihl, George, 78
Rippelmeyer, Lynn, 147
Robert Graham Competition, 194
Robertson Aircraft Corporation (RACorp), 66, 69
Robertson, Sir Macpherson, 47
Robertson, William and Frank, 69
Rodgers, Calbraith, 14
Rohrbach, Dr Adolph, 107
Rolls, Charles Stewart, 13
Rolls-Royce aero engines, 88, 108, 121
   Avon, 90
   Eagle 360hp, 18
   Nene, 87
   R racing, 36
   RB.80 Conway, 94
   RB.108 turbojet, 33
Rome-Fiumicino Airport attacks and hijackings, 216–17
Roosevelt Field, Long Island, 138
Roosevelt, Franklin D., 71, 74, 134, 136, 205
Roosevelt, Theodore, 70
Roscosmos, 195
   Federal Space Agency, 195
   Russian Aviation and Space Agency and Federal Space Agency, 195
   State Space Corporation, 195
   United Rocket and Space Corporation, 195
Ross Howard, Jean, 142–43
Roussos, Demis,
Royal Aero Club (of Great Britain), 127, 158, 181
Royal Aeronautical Society (RAeS), 194, 233
Royal Aircraft Establishment (RAE), 37, 111
Royal Aircraft Factory, 51, 170
Royal Navy Air Service (RNAS), 19, 35
Royston Instruments and Midas 270 FDR, 211
Ruest, Généreux and CPC DC-3 bombing, 219
runway construction and maintenance, 33, 56, 93, 149–50, 153, 225
Rust, Mathias and flight to Moscow, 175
Rutan model 68 Amsoil Racer, 193
Rutan Voyager, 147, 193
Rutan, Burt, 193

Rutan, Dick, 147, 193
Ryan Aeronautical Company, 65
　Ryan M-1 mail and passenger plane, 65, 68
　Ryan-Standard J-1 biplane, 65
Ryan Airline Company (RAC), 65
Ryan Field, San Diego, 138
Ryan, James, 65, 211–12
　Ryan FDR, 211–12
Ryan, Tubal Claude *also* Ryan NYP *aka Spirit of St. Louis*, 65–66
Ryanair (RYR), 84. 117, 120

Sabena (SAB) (Societé anonyme belge d'Exploitation de la Navigation aérienne), 60–61, 123
sabotage, 212, 218, 228, 230¡1
Sacchi, Louise, 144
Sadi-Lacointe, Joseph, 42
Saint-Exupéry, Antoine, 57
Salmson 2 Limousine model, 57
Santa Monica Flyers, 143
Santos-Dumont 14-bis biplane, 125
Santos-Dumont, Alberto, 125, 190
São Paulo Congonhas Airport *aka* Campo da VASP, 77
Sarabia, Francisco, 39
Savitskaya, Svetlana, 147
Savoia-Marchetti, 36
　S.55, 36
　S.64, 36
　SM.73 Trimotor, 60
　SM.80, 36
　SM.83, 61
Scalewings, 194
Scandinavian Airlines (System) (SAS), 84, 147, 122, 227
Schlesinger African Air Race, 47, 94
Schlesinger, Isidore, 47
Schneider Trophy, 32–36, 42, 114, 162
Schneider, Jacques, 42
Schönefeld Airport East Berlin, 91
Scott, Charles, 47–48, 94, 140
Scott, George, 18
Scott, William, 215
Scottish Airways Ltd *see also* British Airways Ltd (BAL), 56, 95
Scott-Paine, Hubert, 35

INDEX

Sebastian, Dorothy, 113
Sebastio, Silva (governor São Tomé), 209
Second World War, 23, 25–30, 33–34, 42, 44–45, 48, 55, 61–62, 65, 72, 75–77, 79, 81, 89, 92–95, 99, 101, 103, 107, 114–15, 120, 122, 134, 149, 151–53, 157, 159–61, 168, 171–73, 179, 183–84, 186, 191, 194, 198, 203, 211–12, 223
Seilkopf, Heinrich, 157
Selfridge, Thomas, 12
Seversky SEV-2S, 138
Servicio Aéreo Colombiano (SACO), 76
Sha'arawi, Huda, 141
Sharm El Sheikh International Airport, 221
Shaw, Jerry, 61
Shell 3-Kilometer Speed Dash, 38
Sherman Fairchild, 78
   Fairchild FC-2, 78
Shiers, Walter, 19, 47
Sholto Douglas, William, 85–95
Short (Brothers plc), 33–35, 199
   Crusader, 32
   Mayo Composite, 32–33
   S.1 Cockle, 32
   S.21 flying boat, 33
   S.20 Mercury Seaplane, 33
   S.25 Sandringham, 89
   S.26 G-class flying boat, 33
   S.30 Empire flying boat, 55, 199, 202–3
   S.343 glider, 33
   S.38 seaplane, 35
   SA.5 glider, 33
   SA.6 Sealand, 33
   SC.1 (VTOL), 33
   Solent, 85
   Sporting Type, 32
   Sunderland, 200
Short, Eustace, 32
Short, Horace, 32
Short, Oswald, 32
Shuttle Landing Facility (SLF) (Kennedy Space Center), 167
Siberian Airlines (SBI), 208
Siddeley Tiger aero engines, 25
Sikorsky, 16, 172

Ilya Muromets, 16
  R-4 rotorcraft, 172
  S-21 Russky Vityaz, 16
  S-42B, 73
  S-64 Skycrane, 172
  VS-44 flying boat, 100, 152, 205
Sikorsky, Igor, 16, 49–50, 172
Sinai bombing, of Kogalymavia Meroject (KGL) Airbus A321, 221
Singapore Airlines (SIA), 110, 122
Sixth International Conference of American States *also* Havana Conference (1928), 22
Skelton, Betty, 179–80
  and Pitts Special *Little Stinker*, 180
Skiles, Jeffrey *see also* Hudson River emergency landing, 224
Sliney, Ben, 218
Slingsby Sky sailplane, 186
Smith, Joan, 145
Smith, Lowell, 135
Snook, Mary Anita 'Neta', 129–30
Sociedad Colombo Alemana de Transportes Aéreos (SCADTA), 72, 75–76
Società Anonima Costruzioni Aeronautiche Savoia *see* Savoia-Marchetti
Società de Agostini e Caproni,
Società Idrovolanti Alta Italia (SIAI) Seaplane Company *see* Savoia-Marchetti
Società Stipa-Caproni, 87
Société des Lignes Latécoère (SLL), 57
Société Générale des Transports Aériens (SGTA) *see also* Air France, 60
Société nationale des chemins de fer français (SNCF), 99
Soekarno-Hatta International Airport, Jakarta, 154
Southern Railways (SR), 52, 55
Southern, Bill, 129
Southwest Aviation Corporation (SWAC), 142
Soviet space programme, 146, 194–95
Spaatz, Carl, 72
Space Shuttle, 147, 195–96
  Challenger disaster, 196
  Columbia disaster, 196
  Constellation programme, 196
SpaceShipOne, 193
SpaceShipTwo, 193
SpaceX (Space Exploration Technologies Corp, 196–98
  Starship, 197
  Starship Human Landing System (HLS), 234

Spartan Aircraft Company, 55
    Cruiser Trimotor, 55–56
Spartan Airlines (SAL) see also British Airways Limited (BAL), 55
SS *Ionia*, 17
Stamper, Malcolm, 108
Standard J-1 (Ryan-Standard), 65
Standard JR1B mail biplane, 64
Standard Oil Company, 68, 134
Star Alliance, 122–23
Stead, Bill, 44
Stella aero club (women only), 125
Stethem, Robert (USN), 217
Stimpson, Steve, 161
Stinson (aircraft company), 69, 76
    SR-9B monoplane, 142
Stipa, Luigi, 87
Stipa-Caproni, 87
Stits, Ray, 191–92
    Stits SA-2A Sky Baby biplane, 191 –92
    Stits SA-3A Playboy, 191
Stultz, Wilmer, 130
Sud Aviation, 97
    SE 210 Caravelle, 89, 104, 220
Sud-Est SE-161 see also Bloch MB.161, 30
Sullenberger Chesley 'Sully' see also Hudson River emergency landing, 224
Supermarine (Supermarine Aviation Works (Vickers)), 34–36, 52, 114
    Baby, 35
    P.B.25, 35
    S.4, 35
    S.5, 35
    S.6B, 36
    Sea Eagle, 52
    Sea Lion, 35
    Sea Lion II, 35
Supermarine Aviation Ltd see Supermarine
Supersonic Transport Aircraft Committee (STAC), 111
Sutter, Joseph, 108
Svensk Interkontinental Lufttrafik AB (SILA) see also Scandinavian Airlines (SAS), 84, 206
Swallow J-5 biplane, 66
Swissair (SWR), 61, 123, 161, 205–6, 214, 220
SWR Convair CV-990-30A-6, destruction of, at Würenlingen, 220

Syndicat national d'Étude des Transports Aériens (SNETA) *see also* Sabena, 59–60
Syndicato Condor, 76
Szybowcowy Zakład Doświadczalny sailplanes, 186
   SZD-19 Zefir, 186
   SZD-24 Foka, 186

Tait, Maud, 38
Talon supersonic trainer, 139
Tank, Kurt, 31
Tasman Empire Airways Limited (TEAL), 203
Taylor, Clarence and Gordon, 176
Taylor, John, 192
   Taylor J.T.1, 192
   Taylor J.T.2 Titch, 192
Taylorcraft Aviation, 142, 176
Technical Standards and Procedures (in ICAO), 81
Tereshkova, Valentina, 194
terrorism, 112, 123, 212–14, 216, 218, 220–21, 226, 231
Texas International Airlines, 147
Textron, 176
Thaden, Louise, 44, 113, 134–35, 142
Thai Airways International (THA), 122
'The Incredibles' (747 team), 109
The Times, 200
Thomas, Carter, 101
Thomas, Charles Sparks, 101
Thompson Trophy, 38–39, 45
Thompson, Charles E., 45
Thomson Airways (TOM), 122
Tillinghast, Charles Jr, 101
TNT, 90
Traeger, Alfred, 170
Trans Australia Airlines (TAA), 86, 89–90
Trans World Airlines (TWA), 28, 67, 101–3, 116, 123, 205, 213–14, 227
Transcontinental Air Transport (TAT), 68–69
Transcontinental and Western Air (TWA) *see also* Trans World Airlines (TWA), 69–72, 85, 94
Transpacific Route Case, 102
Transport Canada, 213
Transportes Aereos del Continente Americano (TACA), 77
Transports Aériens Intercontinentaux (TAI) *see also* Union de Transports

Aériens (UTA), 99
Travel Air Factory, 134
  Travel Air 6000, 67
  Travel Air Type R Mystery Ship, 134
Treaty of Versailles, 23, 30, 58–59, 162
  and restriction of German aviation industry, 23, 30, 58–59, 162
Trippe, Juan, 66, 72, 78
Tryggve Gran, Jens, 15
Tsiolkovsky, Konstantin, 194
Tula Advanced Flying School, 145
Tupolev, 92
  ANT-20, 64
  ANT-37bis, 146
  Tu-16, 93
  Tu-104, 92–93
  Tu-114, 93
  Tu-124, 93
  Tu-144, 87, 111, 233
  Tu-154, 208
Tupolev Tu-144 crash, 111
Turner, Roscoe, 40, 44

UEC-Aviadvigatel aero engines, 88
U-Fly Alliance, 123
Ukraine International Airlines (AUI), 208
Ukrpovitroshliakh (Ukrainian airline) *see also* Dobrolyot, 63
Ulm, Charles, 72
Ultra-Light Aircraft Association (ULAA) (UK), 192
Union Aéromaritime de Transport (UAT) *see also* Union de Transports Aériens (UTA), 99
Union de Transports Aériens (UTA), 99
United Aircraft and Transport Corporation (UATC), 70
United Airlines (UAL), 66, 102–3, 122, 157, 161, 218
United Airlines Ltd (UK) (UALUK) *see also* British Airways Limited (BAL), 55
United Nations Aviation Standards for Peacekeeping and Humanitarian Air Transport Operations (UNAVSTADS), 170
United Nations Humanitarian Air Service (UNHAS), 169–70
United Nations Office for Outer Space Affairs (UNOOSA), 194
United States Air Force (USAF), 103, 135, 138, 188, 215
United States Air Force Reserve, 139
United States Army Air Forces (USAAF), 95, 103, 205, 211
US Army, 12, 19, 66, 156, 159

US Army Air Corps (USAAC), 72, 137–38, 203
US Navy (USN), 16–17, 19, 217
US Post Office, 64, 66–67
   Contract Air Mail Routes (CAMs), 66–71
USS *Aroostook*, 16

Valier, Max, 194
Value Alliance, 123
Van Vlissingen, Frits Fentener, 61
Van Zanten, Jacob Veldhuyzen, 226
Van's Aircraft, 192
  RV-14 tandem-seat, 192
VanGrunsven, Richard, 192
Vanilla Alliance, 123
VariEze, 193
Varig (Sociedade Anônima Empresa de Viação Aérea Rio-Grandense), 76
VariViggen, 193
Varney Airlines (VAL), 66, 69
Varney, Walter, 66, 103
Védrines, Jules, 15
Viação Aérea São Paulo (VASP), 77
Vial, Jacques, 126
Vickers Ltd, 35
  Vimy Commercial, 19, 23, 47, 51
  Vimy VI bomber, 18
Vickers-Armstrongs *see also* Supermarine, 35, 86, 97
  Vickers Vanguard. 96
  Vickers VC10, 93, 95, 214
  Vickers Viking, 86, 95–96
  Vickers Viscount, 86, 89–90, 96, 98–99
  Viscount V630 (prototype), 96
Virgin Atlantic (VIR), 120, 123, 193
Virgin Galactic, 193
Virgin Holidays, 120
Voisin, Charles, 125–26
Voisin, Gabriel, 12, 125
Von Braun, Wernher, 194
Von Kármán Theodore, 197
Von Krohn, Hellmuth, 75
Von Opel, Fritz, 194
Von Zeppelin, Count Ferdinand Graf, 49, 58
Voss, Albert, 230–31

Voyenno-morskoy flot SSSR (VMF) (Soviet Navy), 201

Waddell, Jack, 109
Wallace, Dwight, 174
Wang Po-Chun, 73
War Assets Administration (WAA), 82
War Training Service (WTS), 203
Warner, James, 72
Warren, David, 211
Washington Air Derby Association Trophy, 143
Weber, Johanna, 111
Wedell, James, 40–41
Wedell-Williams Air Service Corporation, 39–40
   Model 22 'We-Will Jnr', 40
   Model 44 'We-Will', 40–41
Western Air Express (WAE), 27, 66, 69
Wheeler Field, Honolulu, 131
Whirly-Girls, 143–44
Whitchurch Airport, Bristol, 55
White Knight One and Two, 193
Whitteneck, Clem, 180
Whittle, Frank, 86
Widerøe, Turi, 147
Williams, Harry, 40
Wills, Philip, 185
Wills, Williams, 133
Wilson, Woodrow, 135
Winer, Paul, 76
Witt, Oliver, 182
Women Advisory Council (FAA), 142
Women Airforce Service Pilots (WASP), 39, 136, 139, 142, 201, 204
Women's Aerial Derby, 44
Women's International Association of Aeronautics, 143
Women's Advisory Committee on Aeronautics, 134
Women's Auxiliary Ferrying Squadron (WAFS) (US), 139, 204
Women's Engineering Society Council, 127
Women's Flying Training Detachment (WFTD) (US), 139, 203–4
Works Progress Administration (in New Deal), 134
World Aerobatic Championships, 146–47, 180
World Bank, 49
World Championships for Space Models, 182
World Food Programme, 169

Wright brothers, 11–12, 50, 158
Wright aero engines,
Wright Company *also* Wright Aeronautical, 33, 67
   Wright Biplane, 14
   Wright Flyer, 11, 13, 16, 193
Wright Cyclone, 25
Wright Whirlwind rotary engine, 38
Wright, Orville, 11, 149
Wright, Wilbur, 11, 13
Wygle, Brien, 2

Yakovlev, 147
   Yak-18, 147
   Yak-32, 146
   Yak-40, 93
Yeager, Jeana, 147
Yerex, Lowell, 77
Yost, Ed, 184

Zakaria airline, 63
Zara, Antonio, 216
Zeppelin, 23, 58, 106–7, 113, 161
   Staaken E-4/20 monoplane, 107
Zeppelin Hydrogen and Oxygen Company (ZEWAS), 58
Zeppelin-Hallenbau GmbH, 58
Zhou Enlai, bombing assassination attempt by Kuomintang (KMT), 219
Zindel, Ernst, 61